# Plant Disease
AN ADVANCED TREATISE

## VOLUME IV
### How Pathogens Induce Disease

# Plant Disease

## AN ADVANCED TREATISE

### VOLUME IV

### How Pathogens Induce Disease

*Edited by*

## JAMES G. HORSFALL
The Connecticut Agricultural
Experiment Station
New Haven, Connecticut

## ELLIS B. COWLING
Department of Plant Pathology
and School of Forest Resources
North Carolina State University
Raleigh, North Carolina

ACADEMIC PRESS    New York    San Francisco    London    1979

A Subsidiary of Harcourt Brace Jovanovich, Publishers

ACADEMIC PRESS, INC.
111 Fifth Avenue, New York, New York 10003

*United Kingdom Edition published by*
ACADEMIC PRESS, INC. (LONDON) LTD.
24/28 Oval Road, London NW1 7DX

**Library of Congress Cataloging in Publication Data**
Main entry under title:

Plant Disease.

    Includes bibliographies and indexes.
    CONTENTS: v. 1. How disease is managed.——v. 2.
How disease develops in populations.——v. 3. How plants
suffer from disease.——v. 4. How pathogens induce
disease.
    1. Plant diseases. I. Horsfall, James Gordon,
1905–ﾠﾠII. Cowling, Ellis Brevier, Date
SB731.P64ﾠﾠﾠﾠﾠﾠ632ﾠﾠﾠﾠﾠﾠﾠﾠ76–42973
ISBN 0–12–356404–2 (v. 4)

PRINTED IN THE UNITED STATES OF AMERICA

79 80 81 82ﾠﾠﾠ9 8 7 6 5 4 3 2 1

To Cynthia Wescott,
the prototype practitioner of plant pathology,
and all the others
who have made plant pathology useful.

# Contents

## CHAPTER 1  PROLOGUE: HOW PATHOGENS INDUCE DISEASE
### ELLIS B. COWLING AND JAMES G. HORSFALL

## CHAPTER 2  THE EVOLUTION OF PARASITIC FITNESS
### R. R. NELSON

## CHAPTER 3   THE ENERGETICS OF PARASITISM, PATHOGENISM, AND RESISTANCE IN PLANT DISEASE

### TADASHI ASAHI, MINEO KOJIMA, AND TSUNE KOSUGE

## CHAPTER 4   PREDISPOSITION BY THE ENVIRONMENT

### JOHN COLHOUN

## CHAPTER 5   EXTERNAL SYNERGISMS AMONG ORGANISMS INDUCING DISEASE

### COLIN H. DICKINSON

## CHAPTER 6   INTERNAL SYNERGISMS AMONG ORGANISMS INDUCING DISEASE

### N. T. POWELL

## CHAPTER 7   HOW THE DEFENSES ARE BREACHED
### R. L. DODMAN

## CHAPTER 8   HOW THE BEACHHEAD IS WIDENED
### RICHARD D. DURBIN

## CHAPTER 9   HOW FUNGI INDUCE DISEASE
### PAUL H. WILLIAMS

## CHAPTER 10   HOW BACTERIA INDUCE DISEASE
### ARTHUR KELMAN

# CHAPTER 11   HOW MYCOPLASMAS AND RICKETTSIAS INDUCE DISEASE
## KARL MARAMOROSCH

# CHAPTER 12   HOW NEMATODES INDUCE DISEASE
## VICTOR H. DROPKIN

# CHAPTER 13   HOW INSECTS INDUCE DISEASE
## DALE M. NORRIS

# CHAPTER 14   HOW VIRUSES AND VIROIDS INDUCE DISEASE
## MILTON ZAITLIN

## CHAPTER 15   HOW AIR POLLUTANTS INDUCE DISEASE

### EVA J. PELL

## CHAPTER 16   HOW PARASITIC SEED PLANTS INDUCE DISEASE IN OTHER PLANTS

### DONALD M. KNUTSON

## CHAPTER 17   ALLELOPATHY

### RICHARD F. FISHER

# CHAPTER 18   SELF-INDUCED DISEASE
## H. HOESTRA

# CHAPTER 19   IATROGENIC DISEASE: MECHANISMS OF ACTION
## JAMES G. HORSFALL

# CHAPTER 20   MYCOTOXINS AND THEIR MEDICAL AND VETERINARY EFFECTS
## W. F. O. MARASAS AND S. J. van RENSBURG

# CHAPTER 21   THE EPIDEMIOLOGY AND MANAGEMENT OF AFLATOXINS AND OTHER MYCOTOXINS
## ROGER K. JONES

## CHAPTER 22 THE EFFECTS OF PLANT PARASITIC AND OTHER FUNGI ON MAN

### C. M. CHRISTENSEN

# List of Contributors

*Numbers in parentheses indicate the pages on which the authors' contributions begin.*

TADASHI ASAHI (47), Laboratory of Biochemistry, Faculty of Agriculture, Nagoya University, Nagoya 464, Japan

C. M. CHRISTENSEN (393), Plant Pathology, University of Minnesota, St. Paul, Minnesota 55108

JOHN COLHOUN (75), Cryptogamic Botany Laboratories, University of Manchester, Manchester M13 9PL, England

ELLIS B. COWLING (1), Department of Plant Pathology and School of Forest Resources, North Carolina State University, Raleigh, North Carolina 27650

COLIN H. DICKINSON (97), Department of Plant Biology, The University, Newcastle-upon-Tyne NE1 7RU, England

R. L. DODMAN (135), Plant Pathology Branch, Queensland Wheat Research Institute, Department of Primary Industries, Toowoomba, Queensland, Australia 4350

VICTOR H. DROPKIN (219), Department of Plant Pathology, University of Missouri–Columbia, Columbia, Missouri 65211

RICHARD D. DURBIN (155), USDA, SEA, AR, NCR, Plant Disease Resistance Research Unit, Department of Plant Pathology, University of Wisconsin–Madison, Madison, Wisconsin 53706

RICHARD F. FISHER (313), School of Forest Resources and Conservation, University of Florida, Gainesville, Florida

H. HOESTRA (331), Department of Nematology, Agricultural University, Wageningen, The Netherlands

JAMES G. HORSFALL (1, 343), The Connecticut Agricultural Experiment Station, New Haven, Connecticut 06504

ROGER K. JONES (381), Department of Plant Pathology, North Carolina State University, Raleigh, North Carolina 27650

ARTHUR KELMAN (181), Department of Plant Pathology, University of Wisconsin–Madison, Madison, Wisconsin 53706

DONALD M. KNUTSON (293), Pacific Northwest Forest and Range Experiment Station, U.S. Forest Service, Corvallis, Oregon 97330

MINEO KOJIMA (47), Institute for Biochemical Regulation, Faculty of Agriculture, Nagoya University, Nagoya 464, Japan

TSUNE KOSUGE (47), Department of Plant Pathology, University of California, Davis, California 95616

KARL MARAMOROSCH (203), Waksman Institute of Microbiology, Rutgers, The State University, Piscataway, New Jersey 08854

W. F. O. MARASAS (357), National Research Institute for Nutritional Diseases, South African Medical Research Council, Tygerberg 7505, South Africa

R. R. NELSON (23), Department of Plant Pathology, Pennsylvania State University, University Park, Pensylvania 16802

DALE M. NORRIS (239), Department of Entomology, University of Wisconsin–Madison, Madison, Wisconsin 53706

EVA J. PELL (273), Department of Plant Pathology, Pennsylvania State University, University Park, Pennsylvania 16802

N. T. POWELL (113), Department of Plant Pathology, North Carolina State University, Raleigh, North Carolina 27650.

S. J. van RENSBURG (357), National Research Institute for Nutritional Diseases, South African Medical Research Council, Tygerberg 7505, South Africa

PAUL H. WILLIAMS (163), Department of Plant Pathology, University of Wisconsin–Madison, Madison, Wisconsin 53706

MILTON ZAITLIN (257), Department of Plant Pathology, Cornell University, Ithaca, New York 14853

# Preface

*Every great advance in science issues from a new audacity of imagination.*

John Dewey

A steady stream of audacious ideas has brought plant pathology to its present stage of sophistication. A continuing stream will be required to fill the persistent gaps in our knowledge of disease processes. Our hope in these five volumes is to stimulate both the abundance and the audacity of imaginative ideas about the art and the science of plant disease.

Volumes III to V of this treatise deal with the phenomena of disease in individual plants. In Volume III the focus was on suffering that plants experience when their vital functions are disrupted.

This present volume deals, as none of the others do, with the mechanisms by which hungry parasites and irritating pathogens induce the various types of dysfunction described in Volume III. How do fungi, bacteria, nematodes, mycoplasmas, rickettsias, insects, parasitic seed plants, and air pollutants induce disease? What unique mechanical, chemical, physical, and regulatory weapons are used by each type of pathogen? How do they penetrate their host? How is each one disseminated? What are the unique attributes of pathogenesis induced by each type? Volume IV also sets the stage for the discussion in Volume V as to how plants defend themselves against all the different kinds of pathogens described in Volume IV.

In Volume III the focus was on the suffering of the plant. In Volume IV the focus shifts to the attack by the pathogen. In Volume V the focus shifts again—to defense by the plant. All these shifts and differences in perspective have been designed to stimulate new and audacious ideas about all the manifold concepts of disease: pathogenesis and parasitism, synergism and antagonism, virulence and avirulence, resistance and susceptibility, compatibility and incompatibility, morphological and anatomical changes, physiological and genetic changes, phytotoxins and pathotoxins, enzymes and growth regulators, energetics and dispersal, coevolution and competition, coexistence and stress, hypovirulence and hyperparasitism, hypersensitivity and tolerance, cross-protection and immunity, etiology and necrology, senescence and death, and the concepts of single, multiple, and sequential induction.

This present volume, like its predecessors, has been designed for the advanced researchers in our profession—those who will add the audacious ideas on which tomorrow's new concepts will depend. To make it so, we have enlisted the creative talent of 23 of the most imaginative plant pathology researchers from all parts of the world. They have written imaginatively and provocatively to make us think. We have been wonderfully stimulated by their ideas. We hope you will be too!

*James G. Horsfall*
*Ellis B. Cowling*

# Contents of Other Volumes

# Tentative Contents of Other Volumes

*Chapter 1*

# Prologue: How Pathogens Induce Disease

ELLIS B. COWLING AND JAMES G. HORSFALL

## I. INTRODUCTION

With this chapter we welcome you to the fourth of five volumes in "Plant Disease: An Advanced Treatise." Here we shall deal with the mechanisms by which pathogens induce disease. In Volume III we were concerned with plants that are sick—how they function poorly, grow poorly, yield poorly, and sometimes die from lethal attacks of pathogens. In this volume we discuss how different types of pathogens induce all the various kinds of suffering described in Volume III. Volume IV also sets the stage for the discussion in Volume V about how plants defend themselves against various types of pathogens.

1

PLANT DISEASE, VOL. IV

## II. THE BATTLEFIELD OF ATTACK AND DEFENSE

Volumes IV and V in this treatise deal, as none of the earlier volumes do, with the fascinating warfare that occurs when the pathogen attacks the host and the host defends itself. As theoreticians, we can sit back and see that attack by the pathogen (Volume IV) and defense by the host (Volume V) are clearly defined concepts: The pathogen attacks; the host defends. We recognize too that defense often is not passive. We recognize a biological truth in the old football adage, "A good offense is a good defense." In this sense, the host attacks the pathogen, and so the pathogen must defend itself too. These are clear-cut concepts.

When we move from being theoreticians to being experimentalists in the real world, however, we find ourselves in the middle of the swirling confusion of the battle. What is host? What is pathogen? What is offense? What is defense? The answers are not as clear as they appeared to be when we were theoreticians. For example, take the carbon compounds and minerals that flow into the battle zone of a lesion on a leaf. Are they merely flowing down hill to where the pathogen soaks them up, or are they deployed there by the host as part of the defense? Or consider the increased respiration in disease lesions. Does it provide the energy for the pathogen to battle the host, for the host to defend itself, or perhaps both? Or, finally, consider the unique compounds that are found in the battle zone. Are they made by the pathogen to help fight its way in, or by the host to keep the pathogen out? Or are they made by the host and then altered by the pathogen, or vice versa? Are they liberated by the pathogen as a chemical smoke screen to confuse the metabolic machinery of defense in the host? Or are they irrelevant, just battlefield debris—a residual evidence of disrupted metabolism?

The challenge for each of us as authors in Volumes IV and V is to stand back from the swirling confusion of the battlefield far enough to distinguish the phenomena of offense from the phenomena of defense. Clearly many of the experimentalists we cite have experienced grave difficulty here. It surely is not easy. We have tried, in the design and in the development of these two volumes, to suggest new theories and new experimental approaches. We have sought to make more clear the distinctions between offense by the pathogen and defense by the host and thus to illuminate the dynamics of the battle between the pathogen and the host.

## III. THE PERSPECTIVE OF THE PATHOGEN
## IN PLANT DISEASE

The perspective of this volume is the perspective of the pathogen. In fact, we suggested to authors of Volume IV that they might profit from learning to "think like a pathogen." By the same token, we suggested that Volume V authors might "think like a plant."

We recognize a bit of danger in this advice. But, so long as we, the authors, and you, the readers, remember that we are pathologists and not pathogens, it can be a useful intellectual exercise to apply what we know of disease in attempts to answer such questions as the following:

If I were a pathogen, how would I attack a host? Alone? Simultaneously with other pathogens? With assistance from a mobile vector?

Would it be better to penetrate the host, or to operate from the outside as an exopathogen? If the latter, how will I obtain enough energy to survive?

If I penetrate the host, shall I do this passively through wounds or with my own chemical or mechanical power? What are wounds like? How much power will be needed to penetrate? Shall I wait for a particular set of predisposing environmental conditions before penetrating the host? If I wait, what shall I do in the meantime?

Once inside the host, how shall I behave—as a congenial but hungry parasite or as a vigorously irritating and disruptive pathogen? Is it easier to succeed as an obligate parasite or as a facultative saprophyte?

What offensive weapons shall I use to gather the energy that I need to live? Toxins? If so, what kind would be best in this particular host—a general metabolic inhibitor or a host-specific one? Would extracellular enzymes be useful in attacking this host? If so, what enzyme(s)? Maybe it would be better just to activate the catabolic enzymes of the host and act like a virus—simply take over the machinery in the host rather than synthesize all those complex proteins from scratch? That would save a lot of energy. If I use growth regulators, what will my added cytokinin do to the auxin–cytokinin ratio in the host endodermis? Maybe a sequence of weapons would be better—pectinase to get in, growth regulators during early disease development, and hold off on the toxin until colonization is almost complete.

Is it better to kill the host cells first and then digest the contents or use haustoria and be a little more sophisticated about this? Will whole mitochondria go through a stylet? If not, they will have to be broken first.

How shall I cope with the defenses of the host? Those papillae are tough! What shall I do when a lysosome bursts when I first penetrate the epidermis? How many phytoalexins do carrots have? Is the pith of corn lignified in mid-August? Wouldn't it be better to wait until night-fall, when the Hill reaction is shut down for the day?

Is there any way to tell in advance if a given host already is infected? Would it be possible to get to the sporulation stage of pathogenesis in 6 days instead of 10? How much energy would that extra speed cost?

The answers to many such questions are available in the various chapters of this volume. Before we take up these specific matters, how-ever, some general principles of pathology and some definitions and perspectives are in order. We will begin with a conceptual definition of what a pathogen is and what it does.

## IV. WHAT IS A PATHOGEN?

Simply stated, a pathogen is an inducer of disease. Disease results whenever a vital process in a host plant is disrupted continuously. Disease is particularly serious when the coordinated capture and utili-zation of energy are impaired (see Volume III, Chapters 3 and 4). Thus, a pathogen may be defined as any factor that induces the unrelieved impairment or disruption of one or more vital functions in a plant. Some functions such as absorption of nutrients are highly localized; others such as respiration occur in all living tissues; still others such as photosynthesis and photorespiration are extensively coordinated. Thus, the effects of pathogens can range from very localized to very general. Pathogens can induce disease in any one or in many parts of a plant, from the smallest organelle and obscure metabolic pathway, to whole cells, tissues, organs, or, for that matter, the entire plant.

Injury sometimes is confused with disease, and injurious agents some-times are confused with pathogens. But this confusion can be avoided. Injury results when plants are disrupted momentarily, as by a mowing machine or a grasshopper. In contrast, disease results when plants are subject to sustained impairment or disruption (or, as some authors prefer, "continuous irritation") by a pathogen. Examples include wilt-inducing bacteria and yellows-inducing viruses.

Some pathologists prefer to restrict the term "pathogen" to the living inducers of disease (which they usually define as including viruses and possibly viroids). But diseases induced by abiotic factors (for example, air pollutants and drought) have many features in common with those induced by biotic pathogens (such as toxinogenic fungi and wilt bacteria). For this reason, we prefer to use the term "pathogen" to denote *any* inducer of disease.

If a distinction between living and nonliving inducers is needed, we prefer the terms "biotic pathogen" and "abiotic pathogen." Most biotic and many abiotic pathogens penetrate their hosts in order to induce disease. But some pathogens exist only outside their hosts; they are called exopathogens (Woltz, 1978).

Thus, we offer our current definition of a pathogen: any biotic or abiotic factor that induces the sustained impairment of one or more vital processes in a plant or in any of its parts. We recognize that pathogens may act internally or externally and that those which interfere with the coordinated utilization of energy are particularly serious threats to the health and survival of their hosts (Tansley, 1929).

We mention "current" in offering this definition because we recognize that most definitions should be provisional. Our present thinking about pathogens and their modes of action is conditioned by what we know now. In another year, more may be known and a new definition may be necessary. In this connection we reemphasize the statement on pages 16–17 of Volume I: "Truth is a perception of reality which is consistent with all available evidence and contradicted by none. Thus, it seems almost certain that our conceptions of truth will continue to change as we continue to expand the body of relevant evidence in our field."

## A. Biotic Pathogens

The biotic inducers of disease are many and varied. Fungi are the most abundant. Thus, it is not surprising that they were the first pathogens to be studied in detail (see this volume, Chapter 9). After the fungi came the bacteria (this volume, Chapter 10), and soon after that nematodes were shown to induce disease in plants (this volume, Chapter 12). Much more recently, mycoplasmas and rickettsias were added (this volume, Chapter 11).

Mistletoes were recognized as parasites of plants long before fungi (about A.D. 1200), but the notion that they were also pathogens is only about 40 years old. We will come back to this distinction later, but suffice it to say that many parasitic and pathogenic seed plants are now recognized (see this volume, Chapter 16). The chemical warfare that occurs among higher plants (allelopathy) is a new dimension of plant pathology but an old phenomenon among plants (see this volume, Chapters 17 and 18).

Some insects (such as grasshoppers and cutworms) cause injury to plants, and others are vectors for plant pathogens. But many insects are pathogens in their own right; they cause disease in the same sense that nematodes, fungi, and bacteria do! (see this volume, Chapter 13). Entomologists take note! Gall-forming insects induce abnormal growth. Mites, lace bugs, and webworms interfere with photosynthesis; they induce chlorosis and necrosis of foliage and stem tissues. Bark beetles, weevils, wireworms, and grubs interfere with transport of food and water in stems and roots. Seed and cone insects interfere with reproduction. Certain weevils affect flowering parts or kill young seedlings. Wood-boring insects interfere with xylem function and affect the supply of nutrients and water to stems and foliage. Aphids, scale insects, and thrips induce cellular dysfunction.

It is interesting to speculate about why insects have not been recognized as plant pathogens. In many countries, entomology is considered a discipline distinct from plant pathology. This pattern of declustering biologists has inhibited interactions between pathologists and entomologists for decades. Most plant pathologists are botanists concerned primarily with the biology of plant hosts, whereas entomologists are zoologists concerned primarily with the biology of insects. Whatever the whole reasons for reluctance, we believe it is time for a change in the attitude of both plant pathologists and entomologists toward insects and their role in plant disease. As discussed more fully in Chapter 13 of this volume, cooperation between scientists in these two fields offers great promise of mutually productive inquiry. Insects that induce sustained disruption of vital functions in plants are pathogens and will be treated as such in this treatise.

## B. Viruses, Viroids, and Plasmids

Some plant pathologists include viruses, viroids, and plasmids among the biotic pathogens; others consider them abiotic pathogens; some do not consider plasmids pathogens at all. Because the biochemical nature, mode of reproduction, and mechanisms of pathogenesis of all three entities are so similar and so unique, however, we believe they deserve assignment to a special class of their own.

Viruses contain two major constituents—a nucleic acid core and a protein coat. Viroids are particles of naked nucleic acid; they lack a protein coat. Plasmids (and episomes) are extrachromosomal fragments of DNA found in bacterial cells. Recently a plasmid in *Agrobacterium tumafacians* was demonstrated to be responsible for the ability of this bacterium to induce tumors in plants. The plasmid containing the "tumorogenic code" apparently can be transferred into cells of the plant host by the bacterium. These two observations are leading to the conclusion that plasmids should be regarded as pathogens and their bacterial hosts as vectors (see Volume III, Chapter 9). Similar observations of cytoplasmically transmissible factors that condition virulence in *Helminthosporium* (Lindberg, 1971) and *Endothia* (Grenete and Sauret, 1969) suggest that viruses, viroids, plasmids, and episomes may represent parts of a continuum of particles that inhabit plants and pathogenic microorganisms, each with important and distinctive roles in plant disease.

All of these entities include nucleic acid. All lack metabolic capacity of their own. They all reproduce and cause disease by diverting a part of the metabolic machinery of host cells to synthesis of virus, viroid, plasmid, or episome particles instead of host nucleic acid and protein. Neither biotic nor abiotic pathogens can do that! Viruses, viroids, and plasmids are indeed unique, as we shall learn in Chapters 10 and 14 of this volume.

## C. Abiotic Pathogens

The nonliving inducers of disease in plants are also many and varied. These pathogens include the following:

1. Toxic air pollutants. These substances are transferred to plant surfaces by deposition processes including absorption and adsorption of toxic gases, impaction of injurious aerosol particles ($> 3$ $\mu$m in diameter), and gravitational settling of poisonous particulate matter ($> 3 \mu$m) (see this volume, Chapter 15).

2. Toxic rains and snows. These pollutants include salt spray from oceans as well as strong mineral acids and other injurious substances resulting from combustion, industrial, and waste-disposal processes (see this volume, Chapter 15).

3. Toxic substances in soil. These injurious materials include toxic metal ions such as aluminum, salts in irrigation water, and humic acids (see Volume V).

4. Nutrient imbalances and deficiencies. These pathogens result from shortages of nutrients in soil, inadequate development of root symbionts including nitrogen-fixing organisms and mycorrhizae, or excessive leaching of nutrients from foliar organs (see Volume III, Chapters 7 and 11).

5. Extremes of temperature. These pathogens include heat, chilling, and frost.

6. Extremes of water availability. Shortage of water leads to wilt;

standing water leads to oxygen deprivation in roots, stems, and sometimes foliage; and excessive or corrosive rains lead to loss of nutrients
by leaching or damage to foliar surfaces by erosion of cuticular waxes
(see Volume III, Chapters 2, 5, and 6).

## D. The Distinction between Hungry Parasites and Irritating Pathogens

Nature has evolved a magnificent array of relationships and interactions among organisms. This array represents a continuum which is
difficult to comprehend in its entirety. In order to increase our understanding of such complex relationships, scientists find it useful to classify
things. In essence, we divide nature's continuum into components and
divisions within which the entities are sufficiently similar and between
which the entities are sufficiently different that comprehension of the
whole is easier to achieve. This process is sometimes called pigeonholing. Pigeonholing is a well-developed art in plant pathology. It
must be very amusing to God who created the continuum to observe
us scientists as we divide the continuum into arbitrary segments and
then spend much of our time arguing about the borderlines between
the arbitrary segments.

Four particular sets of pigeonholes have proved very helpful in
plant pathology. They involve the following four series of terms: (1)
mutualism, commensalism, pathosism, saprophytism; (2) predator, pathogen, parasite, perthophyte, saprophyte; (3) obligate parasite, faculative
saprophyte, faculative parasite, obligate saprophyte; and (4) obligate
biotrophism, faculative necrotrophism, faculative biotrophism, obligate
necrotrophism.

In Chapter 3 of Volume III, Bateman has defined and illuminated
the usefulness of many descriptive categories of organisms and relationships. We do not intend to repeat his discussion here, but there are a
few comments we should like to make about some of these terms.

With understandable trepidation in the original treatise, Horsfall and
Dimond (1960) proposed their famous "mother-in-law analogy" to
explain the essential difference between parasites and pathogens. This
analogy is so useful that we decided to risk parental and feminist
sensitivity once again and repeat the analogy in the Prologue of Volume
III. In perfecting the design for this new treatise, we have found it
useful to use the terms "hungry parasites" and "irritating pathogens" to
remind our fellow authors of these same essential distinctions.

Parasitism is a process by which energy and substance for growth and
development of a parasite are obtained at the expense of the host. Thus,

the terms "hungry parasite" and "nourishing host" make for a useful description of host–parasite relationships.

Pathogenesis is a process by which the vital functions of a plant are impaired or disrupted by the continuous irritation of a pathogen. In this case the terms "irritating pathogen" and "impaired" or "disrupted host" also provide a useful description of reality.

Bateman and others also make the point that universal use of the term "host" for the plant affected by both parasites and pathogens has led to some confusion. He states on page 64, in Volume III.

> It should be clear that not all parasites are pathogens and not all pathogens are parasites. Pathogens may be biotic or abiotic. Biotic pathogens may be parasitic or nonparasitic. An organism that is susceptible to disease is designated a suscept and may or may not be a host. The term suscept is useful since the pathogen may not be a parasite. We thus have the couplets host–parasite and suscept–pathogen. The couplet host–pathogen is correct only if the pathogen is a parasite. In all cases where disease is involved the suscept–pathogen couplet is correct and has specific meaning.

We agree with the logic of these defined couplets but find it difficult to embrace the proposal to use the word "suscept." "Host" is a well-known term both inside and outside plant pathology. "Suscept" is hardly a euphonious word. We agree that "suscept" can be useful for communication within plant pathology, but we are concerned that its use inside will impede our communication outside plant pathology. Thus, we believe that the term "host" will have to suffice for both parasites and pathogens.

Another useful distinction involves an intermediate form between parasites and saprophytes. Perthophytes are organisms that first kill the cells of a host and then derive nourishment from the dead host cells. This is a useful distinction because the mechanisms of pathogenesis are quite different although they may involve many common components.

## V. THE CONCEPT OF INDUCTION

The title of this volume is "How Pathogens Induce Disease." We chose the verb "induce" with great care. We did not choose "cause." We did not choose "incite." We chose "induce" because the concept of induction more exactly fits the reality of what pathogens do to their host plants than any alternative word.

It is misleading to say that pathogens *cause* disease, because it is a rare pathogen indeed whose capacity to affect the host is not limited or increased by environmental factors. It is also misleading to say that pathogens *incite* disease. As discussed more fully by Bateman (Volume III, Chapter 3,) the term "incitant" or its equivalent was suggested by Link (1933) to call attention to the multiplicity of environmental factors that affect the susceptibility of a host to a pahogen. Following Link's lead, L. R. Jones, J. C. Walker, and others of the Wisconsin School of Plant Pathology called all pathogens "incitants" to emphasize the importance of the weather in the complex web of causality. This term helped our thinking about the importance of the environment, but it muddied the waters of our thinking about the processes of infection and pathogenesis.

While a riot in the street will continue without the original incitant, we know of no plant disease that will continue without the pathogen. In fact, the term "incitant" is antithetical to the very concept of "continuous irritation" in the definition of disease (see Volume III Chapter 3) and the concept of sustained disruption in the definition of pathogens as discussed in the preceding paragraphs of this chapter.

We prefer the term "induced" because it recognizes both the importance of environmental factors and the importance of pathogens as major determinants of disease. Disease is induced by pathogens, not caused or incited by them.

## VI. COMMUNICATION ABOUT PATHOGENIC AND ENVIRONMENTAL STRESS FACTORS IN PLANTS

Many plant physiologists use the term "stress factor" to identify any factor or condition that limits the normal growth and development of plants. Plants respond to stress by making a variety of adaptations by which they attempt to cope with, or compensate for, the negative influence of the applied stress. Thus, one might say that plants struggle when they are stressed to keep on with their various functions despite the inhibiting influences of the stresses to which they are subjected.

Plant physiologists recognize both environmental and pathogenic factors as causes of stress and use the terms "environmental stress factors" and "biotic stress factors." The former are clearly in the domain of plant physiology; the latter are in the domain of plant pathology. The terms "environmental" and "biotic stress factors," as used by plant physiologists, are essentially equivalent in meaning to the terms "abiotic pathogens" and "biotic pathogens," as used in plant pathology.

In recent years, some plant pathologists have adopted the terminology of plant physiology in writing about plant disease. Others use the terminology of both fields interchangeably. For example, it does not matter whether one says, "Growth of the plant was limited by both environmental and biotic stress factors," or "Growth of the host was limited by both abiotic, and biotic pathogens," or "Growth of the plant was limited by both environmental stress factors and biotic pathogens." In fact, the last sentence, using the mixed terminology, is probably superior for communication with both plant pathologists and physiologists.

## VII. THE STIMULATORY AND INHIBITORY INFLUENCES OF KOCH'S POSTULATES

As discussed more fully in Chapter 2 of Volume I, a long and difficult intellectual struggle was necessary before the germ theory of disease could break the strangling chains of ignorance and superstition that for so many centuries blocked productive thought about plant disease. The combined genius of Tillet, Prèvost, Pasture, Koch, and de Bary was necessary before fungi finally were recognized as inducers of disease rather than merely excresences of diseased tissue.

Koch's rules for proof of causality have been wonderfully beneficial to our science. In fact, they came to dominate the way that many plant pathologists thought and conducted our experiments. This domination was beneficial in many ways, but it also had some detrimental side effects.

On the positive side, the postulates gave us an elegant model for experimental analysis of etiology. This model was applied to determine and verify thousands of specific plant–pathogen relationships. The inherent logic of the postulates and the strength of evidence that they require also established a very high standard of rigor in experimentation that benefits plant pathology to this very day.

On the negative side, however, Koch's postulates also led to two serious biases in plant pathology that also persist today. These biases suggest that (1) most plant diseases are induced by single pathogens working alone and (2) biotic pathogens (especially those amenable to analysis by Koch's postulates) are more important to the discipline of plant pathology than abiotic pathogens.

The constraint of the first bias was so powerful that it was not until the early 1930's (nearly 80 years after de Bary's proof that fungi could induce disease!) that plant pathologists generally began to recognize diseases induced by multiple pathogens acting together or sequentially. If you ask Ph.D. candidates in plant pathology to describe how they would determine the etiology of an unknown disease, the answers rarely go beyond Koch's postulates. And yet Koch's rules of proof must be extensively adapted for the analysis of multiple pathogen complexes (Wallace, 1978; see also this volume, Chapters 5, 6, and 13.) The notion that biotic pathogens are the only "real" pathogens continues to haunt those who study disease induced by air pollutants, frost, heat, insects, and nutrient deficiences. Fortunately, experimental models for analysis of diseases of complex etiology are now available (Wallace, 1978 see also this volume, Chapter 6); protoplast culture methods for viruses have been developed (Takebe, 1975); media are available on which to culture many "obligate parasites" (Scott and Maclean, 1969); a handbook of experimental procedures for analysis of dose–response relationships for air pollutants has been prepared recently (Heck *et al.*, 1979); and insects are now being recognized as pathogens too (see this volume, Chapter 13).

## VIII. THE TACTICAL WEAPONS OF OFFENSE

What are the offensive weapons by which pathogens attack their hosts? Let us consider four general and several lesser categories of weapons.

### A. Mechanical Force

Many pathogens have developed mechanical devices to aid in penetrating their hosts. Nematodes have swords that we call stylets. They are thin, hard, and sharply pointed; nematodes drive their stylets into their hosts with powerful muscles in the head of the nematode. The stylets are usually hollow so they can be used to inject injurious substances directly into the cells of the host. Furthermore, the stylet in many nematodes is connected to a powerful esophogeal pump with which the nematode can also suck out the contents of host cells.

Many pathogenic insects and mites have similar swords that we call stylets or proboscises, depending on the group of insects involved. Some of these swords are amazingly long, flexible, and remotely controllable. They can be inserted into host cells, bend around corners, and be thrust into one cell after another! There is some debate among entomologists about whether aphids actually suck out the contents of host cells or depend on the internal turgor of host cells to move host fluids into the esophagus of the insect. Aphids are depicted by Blackman (1966) as having both a sucking pump and a separate salivary pump for injection of substance into host cells.

As every plant pathologist knows, many fungi can form appresoria, which provide anchorage for infection pegs that can be used to pierce host cell walls. As far as we know, viruses, mycoplasmas, and rickettsia all enter their hosts passively, without mechanical force of their own. Many of these pathogens depend on animal vectors.

## B. Chemical Weapons

The chemical weapons with which biotic pathogens attack their hosts are legion. They include three major types of substances: toxins, extracellular enzymes, and growth regulatory substances. We shall discuss each in turn.

Toxins are nonenzymatic metabolities produced by biotic pathogens; they induce disruption of normal functions in living tissues of the host. The toxins can be general metabolic poisons such as the juglone secreted by walnut trees or highly host specific such as "victorin," produced by *Helminthosporium victoriae*. In some cases we know only that toxins are produced by a given pathogen. In other cases we know their chemical structure in detail. Some pathogens also secrete substances that activate specific lytic enzymes of host cells. Some abiotic pathogens (such as air pollutants) are themselves toxins that also can be general metabolic poisons or relatively host specific (see this volume, Chapter 15).

Extracellular enzymes of many sorts are secreted by fungi, bacteria, nematodes, mycoplasmas, and rickettsias. They may also be secreted by parasitic and pathogenic seed plants. Cutinolytic enzymes aid in penetration of the waxy surface of leaves (van den Ende and Linskens, 1974). Pectinases and other hemicellulases, as well as celluloytic and lignin-degrading enzymes, assist with cell wall penetration and dissolution of intercellular substances as in the case of the soft rot of vegetables or the decay of wood (see Volume III, Chapter 13). Proteolytic and lipid-degrading enzymes disrupt (and probably digest) the lipoprotein constituents of cell membranes. Nucleases cause the solubilization of cytoplasmic and nuclear DNA and RNA. Oxidative enzymes of many types induce changes in many different metabolites; catalase, peroxidase, tyrosinase, phenol oxidases, cytochrome and ascorbic acid oxidases, and a host of dehydrogenases are produced by many pathogens (see Volume III, Chapter 16 and this volume, Chapters 9–13).

The growth regulatory substances produced by plant pathogens include all of the hormones known to be produced in host tissues. These include indoleacetic acid and other auxins, gibberellins, cytokinins, abscissic acid, and ethylene. Many fungi, insects, bacteria, nematodes, parasitic seed plants, mycoplasmas, and rickettsias have been shown to produce growth regulatory substances or to cause impairment of normal growth functions in plants (see Volume III, Chapters 8–10 and this volume, Chapters 9–13, and 16).

## C. Diversion of Biosynthetic Machinery

As indicated earlier, viruses, viroids, and plasmids all have a unique mode of pathogenesis. They simply take over the nucleic acid- and protein-synthesizing machinery of host cells and divert this machinery to the synthesis of virus, viroid, and plasmid particles instead of host protein and nucleic acid (see this volume, Chapter 14).

## D. Physical Weapons

Some pathogens attack plants with physical weapons. For example, acid rains erode the cuticular waxes on plant surfaces. Wilt bacteria and fungi plug up xylem vessels. Sooty molds physically shade green leaves. High temperatures cause desiccation of host cells or destroy cellular or metabolic functions. Drought lowers the water potential of soil below the limits for uptake of water by plants. Cold causes formation of ice crystals, which disrupt cell membranes and organelles. (See Volume III, Chapters 2, 4, and 5, and this volume, Chapters 9, 10, and 15.)

## IX. SOME HIGHLIGHTS OF VOLUME IV

This volume begins with a series of theoretical and analytical chapters on the evolution and energetics of pathogens, predisposition phenomena, multiple pathogen interactions, and the penetration and colonization of hosts by pathogens. These first eight chapters set the stage for the more detailed discussions in Chapters 9–19 about how different types of pathogens induce disease. The volume closes with three chapters about how diseased plants and plant pathogens affect livestock and man.

In Chapter 2, Nelson describes the evolution and nature of parasitic fitness. The processes of parasitism and pathogenesis are important forces in the evolution of plants as well as their parasites and pathogens. This chapter describes how plants, parasites, and pathogens have coevolved and predicts how they may be expected to coevolve in the future, partly influenced by man.

In Chapter 3 we consider the energetics of parasitism, pathogenesis and resistance in plant disease. In this chapter, Asahi, Kojima, and Kosuge have developed a comparative analysis of the energy budget of pathogens, parasites, perthophytes, and saprophytes.

The energy and mechanisms of offense required by pathogens for successful parasitism or pathogenesis are altered by environmental factors such as temperature, humidity, leaf wetness, light, pH, and mineral nutrition. In Chapter 4, Colhoun describes how various predisposing environmental factors affect the likelihood of infection and the progress of pathogenesis.

Biotic and abiotic pathogens often interact synergistically with other pathogens as they cause disease. These interactions can occur both outside the host plant prior to infection (Chapter 5) or inside the host plant following infection (Chapter 6). For example, soil-inhabiting pathogens often are aided by other organisms in the rhizosphere. Some organisms aid in transport, others in penetration. Some do both. These symbiotic relationships vary from obligatory vector relationships through neutral symbiosis to positive symbiosis with other organisms. In Chapter 5, Dickinson describes external synergisms between organisms that induce disease.

It is rare indeed that single pathogens induce disease in a given host in the absence of other pathogens. Just as a man with a viral infection of the throat is more susceptible to bacterial pneumonia, so it is with plants; pathogens often act together to cause disease in the same host. Multiple pathogen interactions occur when any two or more pathogens act together, simultaneously or sequentially, in the same or in separate courts of infection. These interactions occur between all classes of pathogens. The specific permutations and combinations are endless. In Chapter 6, Powell summarizes these general principles to help us understand the concept of multiple-pathogen etiology.

How do pathogens penetrate their hosts? What offensive weapons do they use—massive numbers of propagules, stomatal entry, appresoria, infection pegs, toxins? Each mechanism of defense by the host (Volume V) can be overcome by specific offensive weapons of the pathogen. In Chapter 7, Dodman describes how these defenses are breached when pathogens penetrate their hosts.

The progressive expansion of disease lesions is a special phase of pathogenesis. What mechanisms of offense are involved in each stage of lesion expansion? How are zonate lesions and green islands induced? How do pathogens regulate their offense to cope with the daily and seasonal changes in the mechanisms and energy of defense? Durbin stimulates our curiosity about these questions in Chapter 8.

Chapters 9–19 are a series of related chapters about the unique features and special adaptations of each major group of pathogens. In Chapter 9, Williams draws an analogy between the dialogue between protagonists in a drama and the interactions between host and pathogen(s) during disease. He also describes the arsenal of structural, enzymatic, toxic, regulatory, metabolic, and other weapons used by fungi as they induce disease.

Similarly, in Chapter 10, Kelman discusses four major mechanisms by which bacteria induce disease—tissue disintegration, interference with water movement, interference with metabolism by toxins, and induction of abnormal growth. He also has a special section on future needs in research on how bacteria induce disease.

Mycoplasmas and rickettsias are a unique group of pathogens. In Chapter 11, Maramorosch describes the special relationships these pathogens have developed with insects and viruses, their effects on growth and differentiation of their plant hosts, and the origin and evolutionary development of these unique pathogens.

Nematodes and insects are the major animal pathogens of plants. Nematodes have various habits of life—surface feeders, ectoparasitic forms, migratory forms, sessile forms. In Chapter 12, Dropkin discusses the mechanisms used by all these different types of nematodes in altering cell walls and photosynthesis, regulating growth and the permeability of membranes, and altering the water status of plants. Similarly, in Chapter 13, Norris introduces the myriad types of insects that induce disease in plants and the special anatomical, chemical, and symbiotic adaptations insects have made to a pathogenic and/or parasitic habit of development.

As discussed earlier in this Prologue, viruses, viroids, and plasmids cause disease by taking over the synthetic machinery of host cells. Some of these pathogens are uniquely adapted to persist and even reproduce in their insect or fungus vectors as well as their host plants. Zaitlin introduces all of these unique features and adaptations and calls attention to the great contrast in our understanding of how these pathogens multiply and how they induce disease in their hosts.

In recent years injurious episodes of toxic air pollution have become more and more common in many parts of the world. Air pollutants are both pathogens and toxins. Eva Pell tells us how these poisons induce disease in Chapter 15.

Parasitic and pathogenic seed plants include both true and dwarf mistletoes, dodder, beechdrop, witchweed, and various species of *Seymeria*. Most of these plants merely inhibit growth but others induce mortality, decrease production of seeds, induce abnormal growth, and predispose plants to damage by insects or other pathogens. How do they do all these things? Knutson stimulates our curiosity about such questions in Chapter 16.

Allelopathy is a new dimension of plant pathology but an old aspect of plant disease. In Chapter 17, Fisher describes what is known about how plants induce disease in other plants with toxic chemicals. He also outlines the things that must be learned about this important aspect of plant disease.

In some cases plants induce disease in subsequent generations of the same plant. This "sickness" phenomenon is known in peas, clover, alfalfa, flax, peaches, and the creosote bush. Hoestra describes in Chapter 18 the role of toxic residues in plant refuse and other explanations of this phenomenon.

In Chapter 19, Horsfall describes a series of iatrogenic diseases and the mechanisms by which they are induced. These diseases result from the unintended side effects of chemicals prescribed by plant pathologists to protect plants against other diseases.

Chapters 20 and 22 of this volume deal with how diseased plants and plant pathogens affect livestock and human beings. In Chapter 20, Marasas and Rensburg describe mycotoxins and their medical and veterinary effects. The toxic substances that accumulate in food and feed crops can have serious and sometimes disastrous effects. What substances, organisms and environmental factors are involved in the field and in storage?

In Chapter 21, Jones discusses the epidemiology and management of aflatoxins and other mycotoxins.

In the final chapter of this volume, Christensen discusses the fascinating effects of ergotism, poisonous mushrooms, allergies, and other direct effects of pathogens on man.


## X. AN OVERVIEW OF THE TREATISE

Each volume of this treatise must stand alone. For this reason, we have included in this section a brief rationale for the treatise as a whole.

This treatise is designed for advanced researchers in plant pathology, whatever their specialty and status may be—from applied mycologist to virologist and from graduate student to Nobel Laureate. We hope it will broaden their view, stimulate their thinking, help them to synthesize still newer ideas, and to relate the previously unrelated. We hope the treatise will stretch the minds of its readers. To do so it must be comprehensive and timely, provocative and forward looking, practical and theoretical in outlook, and well balanced in its coverage.

We chose to call this treatise "Plant Disease," not "Plant Pathology." The term "plant pathology" means the study of suffering plants. Study is something man does. Man may suffer when disease hits his crops, but his suffering is secondhand. It is the plant that is sick, not the man. We seek to understand disease as plants experience it and thus to make this treatise plant centered rather than man centered.

To look at disease as plants do, plant pathologists must learn enough to predict the reactions of plants to pathogens. That is a tall order! It is difficult enough to understand how healthy plants grow. It is even more difficult to understand how plants behave when they are sick. The acid test of our understanding of disease processes is our ability to predict the progress of disease both in individual plants and in whole populations in the field.

Many earlier books about disease in plants have been given titles with plural subjects such as "Manual of Plant Diseases" (Sorauer, 1914; Heald, 1933), "Recent Advances in Plant Viruses" (Smith, 1933), "Plant Diseases and Their Control" (Simmonds, 1938), or "Pathology of Trees and Shrubs" (Peace, 1962). Use of the plural is understandable; the total number of diseases is as astronomical as the national debt.

It is impossible to learn all the diseases of plants and probably foolish to try. Even to learn all the diseases of one plant is a herculean task. So we designed this treatise to emphasize the commonalities of disease in plants—the unifying principles and concepts that will integrate our thinking. This rationale was used in the design of the original treatise. It emphasizes the common features of disease rather than the diversifying factors that tend to fragment our thinking. Thus, as a symbol of our desire for synthesis and unification, we left the "s" off diseases and called this treatise just "Plant Disease." *

Given five volumes in which to set out the art and science of plant disease, we found that the subject could be divided readily into the required five parts:

    I. How Disease Is Managed
    II. How Disease Develops in Populations
    III. How Plants Suffer from Disease
    IV. How Pathogens Induce Disease
    V. How Plants Defend Themselves

When the original treatise was designed in the late 1950's, plant pathology was reaching for maturity as a science, and basic research was coming into its own. For that reason the original treatise was organized around the scientific foundations of disease processes. During the last decade, however, society has called on plant pathology to demonstrate its usefulness in a world of worsening hunger.

---

* After our decision was made, we found that Stevens (1974) also used the title "Plant Disease" for his recent introductory text.

This treatise was conceived in 1974, the year when the world passed another great milestone along the road to global starvation. In that year the world population reached 4 billion people, enough to form a column marching 30 wide and a meter apart around the equator. It scared the wits out of us!

The growing urgency of the world food problem made us decide that it was timely, even urgent, that this treatise begin with the art of disease management and go on to the science of plant disease. In this way it could also relate the basics of our science more effectively to its usefulness to society.

After long and earnest debate we decided to move from the general to the specific—to put management in Volume I and epidemiology in Volume II. This sequence provides the strongest possible foundation for understanding both the theory of disease management and the dynamic nature of disease in Volumes III–V.

Volume I is not a cookbook about how to control specific diseases. Rather, it is a theoretical and philosophical treatment of the principles of managing disease, by altering the genes of plants, by changing the associated microbiota, by selecting or altering the environment, by using chemicals. Since the first volume sets the stage for the others, Volume I also contains chapters on the profession of plant pathology, its sociology, and how it works to benefit society.

After management in Volume I comes epidemiology in Volume II: "How Disease Develops in Populations." Since 1960 explosive progress has been made in understanding epidemics of plant disease. The latest explosion has come in the mathematical analysis of factors that make epidemics increase and subside. This provides the foundation for the emerging new field of theoretical plant epidemiology (Zadoks and Koster, 1976). Volume II also includes analyses of the genetic base of epidemics, the methodology and technology of epidemiological analysis and forecasting, the concepts of inoculum potential and dispersal, the climatology and geography of plant disease, agricultural and forest practices that favor epidemics, and the use of quarantines as a defense against epidemics of introduced disease.

In Volume III, we move from disease in populations to disease in individual plants: "How Plants Suffer from Disease." The early chapters set the stage for all the later chapters in Volume III plus those in IV and V. First, they describe how healthy plants grow. Next comes a modern conceptual theory of how disease develops in plants. Here disease is presented as the end result of all the positive and negative influences of hosts on pathogens and pathogens on hosts. The later chapters describe the many different kinds of impairments that can occur when plants are diseased. Volume III considers various dysfunctions: in the capture and use of energy, in the flow of food, in the regulation of growth, in the processes of symbiosis and reproduction, in intermediary metabolism and mineral nutrition, in the integrity of membranes, and in the engineering mechanics and biological rhythms of plants.

Having set out in Volume III the potential for dysfunction due to disease, Volume IV considers how pathogens induce these various dysfunctions, i.e., "How Pathogens Induce Disease." This volume describes the concepts of single, multiple, and sequential causality and their relationship to stress; the evolution and energetics of parasitism and pathogenism; and the concepts of predisposition, allelopathy, and iatrogenic disease. Next we compare and contrast the unique features of all the various pathogens of plants—fungi, bacteria, insects, mycoplasma, rickettsia, parasitic seed plants, nematodes, viruses, viroids, air pollutants, and so on. How are the effects of these pathogens similar? How are they unique? What offensive weapons does each type of pathogen use to be successful? Finally, we consider the effects of diseased plants and pathogens on livestock and man.

Volume IV sets the stage for Volume V "How Plants Defend Themselves." The chapters of this final volume will be closely linked with the analogous chapters in Volumes III and IV. Volume IV deals with how pathogens thwart the defenses of the host. Volume V describes how plants thwart pathogens.

Plants have many natural enemies, and they have evolved a magnificent array of armaments to keep their enemies out or to minimize the damage they cause once they get in. Some plants escape from disease. Others tolerate disease and grow well in spite of their sickness. Still others have evolved mechanisms to defend against disease, often with great success.

Volume V is designed by analogy to the defense of a medieval castle. The medieval castle had defenses at the perimeter—the outer walls and gates. In case the gates did not hold, it also had internal defenses. There were dynamic defenses triggered by the invader and even defenses triggered by previous invaders. What did the previous invaders teach the occupants of the castle? Another chapter describes how the metabolic resources of the plant are allocated to maintain and repair its defenses. The dynamic competition between the offensive weapons of the pathogen described in Volume IV and the defensive weapons of the host described in Volume V should read like a battle royal. Since the treatise begins with *management* in Volume I, it is fitting that it end with *defense* in Volume V, a major goal of integrated management being the enhancement of natural defenses against disease.

Throughout this treatise the objective remains the same—to understand disease so well that we can manage it effectively.

### References

Blackman, R. (1966). "Aphids." Ginn, London.

Grente, J., and Sauret, S. (1969). L'hypovirulence exclusive, phénomène original en pathologie végétal. *C. R. Acad. Sci., Ser. D* **268**, 2347–2350.

Heald, F. (1933). "Manual of Plant Diseases," 2nd Ed. McGraw-Hill, New York.

Heck, W. W., Krupa, S. V., and Linzon, S. N., eds. (1979). "Handbook of Methodology for the Assessment of Air Pollutant Effects on Vegetation," Tech. Inform. Rept. No. 3. Air Pollut. Control Assoc., Pittsburgh, Pennsylvania.

Horsfall, J. G., and Dimond, A. E., eds. (1960). "Plant Pathology: An Advanced Treatise," Vol. II. Academic Press, New York.

Lindberg, G. D. (1971). Disease-induced toxin production in *Helminthosporium oryzae. Phytopathology* **61**, 420–424.

Link, K. K. (1933). Etiological phytopathology. *Phytopathology* **23**, 843–862.

Peace, T. R. (1962). "Pathology of Trees and Shrubs." Oxford Univ. Press, London and New York.

Scott, K. J., and Maclean, D. J. (1969). Culturing of rust fungi. *Annu. Rev. Phytopathol.* **7**, 123–146.

Simmonds, J. H. (1938). "Plant Diseases and Their Control." David White, Brisbane, Australia.

Smith, K. M. (1933). "Recent Advances in Plant Viruses." McGraw-Hill (Blakiston), New York.

Sorauer, P. (1914). "Manual of Plant Diseases," Vol. I. Record Press, Wilkes-Barre, Pennsylvania.

Stevens, R. B. (1974). "Plant Disease." Ronald Press, New York.

Takebe, I. (1975). The use of protoplasts in plant virology. *Annu. Rev. Phytopathol.* **13**, 105–125.

Tansley, A. G. (1929). "The New Psychology and its Relation to Life." Dodd, Mead, New York.

van den Ende, G., and Linskens, H. F. (1974). Cutinolytic enzymes in relation to pathogenesis. *Annu. Rev. Phytopathol.* **12**, 247–258.

Wallace, H. R. (1978). The diagnosis of plant diseases of complex etiology. *Annu. Rev. Phytopathol.* **16**, 379–402.

Woltz, S. S. (1978). Nonparasitic plant pathogens. *Annu. Rev. Phytopathol.* **16**, 403–430.

Zadoks, J. C., and Koster, L. M. (1976). A historical survey of botanical epidemiology. A sketch of the development of ideas in ecological phytopathology. *Meded. Landbouwhogesch. Wageningen* **76**(12), 1–56.

*Chapter 2*

# The Evolution of Parasitic Fitness

R. R. NELSON

## I. INTRODUCTION

The evolution of parasitic fitness in the fungi is a unique and spectacular biological saga. The myriad events, both subtle and dramatic, that must have taken place to bring them to their current levels of fitness and precision are staggering to contemplate. While my treatment of their story may be less spectacular than the story itself, it is nonetheless a rare and challenging opportunity to speculate about the events that shaped their evolutionary destinies. The editors have permitted me to wax philosophical, an approach that I considered necessary in view of the paucity of knowledge on the evolution of the fungi. My task has been made considerably more difficult by not having been there when it all took place. The fact that no one else was there, however, has a calming effect on my concern for the matter.

The physiological aspects of improved parasitic and pathogenic fitness are so obscure that it is difficult even to treat the subject on a philosophical tone. Whether parasitism, for example, is the result of the acquisition or loss of some physiological ability is unknown. Obligate parasites

23

PLANT DISEASE, VOL. IV

may be so lacking in biochemical equipment that they cannot survive by themselves, or they may represent the ultimate in biochemical sophistication. It would be exciting to clarify the biochemical evolution that contributed to improved fitness, but since I cannot, for this chapter I must be content with the fact that fungal parasies did indeed evolve to their current levels of parasitic and pathogenic fitness.

My comments on the subject will reflect a bit of intuitive reasoning, a dash of logic, some reasonably documented scientific concepts, hopefully enough scientific facts to make the story somewhat plausible, and enough of my personal philosophy to retain and/or create an image.

Parasitic fitness is not exactly a common term in plant pathology if my travels have introduced me to a random sample of my colleagues. Accordingly, some time is devoted to discussing parasitic fitness, what it is, how it can be measured, the relationship, if any, between fitness and genetic complexity, and finally what parasitic fitness means to our discipline and its commitment to managing plant diseases.

My approach to the evolution of fungal fitness has a pronounced genetic flavor. The extraordinary genetic capacities of the fungi have brought them a long way since their origin. One of their most spectacular achievements has been their ability to coevolve with their natural hosts to a mutually acceptable coexistence by virtue of reaching genetic equilibrium. Similarly striking has been their evolution to an incredible level of tolerance toward man and his uncompromising attitudes toward the fungi. Both of these subjects are given thorough treatment.

Plant parasitic fungi will be here long after man has disappeared from this friendly planet. Their enormous capacities to generate improved parasitic fitness guarantee that dire and somber prediction. Yet, having an unbounded faith that there is still some future for us, I end this chapter with some thoughts on the future evolution of parasitic fitness and the role that man may play in its shaping. After all, I have contended for many years that scientists should spend as much or more time looking forward as they do backward.

A few suggested source references are given at the end of this chapter to provide a partial introduction to certain of the matters discussed herein.


## II. WHAT IS PARASITIC FITNESS?

Evolution, simply defined, is the change in the frequency of gene alleles over time. Populations that are the most fit will pass their genes to the greatest number of individuals, thus changing the frequency of alleles in their favor. Populations that are best adjusted to their particular ecological niches will, in time, become the predominant occupants of those niches. Evolution is a dynamic process and results in equally dynamic products.

Biological fitness can be viewed as the relative ability to persist over time. It is the ultimate testimony to success. Biological fitness can be measured in terms of reproductive success, in relative numbers of progeny or propagules, or by the frequency of certain individuals within a population or populations. Any trait that contributes to increased reproductive capacity and/or survival is a biological fitness character, although some are of more obvious import than others. Some biological fitness traits undergo evolutionary change more slowly than others. Dramatic changes in parasitic capacities, such as virulence, occur more rapidly than do dramatic changes in cell wall thickness or spore pigmentation, although all three characters are biological fitness traits in that they contribute to increased survival and reproductive success of parasitic fungi.

Biological fitness is under genetic domain and is without doubt the strongest driving force in evolution. Natural selection for the most fit in any given ecological niche is one of the few irrevocable biological laws.

A definition of parasitic fitness, then, is the relative ability of a parasitic genotype or population to persist successfully over time. Parasitic fitness is a measure of reproductive success and survival.

Parasitic fitness must be viewed in the context of short- versus long-term success. Reproductive success of a parasite in a short-term sense is, in essence, a case of epidemiological fitness. It is a measure of the epidemiological competence of a parasitic genotype relative to another parasitic genotype on a given host genotype under a given set of environmental conditions. Short-term parasitic fitness is a "one-on-one" relationship. The relative parasitic fitness of a genotype will change when it is compared against other parasitic genotypes or host genotypes or when it is assessed under different environmental conditions.

A pathogen genotype exhibiting the greatest short-term parasitic fitness would be that genotype which induces the greatest amount of disease. Thus, the maximum short-term parasitic fitness would be reflected in the apparent infection rate generated by a pathogen genotype on a given host genotype. Measuring parasitic fitness is discussed in a later section.

A pathogen genotype exhibiting maximal short-term parasitic fitness would not exhibit maximal long-term fitness or reproductive success. The ultimate in long-term parasitic fitness should approach or be equal to the ultimate in biological fitness since both reflect the ultimate in reproductive success over a long period of time. Highly fit parasitic genotypes tend to exert selection pressures on host populations for host genotypes able to cope with the parasite. When the host triumphs to the extent that the interaction between host and parasite is an incompatible one, the parasitic fitness of that particular parasitic genotype approaches zero.

Thus, it seems that nature tends to select for parasitic genotypes that are biologically fit to persist. Nature also tends to select for host genotypes that are biologically fit to persist. Highly resistant populations are not the most fit biologically since they tend to place a selection pressure on the parasite for genotypes able to cope with the host.

The ultimate in parasitic fitness, in the sense of long-term reproductive success, is achieved at some expense to short-term epidemiological success. Similarly, biological fitness and the ultimate in parasitic fitness are achieved at some expense to maximal parasitic fitness if the premise is accepted that ultimate fitness is not equatable to maximal fitness. Ultimate parasitic fitness is achieved through genetic equilibrium between parasite and host. How they reach equilibrium and the resultant biological coexistence is the subject of a later section of this chapter.

## A. Some Elements of Parasitic Fitness

A population of a plant parasite that increases in frequency and distribution relative to other populations of the pathogen must exhibit a number of fundamental assets. The population must, of course, possess the necessary gene(s) for pathogenicity which are effective against the current genotype(s) of its host. It also must possess certain other capacities, which can be characterized in a general way as "fitness attributes." Variation and fluctuation within and among populations of plant parasites essentially reflect changes in parasitic fitness. A brief discussion of some of these attributes is relevant to this discussion. Some of the attributes are concerned with the relative ability of a parasitic species to create novel and significant populations. Others are concerned primarily with the relative ability of new or existing populations to become established as dominant members of the species. Populations acquiring improved parasitic capacities or increased virulence would pose relatively minor concerns to man and his crop plants if the populations lacked the necessary fitness to increase in frequency relative to other populations. Fitness is the key to the efficiency of a population to cause epidemics.

## B. Infection Efficiency

The number of successful infections resulting from a given amount of inoculum is a direct measure of the infection efficiency of a population. Although characterized herein as a specific and identifiable fitness trait, infection efficiency is a composite of several more subtle traits including percentage spore germination, survival of germ tubes, success in finding the appropriate entrance site, and successful entrance. Infection efficiency, viewed as a single trait, is relatively easy to measure experimentally for many pathogenic species that induce disease or travel to other plants by means of propagules or spores.

Improved infection efficiency is a clear signal of improved parasitic fitness. As scientists give increasing attention in the future to monitoring improved fitness and the resulting population shifts of plant parasites, the measurement of improved infection efficiency should receive its just due.

## C. Latent Period

The latent period, considered herein as the time required from infection to the subsequent production of inoculum, is perhaps one of the most important fitness attributes of parasitic species with repeating disease cycles during a given season. It is a major key to epidemiological success. The latent period essentially dictates the number of disease cycles that will occur and is instrumental in determining the rate of disease increase on the host.

To a certain extent, the latent period is similar to the aggressiveness of a population, when aggressiveness is viewed as the time aspect of parasitism. Populations inducing similar or greater amounts of disease, but requiring different amounts of time to do so, differ in their latent periods by virtue of differences in their aggressiveness.

Measuring the latent period of fungal parasites that require dew for sporulation is not a simple task. Improved techniques should, however, permit its determination. The latent period for some parasites, such as the mildews and most of the rusts, can be measured with relative ease by visual observation. The erumpent uredial sorus signals the end of the latent period in the rusts as does the formation of conidia by the mildew fungi.

## D. Inoculum Production

Production of inoculum, or sporulation in most cases, is a significant fitness attribute and one of the prime indicators of reproductive success. Improved fitness in this trait is reflected in several ways. The ability to generate inoculum under an expanded set of climatic conditions is an obvious measure of fitness. Populations that can produce inoculum at lower temperatures, for example, should have superior success early in the growing season in temperate regions. Inoculum production with abbreviated dew periods should confer a universal advantage on such fortunate populations.

Greater inoculum production per given area of diseased tissue promotes increased survival and greater disease increase. Repeated sporulation from the pustule or lesion is an obvious asset to fitness. Each of these facets of inoculum production can be assessed readily for most pathogens in programs designed to monitor improved parasitic fitness.

## E. Virulence

Virulence is the relative ability of a parasite genotype to induce a given amount of disease on a given host genotype under a given set of environmental parameters. That virulence is a relative trait precludes the assessment of the virulence of a single isolate. Again, that virulence is relative mandates that no isolate can express zero virulence and, as such, the term "avirulence" is inappropriate and its use should be avoided.

When the relative virulence of parasitic genotypes is measured by the amount of disease generated over time, virulence represents a composite of fitness traits and, in a sense, approximates short-term parasitic fitness. Thus, virulence may be a more general character than previously thought and probably should not be viewed as a specific fitness trait.

## F. Nonparasitic Survival

Nonparasitic survival of nonobligate parasites encompasses all instances in which the parasite is not functioning as an active parasite of its living host.

The ability of a biotic plant pathogen to survive saprophytically in a given locality may be a significant epidemiological factor conditioning the frequency, prevalence, and severity of the disease that it induces. Dependency on introduced inoculum from other areas for primary infection and disease onset influences the regularity with which the disease occurs from year to year as well as the intensity, at least in instances where disease intensity is correlated with time of disease onset.

Different populations of a pathogen exhibit differential survival capacities in their nonparasitic phase. Nonparasitic survival is not correlated with abilities to induce disease in the parasitic phase. Fitness for one asset is independent of fitness for another. Differential survival among populations of plant parasites should be expected in light of their known capacities to vary for almost any identifiable trait.

Populations of nonobligate parasites that exhibit poor survival qualities in nonparasitic situations would have to possess superior fitness in their parasitic phase to remain dominant members of a species.

## III. MEASURING PARASITIC FITNESS

Parasitic fitness can be measured biologically. Data from some of these measurements provide estimates of microevolutionary changes. Surveys designed to monitor the frequency and distribution of races of pathogenic species might lead to the conclusion that the race that occurs with the greatest frequency is momentarily the most fit. Unfortunately, the validity of such an assumption is suspect. Collections (isolates) obtained from race surveys are placed on appropriate indicator or differential hosts, and race identifications are made on the basis of reaction type, pustule type, etc. The only real measure that such an approach provides is an estimate of the current frequency of virulence alleles. It provides no direct measure of changes in parasitic or pathogenic fitness. Consider, for example, race 1 of a pathogenic fungus under regular surveillance. This year's survey revealed that race 1 constituted 10% of the sampled population. Last year, the same race represented 7% of the sampled population. The host genotypes had remained the same. Is the 3% difference between the two years merely a sampling phenomenon, or is it a signal that some subpopulations of race 1 have acquired improved fitness? The current disposition of the survey collections, such as the routine race identification, must beg that important question. If subpopulations of race 1 are acquiring improved parasitic or pathogenic fitness, then race 1 qualifies as a potential threat to its host and to man. The epidemiologically pertinent fitness attributes of populations of race 1 from both surveys should be compared. The relative ability to induce lesions or pustules from given amounts of inoculum, the latent period, and sporulation potentials are among the more logical candidates for comparative examination. An improvement in one or more of these or other fitness traits is a solemn warning of things to come. Failure to monitor changes in parasitic fitness is one of the major reasons why plant breeders have never released a resistant variety *before* the damage was done. Old, presumably innocuous races that acquire suitable parasitic fitness are of as much concern to man as are the new races that fungi generate without predictability.

Two examples should illustrate the case for improved fitness. The 1970 epidemic of southern corn leaf blight was induced by race T of *Helminthosporium maydis*, a relatively unimportant fungus until corn hybrids were mass-produced in Texas male-sterile (Tms) cytoplasm.

The "new" race of *H. maydis* possessed an extraordinary appetite for such hybrids, and millions of acres of corn were "barbecued on the stalk." The threat abated the following year with the return to hybrids with normal cytoplasm which were resistant to all populations of the fungus. My own subsequent research with isolates of the fungus collected from as early as the mid-1950s, and preserved in the form of dried diseased leaf material to preclude genetic change, indicated that race T was not a new race, as presumed, but rather had had its origin many years before. Studies with the earlier isolates of race T revealed that they were epidemiologically unfit to induce epidemic increase of disease. They produced only relatively small amounts of inoculum, required longer dew periods for infection and sporulation, and produced relatively small amounts of the toxin needed for pathogenesis. The original populations of race T either lacked adequate parasitic fitness to decisively attack male-sterile hybrids, or they lost it through a prolonged relationship with normal hybrids whose resistance was achieved by a polygenic defense system acquired from early maize through its long coevolution with the parasite.

Other elements of our research have shown that the relative parasitic fitness of many isolates of race T can be manipulated in either direction by sustained residence on normal or male-sterile hybrids. As acreages of male-sterile hybrids increased in the southern United States, certain subpopulations of race T gradually moved toward improved parasitic fitness by virtue of their increasing parasitic relationship with a more desirable host. Selection for the improved subpopulations was inevitable. The results of their improved fitness are history. Coevolution of *H. maydis* and maize prior to the use of male-sterile hybrids had reached what appeared to be a genetic equilibrium. Man had inadvertently brought that equilibrium with him as he shifted from open-pollinated selections to the modern-day hybrids. They now appear to be at that point again.

Research designed to evaluate the impact of host resistance on relative parasitic fitness has been virtually ignored. Valuable information on the role of man and his crop plants in the microevolution of parasitic fitness should accrue from such endeavors. Within a short time after the 1970 corn blight epidemic and the return to resistant hybrids, race T virtually disappeared. The most common interpretation of this phenomenon has been that the "old" and common race of *H. maydis*, designated as race O, which was detected in extremely low frequencies during the epidemic year, had made a spectacular return to population superiority. This author has an alternative explanation of the apparent population shift from race T to race O. Residence of race T on resistant hybrids has a dramatic effect on the fitness of its populations, particularly on its toxin-producing apparatus, which is the driving force in its attack on Tms hybrids. The toxin machinery is turned off but not lost. Many populations of what now appear to be race O may indeed be slumbering populations of race T. If someone would care to test my hypothesis, let him plant 50,000 hectares of Tms hybrids somewhere in the Deep South. Will the authentic race O rise and be recognized?

There is another aspect of the corn blight story that is relevant to a discussion of parasitic fitness. *Helminthosporium maydis* is basically a warm weather pathogen. Selection within the species for populations better able to persist under warm weather climatic conditions undoubtedly occurred over time in the southern United States. The 1970 epidemic in the Corn Belt and in Pennsylvania had its onset from massive amounts of inoculum produced in the South and carried by warm, moist winds to the North. The populations of race T introduced in 1970 were preserved in diseased leaf material and compared in 1973 for parasitic fitness under a northern climatic regime with a population of race T collected in Pennsylvania that year. There was every reason to believe that the 1973 populations had resided in the state since 1970 since there was no evidence that the race was introduced from the South after 1970. The two populations were compared under field conditions for their relative abilities to induce disease. The 1973 population was substantially more fit to operate under northern climatic regimes. Selection within populations of race T apparently had occurred after 1970 for subpopulations better able to persist in a cooler environment. The rapidity of their adjustments is staggering. Ecological races are prime examples of populations that have adjusted their parasitic fitness to particular ecological niches.

The epidemic of stem rust of wheat in 1953, induced by race 15B of *Puccinia graminis tritici,* is another example illustrating parasitic fitness, although fewer facts are available. Race 15B was first detected in 1939 on or near barberry, its alternate host. It appeared periodically in low frequencies in various parts of the country in annual surveys in subsequent years. Race 15B never posed a serious threat until the epidemic year of 1953. The original population of race 15B had the necessary virulence to attack the currently grown cultivars; it apparently lacked parasitic fitness to become an epidemic parasite. Either the race had a second origin on barberry in a subsequent year, or subpopulations of the race acquired the necessary fitness to become an epidemic parasite. My story would be much more plausible if the latter were proved to be the case.

There is a sequel to the race 15B saga. Breeders almost immediately found resistance to the race in the form of two race-specific genes now known as *Sr6* and *Sr9d*. The problem appeared to be solved when, lo and behold, a new biotype, race 15B-3, was detected in Canada with the necessary virulence to overcome the new resistance. Race 15B-3 has seldom been detected in Canada since its original detection, even though potentially susceptible cultivars were widely grown. The race does occur annually in the southwest portions of the United States. A plausible conclusion may be that the populations of race 15B-3 lack the parasitic fitness to persist under Canadian conditions. Let us hope that it remains that way.

Relative parasitic fitness is not an attribute of a species or a race. The parasitic fitness of a race cannot be measured because all populations, subpopulations, or isolates of a race do not possess equal fitness. Fitness is the expression of the particular capacities of a given genotype under given conditions. There is no one genotype for any parasitic race. The increased frequency of a race occurs when subpopulations or isolates of that race acquire improved fitness followed by natural selection for the most fit.

The ability of a given parasitic genotype and a given host genotype to persist is the product of their interaction in a given environment. Thus, the relative parasitic fitness of two or more isolates of a race or species can be measured only in the context of a given host genotype under a given environment.

The bulk of relative short-term parasitic fitness, in the sense as given herein, can be estimated by comparing the apparent infection rate of two or more isolates on a given host under a given environment. Missing from the estimate, however, are their relative abilities to generate initial inoculum for disease onset. A more complete or precise estimate of relative parasitic fitness could be obtained by comparing final amounts of disease, although the mechanism of experimental design becomes considerably more difficult. It would not be possible to measure relative fitness in those instances when initial inoculum is introduced from an outside source.

Relative short-term parasitic fitness can be estimated also by competition or mixture studies. This approach pairs two identifiable isolates on the same plants or in a single plot of plants. Equal amounts of initial inoculum are applied at a point source, and the isolate recovered with the greatest frequency at the end of the test, usually the end of the season, is considered to have the better short-term parasitic fitness. A similar approach could be taken with nonobligate parasites to determine nonparasitic survival in the absence of their living host. Diseased plant material is allowed to "overwinter," and the relative frequency of the isolates is determined by sampling of the diseased tissue in the spring in the laboratory.

Relative comparisons among isolates of their individual fitness components, such as latent period or sporulation capacity, cannot be used as measurements of relative parasitic fitness. The value of such comparisons lies, conversely, in determining why isolates vary in relative parasitic fitness and in providing estimates of the relative importance of given components in the overall scheme of fitness. Measures of individual fitness components are of consequence in breeding for disease resistance. Assume, for discussion, that the latent period is the one component that exerts the most force on the apparent infection rate. Parasitic genotypes with the shortest latent period would be selected for, all other things being equal. Correspondingly, host genotypes that effect longer latent periods would be most successful in reducing the apparent infection rate. Selection for such genotypes in a breeding program would be a requisite for improved disease resistance. Tools to identify host genotype that affect certain parasitic fitness components are becoming available to those concerned with the kind of disease resistance that reduces the terminal amount of disease to levels that minimize or negate yield loss.

## IV. PARASITIC FITNESS AND GENETIC COMPLEXITY

Some populations of plant parasitic fungi are more fit than other populations. This holds true for unique genotypes of particular races or for a species in general. Such a biological principle stands without challenge. It is of such truth that no lengthy discourse on the matter is necessary.

Why some populations are less fit than others merits some discussion. It is tempting to defuse the topic by simply stating that populations are less fit because they are less fit—that they are less fit because they lack certain vital genes for fitness. Some developments in recent years have caused certain complications in opting for such a convenient dismissal of the subject.

There is a school of thought that believes that nature tends to place some constraints on parasitic populations munching happily toward unbridled virulence. Nature favors the more rational populations who conserve their energies for better endeavors; or so the argument goes. Teleologically, one could question the wisdom of carrying more weapons in the arsenal than are needed for any given confrontation or for any point in time, particularly if that were the only instance in which the weapon were used.

The same school of thought concedes that nature has no control, before the fact, over populations that accumulate more virulence genes than are necessary for survival and reproductive success. What nature can and does do, the school contends, is see to it that they are less fit to succeed than their less precocious relatives—thus the derivation of the axiom that populations with unnecessary genes for virulence are less fit to survive.

There are at least two theories regarding the ultimate fate of unnecessary genes of any kind. One theory presumes that genes that are not called on to perform in a vital function will drop out of the population or remain at best in very low frequencies. The other theory, which I have come to support, contends that the issue is not whether a gene is unnecessary, but whether its presence in a genome is a detriment to the population carrying it. My support for the latter theory could be mounted in several ways. It is my belief, backed by no evidence whatsoever, that most genes activate whatever they activate for more than just a single function. For example, a gene contributing to the production of an enzyme that is necessary for pathogenesis also may contribute to the production of the same or a different enzyme for another purpose.

Two personal observations bring me back to the real world in my support of this theory. *Helminthosporium maydis* is a parasite that induces southern leaf blight of corn. I have isolated *H. maydis* from small foliar leaf spots occurring on species in 55 genera of the Gramineae in my wanderings throughout the world. The fungus appears to pose no pathological threat to species other than corn, although it has the genetic ammunition to parasitize other species in a modest way. It takes but a single gene for pathogenicity for isolates of *H. maydis* to parasitize corn. Different genes are required to parasitize other gramineous species. One of our genetic studies, for example, revealed that isolates of *H. maydis* need 13 different genes to be pathogenic to nine particular gramineous species. Through a series of genetic crosses and selection, a large number of isolates were developed, each of which possessed from 1 to 13 of these pathogenicity genes. The question put to test was, Are isolates of *H. maydis* possessing all 13 genes carrying 12 useless genes with respect to their capacities to induce disease in corn? The answer was a clear and resounding *no*. Isolates carrying increasing numbers of the 13 genes incited increasing amounts of disease (i.e., were more virulent) on corn than did isolates carrying fewer genes. The 12 genes other than the gene needed to parasitize corn were contributing to the intensity of the attack and thus to fitness.

That research gave me an explanation for my second general observation. Until then I was slightly puzzled by the fact that I could frequently collect isolates of *Helminthosporium victoriae*, causal agent of Victoria blight of oats and champion of host-specific toxins, from other gramineous hosts from areas of the world where oat cultivars susceptible to the fungus had never been grown. There were collections of *H. victoriae* from *Agropyron* and *Hordeum* in Scotland, from *Agrostis* in Holland and from *Poa* in Switzerland near the tree line in the Swiss Alps. And there was *Helminthosporium carbonum*, causal agent of a leaf spot of corn and another toxin producer, collected from a tuft of *Poa* just south of the Arctic Ocean. The genes in those two species that functioned in the parasitism of oats and corn were contributing in some way to the general parasitic fitness of the isolates in areas far removed from their favored hosts.

The matter of whether populations carrying unnecessary genes for virulence are less fit to succeed has become something more than a mere exercise in intellectual trivia. It has become a controversial issue in plant pathology, particularly among those who are concerned with managing plant diseases by means of disease resistance. The idea that populations with few (if any) virulence genes are more fit to succeed on hosts with few (if any) resistance genes than are populations with unnecessary virulence genes (in context with the same hosts) has penetrated our thinking on the potential value of gene deployment and on the genetic composition of multilines. Much has been said and written, pro and con, about the hypothesis of stabilizing selection. Some, including myself, have given up all hope of testing the hypothesis, since the built-in escapes preclude any valid test. Nonetheless, the idea of stabilizing selection is used by some who reflect on population dynamics and parasitic fitness. It has been theorized that within all parasitic species there are populations of super races that carry virtually every virulence gene the species has ever manufactured. They lie there in nature as slumbering giants in minute frequency, because their gamut of virulence genes has rendered them less fit to succeed on host populations that are no match for them on a gene-for-gene basis. There are those who believe that a substantial portion of a multiline must consist of known susceptible tissue, so that stabilizing selection can protect the rest of the multiline from the ravages of a complex race. The proponents of deploying different resistance genes in different but contiguous regions believe that complex races will be stabilized against and would be unable to continue their travels to the next region. I do not agree with them, but I am in no position to judge who is right or wrong.

Let us suppose, for the moment, that there are genes that function to induce disease and that there are other genes that condition overall fitness. It is actually not an unreasonable tenet; in fact, there is some evidence for it. Let us further suppose that at least some of these different kinds of genes are inherited independently of one another. Again, this is not an unreasonable tenet; in fact, there is some evidence of it, too. If these suppositions are true, as I believe them to be, then one might expect to find populations with few genes for virulence that are fit and other such populations that are unfit to succeed. One might also expect to find populations with many virulence genes, some or many of which may be unnecessary by our criteria, that are fit and other such populations that are unfit to succeed. There is no reason why all of these combinations of virulence and fitness genes should not exist in natural populations at some point. If this is true, as I believe it is, then I am forced to reject the equation that simplicity equals fitness and return to the tenet that fit populations are fit because they are fit, pathogenic simplicity aside. We should give some thought to these notions. Perhaps one day we may want to monitor populations for their fitness and not just for their complexity.

To set the matter straight, and in conclusion, I am not yet prepared to reject stabilizing selection as a valid concept of classic population genetics. Nor am I yet prepared to reject the belief that populations at the extremes of a normal distribution are, at least momentarily, less fit than populations about the mean. I am fully prepared, however, to reject the assertion that so-called unnecessary virulence genes render populations less fit. No one has ever provided evidence that any given virulence gene is unnecessary to a genome, let alone detrimental. In fact, I like to think that I have accumulated a little evidence to support the opposite contention.

## V. WHENCE COMEST THOU?

There is no real consensus concerning the origin of the fungi. One school of thought supports the contention that they were derived from primitive ancestors that, as someone so nicely put it, "relinquished flagellar motility in exchange for mitosis." The possibility that the true fungi may have arisen from the slime molds has some credence considering certain similarities in the amoeboid movement of zoospores and swarm cells. The flagellated zoospores of the Phycomycetes resemble some Protozoa sufficiently to speculate that the true fungi may have a protozoan heritage or one in common with the Protozoa. Certain similarities between the fungi and the algae, particularly the green, blue–green, and red forms, prompt some evolutionists to trace fungal origin to the most primitive of the green plants.

While it would be a great honor for me to resolve the issue of fungal origin once and for all, this chapter is more concerned with the origin of parasitism and the subsequent evolution of parasitic fitness. Most certainly the fungi arose out of some symbiotic relationship between a host and one or more guests. They certainly did not become true fungi overnight or in one "fell swoop." All forms of symbiotic associations involving some degree of organic union are essentially types of parasitism. It is doubtful that even in cases of mutalistic symbiosis one would find a 50:50 gain–loss relationship.

One theory of symbiosis suggests that there were many cases of parasitic symbiosis during the evolution of primitive life form. The early bacteria engulfed by the primitive amoebas were true parasites until they decided to share the work load and became mitochondria. Ancient spirochetes attached themselves to the amoebas and parasitized the ancestoral protozoans until they ultimately resigned themselves to being flagella. The algae offered chlorophyll to the protozoans in exchange for a better life on their way to becoming higher plants.

I have been taught to believe that the primitive fungi were saprophytes and the parasitism evolved at a later time. No one ever provided any rationale, let alone evidence, why that should have been the case. While the true origin of the fungi remains obscure, I propose that the first fungus may well have been a parasite and an obligate one at that. This should most certainly be true if indeed the fungi had their origin out of some symbiotic association between two or more life forms.

Obligate parasitism is the ultimate example of mandatory symbiosis. The primitive fungi were unable to persist by themselves and thus created a sensitive relationship with their symbiotic hosts. It was a sensitive association because a guest can ill-afford the luxury of being obnoxious.

The fungi did not evolve to being dependent on their hosts; they were born that way. With the eventual development of independent survival, reproductive success, and improved parasitic fitness, their subsequent evolution took them in many directions. They became facultative and obligate again, short-cycled and long-cycled, broadminded and narrow-minded, homothallic and heterothallic, autoecious and hetereoecious, or whatever. The one common denominator was the evolution of parasitic fitness.

The primitive fungi were concerned initially with survival. Concern for improved parasitic fitness would come later. Many of the evolutionary events that contributed to overall fitness in a general way would contribute directly or indirectly to improved parasitic fitness at a later date.

The reproductive success of the primitive fungi probably was dependent on the reproductive success of their hosts. At least it is doubtful that the earliest fungi had the capacities to form spores or reproductive propagules per se. As evolution proceeded, the first spores or propagules, perhaps formed by the rounding of single cells or by budding, were produced within their host. Such a mechanism protected the propagules from desiccation and radiation but provided little, if any, opportunity for release and dissemination.

Reproductive propagules are one of the most important keys to reproductive success and thus to fitness. Early evolutionary events focused on just such success. True spore forms eventually evolved, and they ultimately became external to their hosts but still protected to some degree. Regulated spore production become a reality. Within the short span of 100 million years, more or less, the evolutionary spotlight shone brightly on several significant events. The open sorus, pedicillate spores, and pigmented cell walls paved the way for their safe departure and travel. Spore take off mechanisms evolved long before the Wright brothers.

Spores learned how to germinate but they did not always like what they saw. Regulated spore germination via dormancy protected them during the difficult times, and favorable environmental conditions awakened them for the good times. At last the fungi were on their own, and parasitic fitness became essential.

When the primitive fungi finally learned how to leave their ancestral and primitive symbiotic friends, they were rudely reminded that they did not know how to get back in. They had never learned how to infect because they never had had to. The evolution of parasitism as we think of it today was about to begin. Spectacular mutations and subtle events led to enzymes, toxins, appressoria, infection pegs, and haustoria. Some of the fungi took the easy way in and entered through wounds or natural plant openings.

Saprophytes eventually appeared on the scene when some of the fungi grew tired of the day-to-day confrontations. Facultative parasites could not make up their minds as to how they wanted to live and chose the better of both worlds. New obligate relationships developed when certain of the fungi remembered that the early days really weren't all that bad. They would never have to worry about Social Security, retirement, or Medicare as long as they behaved themselves and resisted the temptation to eat more than their hosts could afford to share.

The fungi are considered simple forms of life, and yet the multitude of events that have contributed to the evolution of parasitic and pathogenic fitness is staggering to contemplate. Citing just a few of them should make our point. How beautifully compensating is the repeating uredial stage of the hetereoecious rusts as they restore their populations to acceptable levels after so many aeciospores had failed to find their alternate host. The aerodynamics of spore takeoff and travel are engineering classics. Their ability to react and adapt rapidly to environmental stress and new ecological niches epitomizes the power of genetic change and selection. The harmony of the dikaryon is to biology what the Mona Lisa is to art. The evolution of tolerance to fungicides gives man food for thought if not for consumption. The greatest monument to a parasitic fungus is an epidemic; the number of monuments is testimony to parasitic and pathogenic fitness.

## VI. WHAT HAST THOU BEEN UP TO?

No single factor has influenced the evolution of parasitic fitness to a greater degree than the hosts that fungi parasitize. The overwhelming significance of that fact dictates that this chapter is concerned with the coevolution of plant parasitic fungi and their hosts. Their coevolution is a unique and spectacular biological phenomenon. The epic struggle for survival and temporary dominance is a Darwinian classic. The nature of their corelationship varied through time from one of utter disdain for their adversary to one of extreme sensitivity to one another. Microevolutionary changes were subtle and virtually unnoticed, or they were massive and abrupt. The ultimate nature of their cohabitation eventually would be orchestrated by man, while earlier they were left alone to resolve the issue between themselves. Both opponents have the unpredictable ability to stockpile genetic ammunition for the future without impairing their current fitness. The continued survival of some fungi was predicated on the continued survival of their hosts, and their evolutionary "rationale" was openly tempered by that cold reality. The fungi and their hosts are indeed the "odd couple" of the biological world. The most spectacular *coup de grace* ever accomplished by parasitic fungi throughout their long evolution was their eventual ability to coexist with their hosts in genetic equilibrium. It was fitness personified. In an atmosphere of relaxed selection pressure, both parasite and host had learned that coexistence was preferable to the alternating thrill of victory and the agonies of defeat. No longer would the improved fitness of one result in the diminished fitness of the other.

Philosophical or teleological as such a speculation may seem, it probably represents the scientific explanation embodied in the coexistence and coevolution of parasites and their hosts in natural epicenters and ecological niches. We can return to the beginning of their affair to learn how they ultimately arrived at coexistence and equilibrium.

Genetic events at the beginning probably were of a relatively simple nature. At least that assumption is easier for me to accept than the notion that massive genetic alterations were needed initially for pathogen virulence or host resistance. It seems reasonably safe to assume further that little genetic variation existed initially within subpopulations of the parasite, since the necessary events had not yet occurred to place substantial selection pressure on the populations. The original populations of the pathogen probably constituted a single, simple race. Coevolution had only begun. The genetic simplicity of the host with respect to resistance genes is another matter, and I prefer not to speculate in depth on the subject. Resistance genes arise by chance events. Genotypes of plant pathogens influence the ultimate frequency of the sustained presence of resistance genes, but not their origin. Nonetheless, the discussion is more readily developed by assuming that the resistance of the host prior to initial exposure to a plant pathogen was either absent or of a relatively simple nature. The validity of that assumption is not pertinent, however.

The initial resistance of a host to a "new" parasite probably was accomplished by a genetic change at a single locus. Plants with that resistance gene probably reacted to the parasite in a hypersensitive manner much like a modern-day cultivar whose resistance is conditioned by a single gene. Subsequently, members of the populations of the parasite evolved and strains pathogenic or virulent to the then-resistant host were selected. Such must have been the case, since both host and parasite still exist.

The process of coevolution proceeded, perhaps stepwise in a gene-for-gene manner or perhaps in another way. Host genotypes with fewer genes for resistance and parasite genotypes with fewer genes for virulence and/or for fitness probably were selected against, and either dropped out of their respective populations or remained in low frequency. It is difficult to imagine that all genotypes that existed throughout the process of coevolution remain a part of the current populations of host and parasite.

How long the process of coevolution has proceeded and what stage of coevolution currently exists are not important. By the time the host and parasite had reached a stage of equilibrate coexistence, each had accumulated a substantial number of resistance genes or virulence genes. The fact that they existed in relative harmony indicates that the last resistance gene thus far incorporated into the host genome did not confer a massive or hypersensitive response against the pathogen. Nor did the last virulence gene pose a serious threat of extinction to the host. There was no substantial selection pressure on either. Each found safety in numbers of genes for resistance or virulence. Genes that once functioned separately with only temporary success now functioned collectively with permanent success.

If the assumption is correct that the long process of coevolution resulted in the ultimate accumulation of many resistance and fitness genes in the host and parasite, genetic probabilities suggest that subsequent changes in the future course of their continuing coevolution will be largely subtle ones. At least it seems unlikely that either opponent will enjoy the massive, albeit temporary, superiority exhibited in the earlier stages of their coevolution. Genetic changes in the parasitic populations may result in subtle improvements in one or more of its fitness attributes. A change could occur, for example, that would reduce the latent period by a day or so. Contrastingly, a small genetic improvement in the resistance of the host could extend the latent period by a day or more. The epidemiological consequences of either change could be to increase or decrease the number of cycles of diseases within a given year.

The fine tuning of fungal parasites that exist in equilibrium with their hosts reflects a balanced compromise between parasite and biological fitness. It was a spectacular accomplishment. The countless subtle genetic adjustments in fitness were monitored through natural selection and witnessed through improved and long-term reproductive success. The placing of self-imposed constraints on parasitic potential was not a simple matter; it contradicted their basic nature and purpose in life.

Parasitic fitness represents a conglomerate of specific attributes, as previously noted. There is a compensating element in the process of reaching a balance between parasitic and biological fitness. A parasitic genotype that expressed a minimal infection efficiency compensated for its poor performance in that regard by maximizing its potential to produce inoculum or to perform better in some other way. An excessive use of genetic energy to perform in a substandard manner is compensated for by a maximal effort in some other trait that is accomplished with a minimal energy involvement. The judicious use of available energy is what parasitic fitness is all about. Parasitic fungi are a remarkable lot, given the hazards of their profession.

## VII. WHITHER GOEST THOU?

While natural populations of hosts learned to live with their parasites, man could not or would not. Therein lies our dilemma.

Early man began to alter the natural coevolution of hosts and parasites from the time he began to domesticate plants for food and fiber. To be sure, his influence was modest by current standards. He grew his plants close together to conserve space and to facilitate his harvest. Wild plants grow randomly. Disease increase is related directly to the distance between a diseased plant and a healthy one. Thus, domesticated plants sustained more disease. When disease was a factor, early man inadvertently practiced the art of selecting for disease resistance by collecting seed to plant the next crop from plants that looked the best or from all plants that survived. That selection practice placed a selection pressure on the parasite, and so on. However, as we have acknowledged, his influence was nominal.

The present level of interference began with the rediscovery of Mendel's principles of genetics. It began with the discovery that plant diseases could be controlled by hereditary means and that disease resistance could be combined with other characters desired in commercial varieties.

The coevolution of most cultivated plants and the fungal parasites has been sternly guided by plant breeders for the past 75 years. What has been accomplished in those years of breeding for disease resistance? Agricultural scientists, of which I am one, have been extraordinarily successful for the most part. They have been lucky and happy to have been so. They have lost some battles, but seldom the war. They have demonstrated superb foresight on many occasions and a total lack of it on others.

Genetic resistance to plant diseases usually has been achieved with the use of one or a very few genes. Disease resistance usually has been developed for protection against a specific race of a pathogen whose presence threatens the current cultivars. The term "race-specific" resistance is used commonly. The increased frequency of virulent races and the concurrent "loss" of resistance of a host cultivar usually is associated with cases in which cultivars have been developed with just such race-specific resistance. It is virtually a "feast or famine" type of resistance.

In all probability, single-gene, race-specific resistance can be overcome by a single-gene change in the parasite. Since virulence against a race-specific resistance gene is often recessive, mutation to recessiveness renders a race virulent against that particular gene. In short, breeding for race-specific resistance has made variation worthwhile to some plant pathogens.

Breeding for race-specific disease resistance and the resulting population shifts of plant pathogens has, among other things, taken the coevolution of host and parasite out of the genetic equilibrium that apparently existed prior to man's intervention. Spectacular shifts in fungal populations and the concurrent rise and fall of cultivars are phenomena akin to those occurrences that I have proposed in the early stages of host–parasite coevolution. In one sense, man is back to base zero on the evolutionary scale. He has often been unable to cope with the extraordinary abilities of parasitic fungi to acquire rapidly the necessary parasitic and pathogenic fitness.

Man's attempt to use genetic resistance has been and still is complicated by the fact that single-gene resistance does not always solve the problem. For example, efforts to control the blast disease of rice or late blight of potatoes with single genes appear to be hopeless. The blast and blight fungi are too variable to contain with a single gene.

Race-specific, single-gene resistance, when effective, usually is identified by a hypersensitive response of the host to the race. The parasite is restricted to minute infection sites and the struggle is terminated. It is a resistance against the successful establishment of the parasite. The potato late blight story again has relevance. In Mexico, where *Solanum* and the blight fungus coevolved and where *Solanum* rarely sustains more than token levels of blight, it has been stated that no tuber-bearing *Solanum* species was immune or hypersensitive to the pathogen when exposed under natural field conditions in that country. Their process of coevolution has taken them to coexistence and genetic equilibrium. Hypersensitivity is not a mark of such a relationship, as we have discussed.

Another kind of disease resistance is available to man. It is the kind of resistance that plants have achieved through their natural coevolution with their pathogens. It is primarily a resistance against epidemic increase of disease. The resistance restricts infection efficiency, latent period, and inoculum production. It is usually conditioned by many genes. It is often termed "non-race-specific" resistance, because presumably the resistance is effective to some degree against most or all races in retarding disease increase.

Man recently has become aware of the potential value of this "rate-reducing" resistance, as viewed in an epidemiological context. In fact, maize already has this resistance to several foliar pathogens; it was extracted from old open-pollinated varieties which also appeared to have evolved to equilibrium with their parasites.

We turn our thoughts once again to relative parasitic fitness—those fitness attributes needed by virulent races to become dominant members of their species. Increased fitness can turn innocuous races into demons, and man has often paid the price. The epidemics of southern corn leaf blight, coffee rust, potato late blight, stem rust of wheat, to mention only a few, testify to the consequences of increased parasitic fitness.

What does the future hold for parasitic fungi and their hosts? Their coevolution will continue to be guided by man as long as man attempts to manage or control plant diseases by disease resistance, and well he should. I contend that *man can effectively manage many major plant diseases by returning his plants and their parasites to genetic equilibrium.* The differences between managing plant diseases and controlling them are substantial. The underlying philosophy of managing plant disease states that plants can live with a certain amount of disease without sustaining economic loss. The threshold level of disease will vary among specific host–parasite systems, but the principle is the same. Modern man's concept of controlling plant disease has been based largely on a search for essentially zero disease through the use of race-specific resistance genes.

Regardless of the kinds of defense mechanisms that man incorporates into his crop cultivars, he must contend with the enormous capacities of plant pathogenic fungi to vary. Genetic changes effective against single-gene resistance can be expected to have massive impacts. Improved virulence or parasitic fitness effective against a multigenic defense system should be subtle if the concept of genetic probabilities is essentially a valid one.

Man typically has viewed the results of parasite evolution after the fact. That is, he has become concerned with the consequences of pathogen evolution only when presumably resistant cultivars are no longer effective against new and predominant fungal populations. One can argue that his rationale has been reasonably sound. How, for example, can one breed against the unknown?

Man can do nothing to prevent the origin of new fungal genotypes; he can only attempt to obviate the consequences of their presence. Future attempts to manage or control plant diseases must include a substantial research effort designed at early detection of fitness in fungal populations. New fungal genotypes initially will be present in low frequencies within the total population. Some time will be required, perhaps years, for new genotypes to become dominant and widespread members of the population. Early detection of significant genotypes with increasing virulence or fitness would provide added time to develop new sources of resistance or to invoke alternative control strategies.

It can be argued, without contest, that many annual race surveys are conducted to monitor the presence of new races and to determine the relative frequency of old ones. And yet the fact remains that no new cultivar has ever been developed and used with resistance to a race *before* that race has caused a problem. Surely, more intense concern about the presence of new and threatening races is necessary as man moves to the future. Techniques are available to identify new races. Better methods are needed to determine when "old" races are acquiring improved parasitic fitness. Recent evidence demonstrates that components of parasitic fitness are highly heritable traits. Natural selection could effectively improve parasitic fitness, and man should attempt to cope with that phenomenon before the fact.

We acknowledged earlier that man apparently cannot control certain diseases of his major food crops by single-gene, race-specific resistance. Those diseases will have to be managed with a multigenetic defense system—a system that I predict would return plants and parasites to genetic equilibrium. If a man is to be the director of their coevolution, let him see to it that they play in harmony.

A final note about disease resistance that is relevant to the story and to the future. The concepts developed about the terms "specific" versus "nonspecific" resistance and "major" versus "minor" genes for disease resistance have led us to conclude by either direct statement, inference, or deduction that certain host genes condition one kind of resistance (i.e., race-specific), while different genes condition the other kind of resistance (i.e., race-nonspecific). I have contended for some years, and support now emerges from other sources, that the same resistance genes govern both kinds of resistance. There are, in fact, no major genes and minor genes. My concept suggests that there are only genes for disease resistance. Race-specific resistance and race-nonspecific resistance are not indications of the action of different genes, but rather are expressions of different actions of the same genes in different genetic backgrounds. So we should back and gather up our worn and spent resistance genes of the past, take our currently successful ones, find some new ones if we can, and build ourselves a genetic pyramid. And for irony's sake, if not for mine, we should name our first cultivar Equilibrium.

## VIII. EPILOGUE

*The difficulty lies, not in the new ideas, but in escaping from the old
ones. . . .*

John Maynard Keynes, 1936

My story is told for what it is worth. My sincere thanks to editors
Horsfall and Cowling for permitting me to share my thoughts with
others in a manner that I preferred. The Editors should be commended
for asking the authors of these chapters to be speculative and challenging
in the treatment of their subjects. There are pitifully few opportunities
in our science to offer provocative and philosophical interpretations of
matters relevant to our science. The science of plant pathology is, by
and large, a conservative discipline and somewhat less than eager to
venture beyond the walls of our traditional framework. Our motto too
often seems to be that it is easier and safer to gather new data rather
than generate new ideas. Our journals are amply stocked with scientific
data, but only rarely can one find a provocative or philosophical interpretation of the data beyond the immediate scope of the vehicle that
generated the data. Editorial policies discourage or reject provocative
thinking. Stern editorial barriers are erected in front of new ideas or
concepts that challenge existing dogma. However, succeeding generations of scientists should be obliged to assess critically the current
knowledge to better guide its future scientific rationale. A conservative
science, when muffled with orthodoxy, will be slow to grow in stature;
it will only stagnate, and may just fade away. I suggest that those who
dictate editorial policies for our many journals everywhere heed these
words and be aware that they reflect the feelings of a growing number
of their colleagues, who are growing restless. If this treatise breathes
new life into our science before rigor mortis is final or before I become
a member of an endangered species, it will have met its obligation.

### Suggested References

Day, P. R., ed. (1976). The genetic basis of epidemics in agriculture. *Ann. N.Y.
    Acad. Sci.* **287**, 1–400.
Nelson, R. R., ed. (1973). "Breeding Plants for Disease Resistance: Concepts and
    Applications." Pennsylvania State Univ. Press, University Park, Pennsylvania.
Robinson, R. A. (1976). "Plant Pathosystems." Springer-Verlag, Berlin and New York.
van der Plank, J. E. (1968). "Disease Resistance in Plants." Academic Press, New
    York.
Wilson, E. O., and Bossert, W. H. (1971). "A Primer of Population Biology." Sinauer
    Associates, Inc., Stamford, Connecticut.

*Chapter 3*

# The Energetics of Parasitism, Pathogenism, and Resistance in Plant Disease

TADASHI ASAHI, MINEO KOJIMA, AND TSUNE KOSUGE

## I. INTRODUCTION

Healthy plants require large amounts of energy for their growth and development. When a hungry parasite or an irritating pathogen attacks a plant, energy is drained off by the invading organism to support its growth. Energy also is needed in infected plants to fight back against invading parasites or pathogens. Thus, energy and its use in infected plants is a singularly important aspect of growth, development, parasitism, pathogenism, and resistance.

With the single exception of parasitic green plants (see Chapter 16 by Knutson, this volume) no pathogen can trap energy from the sun.

PLANT DISEASE, VOL. IV

In general, parasites depend entirely on substances in host cells as sources of energy. For example, virus synthesis utilizes ATP formed by host cells through phosphorylation of ADP. Other pathogens oxidize sugars and probably also organic acids, lipids, and amino acids from host cells to generate energy for their own development.

Infected plants also are compelled to use additional energy for defense against infection. The energy for defense is supplied through catabolism of carbohydrates in most nonphotosynthetic tissues and, in some tissues, probably also through degradation of lipids and amino acids. In leaves and green shoots photosynthesis is the major source. In contrast to saprophytes, pathogenic organisms must compete for energy sources with host cells in order to grow and invade host cells. Perthophytes must kill host cells while enough energy sources remain available in them. Thus, in localized areas in host tissues there can be strong competition between pathogen and host for energy sources. This competition is keen in the case of parasitism, but even more so in the case of pathogenism.

The objective of this chapter is to compare and contrast the biochemical and physiological competition for energy during parasitism and pathogenesis. How do parasites and pathogens do battle with the host and thus obtain the energy and substances needed for their own growth and proliferation? Such an analysis will help us understand how pathogens induce disease (Volume IV) as well as how plants defend themselves (Volume V).

It should be noted that various types of pests may differ markedly in the types of substances that can be used as sources of energy. Chewing insects can directly ingest sizable pieces of host tissue. Sucking insects and nematodes can directly ingest various high molecular weight substances and even intracellular organelles. Bacteria, fungi, and parasitic seed plants cannot take up insoluble materials from host cells; they must either use already soluble cell constituents of low molecular weight (sugars, amino acids, organic acids, etc.), or break down insoluble or high molecular weight substrates (starch, polysaccharides, protein, lipids, lignin, membrane components, etc.) into soluble small degradation products, which can then be taken up by the cell.

Some fungi and parasitic seed plants have specialized absorptive organs (haustoria, endophytic system, etc), whereas others take up soluble (or solubilized) substrates directly through their cell walls. Viruses and viroids have a unique mode of attack; they take over the metabolic machinery of host cells and direct it for synthesis of macromolecules essential for their multiplication.

## II. TYPES AND LOCALIZATION OF ENERGY SOURCES IN HOST CELLS

In examining the competition for energy between pathogens and their hosts it will be useful first to review the sources of energy that are available in host cells and then to consider certain general aspects of energetics from the perspective of pathogens.

### A. Carbohydrates

Plant tissues contain a diverse group of polysaccharides, oligosaccharides, monosaccharides, and derivatives of monosaccharides. From the standpoint of energetics, the most important polysaccharide is starch, the most abundant and widespread reserve polysaccharide in the plant kingdom. It is easily used by either pathogens or hosts after hydrolysis or phospholysis. Nematodes and insects can ingest starch directly whereas utilization by bacteria and fungi occurs after hydrolysis or phospholysis by enzymes present in host cells or excreted by the pathogen. Starch is deposited in plant cells as granular particles surrounded by membranes. Therefore, the accessibility of starch to $\alpha$- or $\beta$-amylase and phosphorylase is regulated by these membranes. Other types of reserve polysaccharides, such as inulin and mannan, are found in some plant tissues. They also are available to pathogens as well as to host cells as sources of energy, but only after hydrolysis by host or pathogen enzymes.

The cell walls of plants contain many different kinds of polysaccharides, including cellulose, various hemicelluloses, and pectin. Since many pathogens secrete enzymes that degrade host cell walls (Bateman and Basham, 1976) and are nonspecific with respect to their carbohydrate requirement for growth (Scott, 1976), it is very likely that a part of the growth and proliferation of many pathogens is achieved by oxidation of these substances. However, cell wall degradation products probably contribute less to the energy supply of pathogens than degradation products of reserve polysaccharides. Pathogens evidently digest cell walls of host cells mainly as a means of gaining access to new host cells. Biehn and Dimond (1971) found that monosaccharides depress polygalacturonase production by pathogens. This suggests a small contribution of degradation products of cell walls to the energy needed for growth and proliferation of pathogens.

Mono- and oligosaccharides, most of which are dissolved in the cytoplasm of plant cells, are easily taken up and utilized as energy sources by pathogens. Foliar pathogens probably depend almost entirely on soluble sugars produced during photosynthesis. Sucrose is the most abundant photosynthetic product in plant cells; its concentration in leaf tissues frequently is about the same as that of monosaccharides, most of which are present in phosphorylated forms. In certain storage tissues, such as sugarcane internodes, sucrose content is extremely high. It should be noted that sucrose can be transported readily from one host cell to another.

A wide variety of $C$-, $S$-, and $O$-glycosides also occur in plant cells. However, it is very doubtful that these glycosides are important energy sources for pathogens.

## B. Organic Acids

Intermediates in the tricarboxylic acid cycle and other organic acids accumulate in most plant cells and are responsible for the acidity of plant cell extracts. These organic acids are usually concentrated in vacuoles, although there are a few exceptional cases in which they are present in both cytoplasm and vacuoles at comparable concentrations (Matile, 1976). Except for malonic acid, an inhibitor of the tricarboxylic acid cycle, the accumulated organic acids are easily utilized as energy sources by pathogens.

In most cases, accumulated organic acids are available to pathogens only after breakage of the vacuole membrane (tonoplast). This breakage can be achieved either by attacking pathogens or through degradative processes associated with natural senescence of host cells. Since vacuoles contain various acid hydrolases (Matile, 1976), breakdown of the tonoplast leads to degradation of many other constituents of host cells. This situation occurs in senescent or dying plant cells and thus would be favorable for both saprophytes and perthophytes. For further discussion of senescence phenomena in plants, see Chapter 18, Volume III.

## C. Lipids

Some plant tissues, such as fatty seeds, contain large amounts of reserve triacylglycerides. These substances are deposited in particles called spherosomes, which are surrounded by a single membrane. In such host cells, pathogens may use the reserve triacylglycerides as an energy source after hydrolysis by lipases present in host cells or excreted by pathogens. Hydrolysis of reserve triacylglycerides by host enzymes can take place without breakdown of the spherosome membrane, because spherosomes themselves, and especially their membranes, contain acid lipases (Ory *et al.*, 1968). Attack by pathogen lipases would require alterations in the structure of the spherosome membrane, however.

In most plant cells, triacylglycerides also are present as constituents of cell membranes. The most abundant and important lipids are phospholipids. Since many pathogens excrete phospholipases (Bateman and Basham, 1976), they could utilize the fatty acids in phospholipid molecules as energy sources. It is more likely, however, that phospholipases function mainly in breaking host cell membranes (including the plasma membrane and the membranes of reserve-containing cell organelles) rather than in producing substrates for energy capture.

### D. Amino Acids

The carbon skeletons of amino acids could be used as sources of energy. Undoubtedly, pathogens take up these compounds after hydrolysis of host protein by hydrolases present in host cells or excreted by the pathogens themselves. It is more likely, however, that parasites use amino acids as building blocks for synthesis of new protein rather than energy production.

### III. ENERGETICS FROM THE PATHOGEN PERSPECTIVE

With very few exceptions, plant pathogens must multiply in the tissues of their hosts before they can express their pathogenicity. Thus, acquisition of the sources of energy discussed in the proceeding paragraphs is essential to the development of pathogens.

From a whole-plant perspective, it would appear that plant pathogens in host tissues exist in an environment abundant in sources of carbon but occasionally deficient in certain other nutrients, such as nitrogen (Levi and Cowling, 1969). At the tissue and cellular level, however, chemical composition differs, and some pathogens are confined to specific types of tissues in which readily utilizable substrates, such as free glucose, occur in low concentration.

Pathogens differ both in capacity to take up carbon sources from the surrounding medium and in efficiency of extracting energy from the carbon compounds they absorb. Their level of metabolic efficiency undoubtedly reflects the nutrient environment in the plant tissues that they normally colonize. Thus, pathogens that colonize xylem elements would exist under carbon limitation, if they depend solely on xylem fluids for nutrients. Such pathogens must have more effective machinery for extracting nutrients from the surrounding medium and must be more efficient in extracting energy from substrates than pathogens that colonize parts of the plants rich in energy sources.

Nutrient environments in host tissue are conducive to both induction and repression of carbohydrate-metabolizing systems in the pathogen. Although the production of these systems has been studied in plant pathogens, most such studies have been done under *in vitro* conditions that have little resemblance to those encountered in plant tissue. It is not surprising that many plant pathogens "change" and lose virulence after prolonged culture in rich media.

The following discussion emphasizes those components of the pathogen's metabolic machinery that contribute to its overall energy-generating efficiency. It will be evident that information on this subject in plant–pathogen interaction is woefully deficient. Furthermore, much of what is said is speculative and extrapolated from knowledge of metabolism of microorganisms that are not plant pathogens.

## A. Mechanisms for Acquisition of Energy Sources

### 1. Periplasmic and Mural Enzymes

Microbial plant pathogens do not take up polymerized forms of carbon but degrade them to simple molecules for transport into the cell. Degradation of polymerized compounds is achieved by extracellular, usually hydrolytic enzymes that either are excreted into the surrounding medium or are embedded in the cell wall matrix or in the periplasmic space (Heppel, 1971; Heppel *et al.*, 1972). The latter type of enzyme seems particularly adapted to function in plant tissue. For example, phosphatases in the extramembrane matrix (periplasmic space) of gram-negative bacteria catalyze hydrolysis of phosphorylated sugars. Such compounds are negatively charged under physiological conditions, and are less easily transported than the free sugar. Enzymes such as invertase also occur in the periplasmic space and catalyze hydrolysis of sucrose to yield glucose and fructose, which are readily taken into the cell by monosaccharide transport systems.

In fungi, the counterparts of bacterial periplasmic enzymes are mural, or wall-bound, and intramural enzymes (Gander, 1974). For example, *Geotrichum candidum* contains wall-bound polygalacturonase (Barash and Klein, 1969). Germinating conidia of fungi such as *Neurospora* contain wall-bound invertase localized at the growing tip of the germ tube, where metabolic activity is most intense and demand for energy sources is greatest. Thus, the enzyme is concentrated in those regions of the hyphae that have the greatest need for carbon sources. If such pathogen enzymes functioned in host tissue, they would be in intimate contact with carbon sources in host tissue and would fulfill an important role of preparing substrates for transport into the cells of the pathogen. Being embedded in the wall matrix of the pathogen, the enzymes escape dilution and possible inactivation in plant tissue. These enzymes may serve other functions; for example, wall-bound hydrolases in infection pegs of germinating rust spores could function during penetration by catalyzing limited hydrolysis of host cuticle and epidermal cell walls much in the same way that bore holes are formed by wood-destroying fungi as they grow from one wood cell to another (Cowling, 1965).

### 2. Solute Transport Systems

The concept of transport-limited growth dictates that development of a pathogen in host tissue can be controlled by the rate of uptake of the most limiting nutrient (Bull and Trinci, 1977). While it has not been possible to determine if such control exists in plant–pathogen interactions, it seems certain that a pathogen in host tissue encounters a variety of compounds that can serve as sources of energy. Control over solute uptake, therefore, is necessary; otherwise unneeded compounds would be taken up by the cell and energy wasted. Such control is provided by two mechanisms: (1) the plasma membrane, which restricts movement in and out of the cell, and (2) a variety of transport systems, which provide selectivity for uptake of nutrients (Saier and Moczydlowski, 1978). Most transport systems have specificity for substrates, are subject to induction and repression much like enzymes, and have differing affinities for substrates. Indeed, affinity for a given substrate may reflect the relative abundance of that compound in the medium (Koch, 1971). For example, some microorganisms have a high-affinity transport system that functions when glucose concentration is low; the high-affinity system is repressed and replaced by a low-affinity transport system in a high-glucose medium. High-affinity transport systems would allow pathogens to scavenge carbon sources from the surrounding medium and therefore would be important to pathogens in plant tissue that are deficient in nutrients. Thus, induction and repression of transport systems, selective uptake of solutes, and affinity of transport systems for substrates not only help regulate the type of energy source utilized by a plant pathogen in host tissue but also determine the capacity of the pathogen to colonize that tissue.

### 3. Macromolecule Synthesis and Turnover

Many microorganisms produce storage forms of carbohydrates or lipids during conditions of carbon excess and utilize them for energy sources when exogenous carbon sources are not available (Dawes and

Senior, 1973). The availability of a stored carbon source make conidia of some fungi nutritionally independent during early stages of germination when host tissue has not yet been penetrated. Some fungi also produce intracellular protease systems that "recycle" unneeded proteins when external sources of nitrogen are limiting (Holzer et al., 1975). The free amino acids generated by this process are used for synthesis of new protein and energy needed to maintain the organism.

In the absence of growth, plant pathogens still require energy for maintenance of the living state. Endogenous energy-reserve polymers such as polysaccharides and lipids are used for maintenance energy when exogenous sources of carbon are inadequate. Pathogens lacking energy-reserve polymers degrade their own protein and RNA for maintenance energy (Dawes and Senior, 1973; Levi and Cowling, 1969). Such mechanisms permit pathogens to survive and grow under conditions of nutrient limitation.

When excess carbon and energy sources are available but other nutrients are limiting, both fungi and bacteria accumulate exopolysaccharides in culture. The nature of the exopolysaccharides generally reflects the nature of the carbon sources and other nutrients available to the organism. The accumulation of extracellular polysaccharides by the pathogen may have unfortunate consequences for the host plant. If accumulation occurs in vessels of the host, water movement will be curtailed and the plant may wilt. Thus, the extracellular polysaccharide may function as a toxin in the host (Strobel, 1977).

### 4. Induction and Repression of Enzymes

Induction and catabolite repression may play important roles in the nutrition of plant pathogens in vivo and may help account for phenomena associated with the so-called high-sugar and low-sugar diseases.

Glucose catabolite repression prevents synthesis of polysaccharide hydrolases. If the hydrolases are involved in the degradation of host tissue, symptoms will be less severe if such enzymes are repressed. Absence of free monosaccharides will allow derepression of production of hydrolases, and tissue destruction should be more severe under low-sugar conditions. However, any interpretation of the high-sugar-low-sugar phenomena must include consideration of the concentration of the individual carbon sources in those tissues supporting growth of the pathogen. For example, ripened fruit may contain high concentrations of carbohydrates such as fructose and glucose. However, xylem fluid, which supports growth of some pathogens, would contain little or no free sugar. In this case, the pathogen may break down vessel walls and surrounding cells for carbon sources. Although induction and catabolite repression are readily demonstrated in culture, evidence that they occur in pathogens in host tissue is largely circumstantial.

Inducible systems in plant pathogens may play dual roles in host–pathogen interactions. For example, invasion of a cyanogenic plant (birdsfoot trefoil) by *Stemphylium loti* causes breakdown of compartmentation and mixes cyanogenic glycosides with enzymes that catalyze their breakdown to yield HCN (Millar and Higgins, 1970). The pathogen is little affected by the poison, because it produces an inducible system that converts HCN to formate and ammonia (Fry and Evans, 1977). Thus, the pathogen not only detoxifies the metabolic poison but also converts it to forms utilizable as sources of carbon and nitrogen. The same fungus also develops cyanide-resistant respiration, which allows generation of energy via the terminal respiration system to continue in the presence of the metabolic poison (Fry and Millar, 1971). Some pathogens also produce inducible polysaccharide-degrading enzymes that not only permit ingress into host tissue but also permit utilization of polysaccharides for carbon and energy sources. Such versatility in plant pathogens reflects coevolutionary processes that have occurred during the long association between the microorganism and the plant (see also Chapter 2, this volume).

### 5. Energy-Yielding Pathways for Pathogens

Since glucose is the most abundant form of carbon in plants, it is not surprising that energy-yielding metabolic pathways in plant pathogens are similar and are designed to utilize glucose or one of its close relatives.

Energy generation in plant pathogens occurs by carbon flow through the Embden–Meyerhof (EM) pathway and the tricarboxylic acid (TCA) cycle, with generation of ATP through oxidative phosphorylation. Some pathogens such as *Agrobacterium tumefaciens* utilize glucose via the Entner-Doudoroff pathway in place of the EM pathway. Utilization of carbon sources other than the hexose sugars occurs by sequences that feed into the EM pathway or the TCA cycle. For example, galacturonate is converted by *A. tumefaciens* to $\alpha$-ketoglutarate, which can be utilized via the TCA cycle (Chang and Feingold, 1970). It is not known if this pathway is a major source of energy for the bacterium in plant tissue.

There is considerable information on regulation of energy-generating sequences in microorganisms not pathogenic to plants (Chapman and Atkinson, 1977; Lehninger, 1973). The principles of metabolic regulation developed from such studies apply to prokaryotic and eukaryotic microorganisms in general. Thus, it can be expected that enzyme induction and repression, feedback inhibition, compartmentation, and energy charge all contribute to metabolic regulation in pathogens in plant–pathogen interactions. It would be interesting to know how the various devices for regulating metabolism in plant pathogens will respond to the nutritional environment created during plant–pathogen interactions.

## B. By-Products of Energy-Generating Sequences

Both plants and pathogens are known to produce unusual compounds that are products of secondary metabolism of carbon sources (Luckner, 1972). These may be exchanged between plant and pathogen and influence development of the other. Among plant pathogens, fungi in particular accumulate high concentrations of unusual metabolites, many of which are toxic to plants. In culture, production of these compounds begins during the late logarithmic phase of growth and reaches a maximum during stationary phase. From the perspective of energetics of growth of the pathogen, once demands for growth are met, carbon sources previously used for energy generation are "shunted" to the production of secondary metabolites (Detroy *et al.*, 1971) (see also Chapter 16, Volume III).

Production of such compounds is rapid if sources of carbon are available and other necessary nutrients are limiting. While the compounds may be nothing more than shunt metabolites for the pathogen, they may have considerable effect on the host. For example, fumaric acid, a normal intermediate in the TCA cycle, is produced by *Rhizopus* species in high concentrations in hulls of developing almond fruits. It is then translocated in toxic concentrations into the fruit-bearing branchlets and causes twig dieback (Mirocha *et al.*, 1961). Also, oxalic acid produced by *Sclerotium rolfsii* functions synergistically with polygalacturonase in maceration of host tissue (Bateman and Beer, 1965).

Secondary metabolites produced by plants may be toxic to plant pathogens. Thus, the outcome of the battle between host plant and pathogen may be decided by secondary or shunt metabolites of the host or the pathogen.

## IV. COMPETITION BETWEEN PATHOGEN AND HOST FOR CARBOHYDRATES IN HOST CELLS

### A. Production of Soluble Sugars in Infected Plant Tissues

Carbohydrate catabolism in host tissues is accelerated during infection by most parasitic and pathogenic microorganisms. The invading organism may stimulate the host to mobilize its reserve polysaccharides and soluble sugars. These energy sources are then depleted by wasteful host respiration, by host defense reactions such as synthesis of phytoalexins, by synthesis of abnormal host metabolites such as polyphenols, and by further growth and development of the parasite or pathogen.

The first step in all these processes is production of soluble sugars such as hexoses and hexose phosphates. In leaves, the continuing capture of energy in photosynthesis complicates the analysis of changes in the rate of soluble sugar production and the liberation of energy in infected tissues. Thus, starch concentration in infected leaves decreases in some cases and increases in other cases (Akai *et al.*, 1958; Inman, 1962; MacDonald and Strobel, 1970; Mirocha and Zaki, 1966). In host cells invaded by pathogens, drastic effects on the photochemical machinery are observed. The capture of energy and assimilation of carbon dioxide both decrease drastically except in the case of certain virus diseases in which photophosphorylation usually increases (see also Chapter 4, Volume III). In nearby cells not yet invaded by the pathogen, starch is decomposed to soluble sugars, which are utilized as energy sources and as materials for growth by both invaded and noninvaded host cells and by the cells of the invading parasite or pathogen. If the photosynthetic capacity of the host cells near the cells invaded by the pathogen remains unchanged or increases, starch concentration may not decrease but rather may increase after infection. We believe that there is no general rule governing the mechanisms of increase in production of soluble sugars in response to infection.

In nonphotosynthetic plant tissues, starch degradation almost invariably increases, and soluble sugars accumulate in response either to infection or to mechanical injury. In infected cells, amylases produced by an invading pathogen could catalyze the conversion of starch to sugars. In mechanically wounded plant tissues, however, stimulation of starch degradation has been ascribed to an increase in either host phosphorylase or host amylase activity (Kahl, 1974; C. Kato and I. Uritani, unpublished data). This activation of host enzymes may also account for the accumulation of soluble sugars in host cells adjacent to those invaded by pathogens. Phosphorylase in plants does not have a regulatory function although it does in animals. Also, the activity of plant phosphorylase is only weakly controlled by some nucleotides (Turner and Turner, 1975). Both $\alpha$- and $\beta$-amylases are well known to have no regulatory properties. Thus, it is reasonable that activation of starch degradation in host cells adjacent to those invaded by pathogens is caused by increases in the concentrations of host phosphorylase and/ or amylase. Alternatively, release of host amylase from isolated compartments within host cells could lead to enhanced starch degradation. $\alpha$-Amylase has been shown to be localized in organelles (Matile, 1976) or membranes (Hirai and Asahi, 1973) in some plant cells. Starch degradation also would be accelerated if the membrane surrounding starch granules were changed so that, the granules were more accessible to phosphorylase or amylase. We suggest that such a change in the membrane of starch granules probably does occur in infected tissues, but so far we have no experimental evidence for this possibility.

Soluble sugars are actively utilized by host cells as sources of energy and precursors of phytoalexins and polyphenols after infection with pathogens. In sweet potato root tissue, soluble sugars do not accumulate but rather decrease in tissues adjacent to those invaded by *Ceratocystis fimbriata* (Kato and Uritani, 1976). In tissues slightly farther away from the sites of infection, soluble reducing sugars (but not sucrose) do accumulate. Thus, starch degradation takes place much more actively than the consumption of soluble sugars in these tissues. We believe that the nutritional environment is very favorable for an invading parasite or pathogen when soluble sugars accumulate in host cells adjacent to those invaded by a pathogen. Under these circumstances, pathogens are able to take up soluble sugars accumulated in host cells immediately after invasion into the cells. Soluble sugars also may be transferred from noninvaded cells to invaded ones. In this respect, it is very interesting that the bean rust pathogen excretes a diffusible amylase activator to accelerate starch degradation in host cells adjacent to those invaded by the pathogen (Schipper and Mirocha, 1969). Much more attention should be given to the comparative rates of production and consumption of soluble sugars in host cells in order to elucidate the availability of energy sources to pathogens.

## B. A Proposed Mechanism for Competition between Pathogen and Host for Soluble Sugars

Although experimental evidence is still sparse, we propose the following mechanism for the competition between pathogen and host for soluble sugars in infected host tissues.

An enhancement in carbohydrate catabolism takes place even in mechanically wounded plant tissues. Thus, in response to any "injury" (whether induced by mechanical means or by an invading parasite or pathogen) carbohydrate catabolism including soluble sugar production from reserve polysaccharides is stimulated in plant cells adjacent to injured cells. The catabolic reactions generate energy for wound-healing reactions as well as for production of abnormal metabolites such as polyphenols. The acceleration in soluble sugar production is brought about by increases in phosphorylase and/or amylase activity in the cells and probably also by changes in the membranes of starch granules that facilitate the attack of enzymes on starch.

In infected tissues, carbohydrate degradation may be accelerated to fuel defense reactions against infection, which need more energy than wound-healing reactions. In addition, much greater amounts of soluble sugars are required to form polyphenols, phytoalexins, and other abnormal metabolites. If such active defense reactions exist in cells, there will be a need for production of soluble sugars at much higher rates in infected tissues than in mechanically wounded tissues. Greatly enhanced starch degradation in infected tissues may be brought about by marked increases in host enzyme activity and by increased exposure of starch granules to such enzyme activity by changes in cellular compartmentation.

When starch degradation proceeds at a higher rate than consumption of soluble sugars by host cells, soluble sugars accumulate in host cells. This situation creates an environment favorable for the growth and proliferation of parasitic, pathogenic, and perthophytic organisms. All three types of attacking organisms would gain an energetic advantage over host cells if they were able to do one or more of the following: (1) secrete diffusible substances that activate enzymes involved in starch degradation, (2) increase the amounts of these enzymes in host cells, (3) induce a change in the membrane of starch granules, (4) inhibit the consumption of soluble sugars by host cells, or (5) stimulate the transfer of soluble sugars from noninfected to infected host cells.

In parasitic and pathogenic and even in perthophytic disease situations, the attacking organism takes advantage of any tendency toward accumulation of soluble sugars in host cells. On the other hand, host cells resist these effects by attempting to maintain a balance between the production and consumption of soluble sugars and thus to minimize the pool of soluble sugars available to the attacking organism.

## V. CHANGE IN ENERGY METABOLISM IN HOST CELLS IN RESPONSE TO INFECTION WITH PATHOGENS

When parasitism and pathogenism are discussed from the standpoint of energetics, changes in the energy metabolism of host cells in response to infection by pathogens must be considered carefully. Three major metabolic functions may be involved; each will be discussed in turn.

### A. Glycolysis

Both mechanical injury and infection of host cells stimulate glycolysis in plant storage tissues (Kahl, 1974). Phosphofructokinase and pyruvate kinase appear to play important roles in the regulation of glycolysis

(Turner and Turner, 1975). Changes in amounts of ATP, ADP, and AMP (or a change in the energy charge) strongly affect the activities of these two enzymes and thus greatly alter the rate of the glycolysis. Since ATP-utilizing metabolism in plant tissues is activated in response to either injury or infection, ATP concentration decreases with concomitant increases in amounts of ADP and AMP. These changes lead glycolysis to operate actively. Glycolysis then returns to normal when ATP-utilizing metabolism is no longer activated.

The rate of glycolysis is in part controlled by changes in the amounts of glycolytic enzymes. Increased glycolysis occurs by such a mechanism in both infected (Lunderstädt, 1964; Scott and Smillie, 1966) and mechanically wounded tissues (Kahl, 1974). Therefore, in such tissue the rate of glycolysis will not return to normal, unless the glycolytic enzymes are either inhibited or inactivated.

## B. Pentose Phosphate Pathway

Allen (1953) emphasized that infection appears to abolish the Pasteur effect. This abolition appears to be due to stimulation of the pentose phosphate pathway, which is much more greatly enhanced than the glycolytic pathway in infected tissues (Daly, 1976; Shaw, 1963; Uritani and Akazawa, 1959). The pentose phosphate pathway produces precursors for the formation of nucleic acids, polyphenols, and lignin. It also reduces $NADP^+$ to NADPH, which is utilized for the reduction of intermediates in anabolic pathways. The pentose phosphate pathway is greatly enhanced in infected tissues, and provides substrates for polyphenols and/or lignin production in these tissues. However, activation of the pentose phosphate pathway would lead to production of glyceradehyde 3-phosphate, which is then metabolized to pyruvate through the later stage of glycolysis. By appropriate shuttle systems NADPH would also be oxidized through the electron transport system in mitochondria for production of ATP from ADP. Consequently, another reason why the pentose phosphate pathway is accelerated in infected tissues may be that substrates for respiration are supplied to mitochondria at increased rates.

If NADPH is actively utilized either for reduction of metabolites or for oxidation by mitochondria, the pentose phosphate pathway would be enhanced. However, the activation of this pathway in infected tissues is ascribed to increases in amounts of the various enzymes involved in the pathway, especially glucose 6-phosphate dehydrogenase and 6-phosphogluconate dehydrogenase (Farkas and Lovrekovich, 1965; Malca and Zscheile, 1964; Scott et al., 1964). Accordingly, we suggest that the pentose phosphate pathway continues at a rapid rate even after defense reactions are completed unless the relevant enzymes are inactivated.

## C. Respiration

### 1. Respiratory Increase in Infected Tissues

An increase in rate of respiration takes place in almost all infected plant tissues. The only well-known exceptions are leaves infected by certain systemic viruses. Figure 1 shows the typical increase—in this case, in sweet potato root tissue infected by *Ceratocystis fimbriata.*

Allen (1953) proposed that the increase might be due to uncoupling of respiration induced by diffusible toxins produced by the invading pathogens. A toxin produced by *Helminthosporium maydis* race T has been reported to cause such uncoupling in host mitochondria (Gengen-

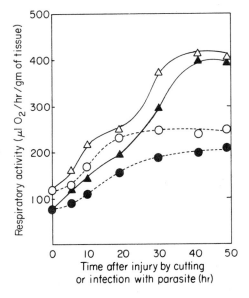

**Fig. 1.** Increase in rate of respiration of sweet potato root tissue in response to mechanical injury or infection with *Ceratocystis fimbriata*. Blocks (about 3 mm in thickness) or sweet potato (*Ipomoea batatas* L. var. Norin No. 1) root tissue were incubated at 30°C in moist chambers with or without inoculation with the fungus. The portion of tissue from 1 to 2 mm below the surface was assayed for respiratory activity in the presence or absence of $10^{-3}$ $M$ cyanide. Solid lines, infected tissue; dashed lines, wounded tissue; open symbols, without cyanide; closed symbols, with cyanide.

back *et al.*, 1973; Koeppe *et al.*, 1973; Miller and Koeppe, 1971), but other toxins have been reported to inhibit respiration in isolated mitochondria (Rudolph, 1976) and could indirectly cause accumulation of soluble sugars in host cells as well as weakening of host resistance. Many toxins are known to attack the plasma membrane or other membranes of host cells and not to induce uncoupling of mitochondria (Scheffer, 1976). Such toxins often increase membrane permeability, although it remains to be determined if the increased permeability brings about enhanced accumulation of soluble sugars in invaded cells.

In both parasitism and pathogenism, an increase in rate of tissue respiration is the most common response of host cells to infection. At least in host cells adjacent to those invaded by pathogens, accelerated oxidative phosphorylation by host cells accompanies increased respiration. Oxidases outside mitochondria do not appear to be involved. In mitochondria isolated from tissues adjacent to those invaded by pathogens, a tight coupling between the electron transport system and the phosphorylating system has been observed (Verleur and Uritani, 1965). Thus, in this case host cells should continue to produce ATP, because any diffusible toxins excreted by the pathogen do not uncouple mitochondria.

## 2. Wound Respiration

In spite of the common increase in rate of respiration in infected tissues, little information is available on the precise mechanisms involved. In analyzing these mechanisms it is helpful to examine the increase in rate of respiration of plant tissues wounded only by cutting. Later we will relate this to infection.

Oxidases outside of the mitochondria, such as ascorbic acid oxidase and polyphenol oxidase, do not participate in wound respiration. Succinate oxidase activity of mitochondrial fractions isolated from wounded plant storage tissues is much higher than that from healthy (or fresh) tissues on the basis of fresh tissue weight (Greksak *et al.*, 1972; Nakano and Asahi, 1970). Furthermore, mitochondria isolated directly from plant storage tissues are highly sensitive to cyanide, while the mitochondria from wounded tissues are resistant to cyanide inhibition (Hackett *et al.*, 1960; Nakano and Asahi, 1970). Increased tissue respiration after wounding is characterized by its high resistance to inhibitors of cytochrome *c* oxidase such as cyanide and carbon monoxide (Ikuma, 1972). Accordingly, wound respiration must be a reflection of an increase in the respiratory rate of mitochondria.

Increased supply of substrates to mitochondria could be responsible for the increase in mitochondrial respiration. More importantly, however, increases in amounts of ADP and/or AMP within cells would result in an enhancement of respiratory rate of the cells through the control of mitochondrial respiration by ADP.

It has been proposed that increased wound respiration could be caused by increases in amounts of ADP and/or AMP in cells through activation of ATP-utilizing metabolism. However, it has been shown that the number of mitochondrial particles increases in wounded storage tissues without accompanying cell division (Asahi et al., 1966; Asahi and Majima, 1969; Lee and Chasson, 1966). As shown in Fig. 1, slices of sweet potato root tissue increase respiratory activity after a lag phase lasting about 10 hours and reach a maximal rate after about 24 hours. The amounts of mitochondrial membrane protein and membrane phospholipid per slice also increase after a lag phase of about 10 hours and attain maxima after about 1 day (Nakamura and Asahi, 1976; Sakano and Asahi, 1971). This indicates that active biogenesis of mitochondria induced in response to injury by cutting also is an important factor in the control of wound respiration.

We conclude that wound respiration is due to both induction of mitochondrial biogenesis and activation of ATP-utilizing metabolism in response to cut injury. The fact that wound respiration is induced at least in part by biogenesis of mitochondria indicates that this phenomenon is really an active and not a passive or inflammatory reaction of plant tissues to mechanical injury. We believe this for the following reasons: (1) Mitochondria possess a very complicated structure. (2) Some proteins such as cytochrome c oxidase and ATPase are composed of subunits synthesized on both cytoplasmic and mitochonrial ribosomes. These proteins are coded by both nuclear and mitochondrial genes (Schatz and Mason, 1974). Furthermore, many other proteins such as the enzymes involved in the TCA cycle are synthesized on cytoplasmic ribosomes under control of nuclear genes and then transferred into mitochondria. (3) Newly synthesized protein is arranged in definite compartments within the mitochondria. Thus, the biogenesis of mitochondria requires cooperative effort between nuclear and mitochondrial genes and the very complicated control system available in whole cells.

As mentioned before, wound respiration is accompanied by development of cyanide-resistant respiration (see Fig. 1). The cyanide-resistant path of electron flow (alternate path) branches at a site near coenzyme Q from the normal and cyanide-sensitive cytochrome path. Evidently, in the alternate path, no phosphorylation occurs with succinate as substrate, and the phosphorylating efficiency decreases to one-third with NADH as substrate. The physiological function of the cyanide-resistant respiration remains speculative. For example, it is possible that the alternate path of electron flow may lead to formation of hydrogen peroxide with superoxide as a possible intermediate, which may be utilized for the production of ethylene (Solomos, 1977).

The inner membrane in newly formed mitochondria is quite different in various properties from the membrane preexisting in tissues before injury by cutting (Nakamura and Asahi, 1976; Sakano and Asahi, 1971). The newly formed membrane contains a very small amount of phospholipid relative to protein as compared with the preexisting one and possesses the alternate path of electron flow, which is resistant to cyanide inhibition. According to a generally accepted model for the structure of biological membranes, that is, the fluid mosaic model proposed by Singer and Nicolson (1972), the fluidity of membrane components is essential to membrane functions. Phospholipid in the mitochondrial inner membrane newly formed after injury by cutting is too sparse to maintain the fluidity of membrane components. We suggest that in such rigid membranes, the alternate path of electron flow operates because of the occurrence of irregular interactions among proteins and a decrease in the freedom of movement of molecules involved in the cytochrome system.

### 3. Pattern of Respiratory Increase in Resistant and Susceptible Tissues

Much effort has been made to find differences in the pattern of increase in rate of tissue respiration between susceptible and resistant disease reactions. No striking differences have been found so far, but the patterns appear to fall into the several classes indicated in Fig. 2 (Akutsu *et al.,* 1966; Millerd and Scott, 1956; Samborski and Shaw, 1956).

Plant tissues may be roughly divided into two classes with respect to the magnitude of increase in rate of tissue respiration in response to either injury by cutting or infection by pathogens. Certain plant tissues (class I in Fig. 2) respond strongly to mechanical injury, and the increased rate of tissue respiration remains unchanged for a long time. In such plant tissues, a further marked increase in rate of tissue respiration is induced in response to infection. An example of this type is shown in Fig. 1 with sweet potato root tissue infected with *Ceratocystis fimbriata.* The increased rate of tissue respiration has a tendency to return to the rate in wounded tissues in resistant combinations (Fig.

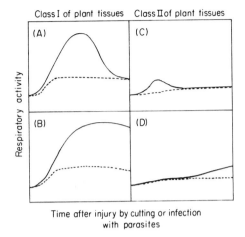

Class I of plant tissues    Class II of plant tissues

**Fig. 2.** Classification of host–pathogen combinations with respect to the pattern of increase in rate of tissue respiration in response to infection. Solid lines, respiratory rate of infected tissues; dashed lines, respiratory rate of wounded tissues; A and C, resistant combinations of hosts and pathogens; B and D, susceptible combinations of hosts and pathogens.

2A), while it remains essentially unchanged in susceptible combinations (Fig. 2B).

In the other class of plant tissues (class II in Fig. 2), tissue respiration is only slightly increased in response to injury by cutting. In susceptible combinations, such tissues recognize infections so leisurely that a slight increase in rate of tissue respiration above wound respiration becomes evident in later stages of disease development and takes place only gradually. In resistant combinations, however, tissue respiration has a tendency to be significantly and rapidly enhanced. In addition, the increased rate of tissue respiration seems to return quickly to the rate observed in wounded tissues.

Of course, there are plant tissues that produce nondistinctive patterns intermediate between the two types discussed above. In general, however, respiration of host tissues tends to be rapidly enhanced and then to be depressed after infection by pathogens in resistant reactions. In susceptible reactions, however, it tends to remain active for a considerable time. In susceptible combinations, host cells may continue active respiration because they lack the ability to quickly complete defense reactions or because the reactions necessary to restore energy metabolism to a normal state may be inhibited by the attacking pathogen. We speculate that biotropic pathogens may take advantage of the enhanced energy metabolism involved in defense reactions.

### 4. Mechanism of Respiratory Increase in Infected Tissues

In all infected plant tissues the rate of tissue respiration increases beyond that observed in mechanically wounded tissues. Is this further increase essentially different from that induced by wounding alone?

Defense reactions against pathogens usually require more energy than wound-healing reactions. Thus, ATP-utilizing metabolism operates more actively in infected tissues than in wounded tissues. In addition, carbohydrate catabolism is accelerated much more by infection than by mechanical injury. Thus, substrates for mitochondrial respiration are supplied to mitochondria at higher rates in infected than in wounded tissues. Consequently, it is probable that the increase in rate of respiration of infected tissues beyond wound respiration is due to accelerated utilization of ATP and/or enhanced supply of substrates to mitochondria.

Mitochondria isolated from sweet potato root tissue adjacent to those invaded by *Ceratocystis fimbriata* are more active in succinate oxidation than those from wounded tissue (Greksak *et al.*, 1972). Cytochrome *c* oxidase activity in infected tissue is also greater than that in wounded tissue (Asahi *et al.*, 1966; Greksak *et al.*, 1972). These results suggest that the respiratory capacity of mitochondria per cell in infected tissues exceeds that in wounded tissues. Accordingly, an increase in respiratory capacity of mitochondria also contributes to a marked enhancement of tissue respiration in infected tissues above wound respiration.

There is no difference in the number of mitochondrial particles per cell between wounded and infected sweet potato root tissues (Greksak *et al.*, 1972). Consequently, in infected tissues, we believe that an increase in the amount of inner membrane material per mitochondrion takes place in addition to the proliferation of mitochondrial particles induced in response to mechanical injury.

When slices of sweet potato root tissue are incubated in the presence of ethylene, the rate of tissue respiration increases beyond the rate of wound respiration (Imaseki *et al.*, 1968a). Furthermore, mitochondrial activity increases in a manner remarkably similar to that observed in infected tissue (T. Asahi, R. Majima, and I. Uritani, unpublished data). Since ethylene is produced actively in infected tissue and only slightly in wounded tissue (Imaseki *et al.*, 1968b,c), increases in mitochondrial activity and prevalence in infected tissue over those in wounded tissue may be caused by ethylene.

## 5. Properties of Mitochondria in Infected Tissues

As shown in Fig. 1, the increased respiration of sweet potato root tissue due to infection by *Ceratocystis fimbriata* is very resistant to cyanide inhibition. Judging from detailed studies on wound respiration described in Section V, C, 2, the mitochondrial inner membrane newly formed in infected tissues is probably so poor in phospholipid that the cyanide-resistant path of electron flow operates in this membrane. Just as proposed for wounded plant tissues, perhaps phospholipid synthesis in host cells is inadequate to supply enough phospholipid to the mitochondria. To test this idea we need detailed biochemical studies on changes in properties of the mitochondrial inner membrane in plant tissues during disease development.

## 6. Conclusions

Activation of the glycolytic and pentose phosphate pathways in infected tissues is brought about by increases in amounts of the necessary enzymes as well as by changes in concentrations of coenzymes. Since synthesis of these enzymes takes time, there is a lag period between establishment of the initial infection and activation of these pathways. It is natural that attention has been paid to the relationship between the length of the lag period and the resistance of host tissue to infection. We believe, however, that much more attention should be given to the time it takes for activated carbohydrate catabolism to return to a normal rate.

If a host plant is to recover from disease and maintain its life under healthy conditions, it would be expected that host metabolism would return from an accelerated to a normal rate after the infected cells have completed their defensive reactions against an invading parasite or pathogen. On the other hand, if metabolic disturbances induced by infection remain uncorrected abnormal metabolites will accumulate in host cells and various regulatory mechanisms needed for the "orderly discharge of energy" (see Chapter 3, Volume III, by Bateman) will be disrupted. The resulting disorderly metabolism of host cells will facilitate a continuing offensive by the attacking pathogen. Thus, in order to understand the mechanisms of offense, experimental studies should be made to determine how quickly host cells recover normal metabolic states after completing their defensive reactions to infection. Recovery from infection will require the destruction of disease-induced enzyme proteins and/or the repair of altered membranes. Uritani (1976) has demonstrated that some enzymes that increase in abundance after mechanical injury or infection are actively decomposed in later stages.

In general, the respiratory rate of infected plant tissues exceeds that of wounded tissues. This further increase in rate may be marked or slight depending on the tissue and the attacking organism involved. In resistant reactions, the rate of respiration quickly returns to that of wounded tissues; in susceptible reactions it remains high. The increased rate of respiration in infected tissues is brought about by an enhancement of the respiratory capacity of mitochondria, activation of ATP-utilizing metabolism, and enhanced supply of substrates to mitochondria. The enhancement in mitochondrial function of infected tissues appears to be due mainly to an increase in the amount of inner membrane per mitochondrion and to active proliferation of mitochondrial particles as observed in mechanically injured tissues.

We believe that the increase in rate of tissue respiration is an active and not a passive or inflammatory reaction of host cells against infection. The production of complicated structures such as mitochondria, which are needed for increased respiratory capacity of host tissue, argues for the increase in rate of respiration as a defensive reaction of host cells to parasitism or pathogenism.

Host cells may need so much energy for defense that they are forced to synthesize the mitochondrial inner membrane in order to accelerate oxidative phosphorylation. The acceleration of tissue respiration in infected tissues appears to be accompanied by accumulation of phospholipid-poor, (i.e., inferior) mitochondrial inner membrane. The membrane newly formed in infected tissues may contain a very small amount of phospholipid relative to protein, so that the alternate rather than the cytochrome path of electron flow seems to operate. It is very likely that hydrogen peroxide and/or superoxide are actively produced in mitochondria in infected tissues. When the inferior mitochondrial inner membrane is not repaired, even after defense reactions have been completed, the alternate path operates very actively, and so ethylene may be produced. If so, host cells would undergo senescence—a situation highly favorable for further attack by parasites or pathogens (see also Chapter 18, Volume III).

Host cells seem to meet with difficulties in phospholipid synthesis during their defense reactions. Not only the mitochondrial inner membrane being formed but also other membranes either being formed or undergoing turnover in infected tissues may be deficient in phospholipid and lacking in fluidity. Thus, we believe that the structure and function of various kinds of membranes in host cells may deteriorate during reactions of susceptible hosts to infection. If these membranes remain unrepaired, host cells will lose vitality, and thus parasites or pathogens could easily invade them. (See also Chapter 15, Volume III).

We suggest that in susceptible host-pathogen combinations, repair of deteriorated membranes may be delayed or does not occur. On the contrary, in resistant combinations, deteriorated membranes in host cells may be repaired rapidly, or serious deterioration in membrane structure does not occur in host cells.

It is possible that some parasites and pathogens take advantage of deterioration in the membrane system of host cells. To test this idea, it will be necessary to determine in resistant host tissues whether serious deterioration occurs in membranes, whether deteriorated membranes are quickly repaired, and whether deteriorated mitochondrial inner membrane is repaired when the rate of tissue respiration returns to that of wound respiration.

## VI. A SPECULATIVE LOOK TO THE FUTURE

As this chapter shows, there are still many unanswered questions concerning the energetics of parasitism and pathogenism. Many reports are available indicating how carbohydrate metabolism, including glucose catabolism and photosythesis, is changed in response to infection. In most cases, these changes have been discussed from the standpoint of their significance in the defense of host cells. Attention should also be directed to the impairment of host cell functions brought about by the changes—particularly those that result in further invasion of host cells. If pathogens stimulate the accumulation of abnormal metabolites in host cells, how important is it for pathogens to take advantage of the accumulation? Do the increased amounts of enzymes involved in carbohydrate metabolism in host cells return to normal after the enhanced carbohydrate metabolism is no longer needed? Is it more or less favorable for a pathogen if the increased amounts of these enzymes remain abnormally high?

Similarly, the increase in rate of respiration in infected tissues should be considered from two standpoints: How is the increase related to defense reactions of the host? Is this increase advantageous or disadvantageous for the host cells? Since detailed studies on the mechanism of respiratory increase in infected tissues are not yet available, we offer the following speculations on the basis of detailed biochemical studies on wound respiration: (1) Deterioration in membranes, including the mitochondrial inner membrane, in host cells is associated with the respiratory increase in infected tissues. (2) Host cells will meet with difficulties in maintaining their life if deterioration in membranes is not repaired after defense reactions have been completed. (3) Pathogens can easily further invade host cells when such deterioration remains unrepaired.

Much experimental effort has been directed to elucidation of metabolic alterations in plant tissues in response to infection and to finding the relationship between metabolic changes and defense reactions. Most of the metabolic alterations seem to be accompanied by changes in amounts of enzymes and/or in prevalence and properties of membranes. It may be unfavorable to host cells that such changes remain unrestored even when defense reactions have terminated. This necessary restoration will require induction of enzyme systems to decompose the unneeded enzyme proteins and/or repair the faulty membranes. We believe that experimentation should be initiated to determine whether deteriorations in metabolic states and membrane systems really occur in infected tissues and whether restoring systems are induced when defense reactions are terminating. Attention should also be given to the mechanism of restoration. We believe that such investigations will provide a profitable new approach in the elucidation of parasitism and pathogenism in plants.

## Acknowledgments

We express our thanks to Professor Ikuzo Uritani for his helpful discussions. We are also indebted to Professor E. B. Cowling and Dr. J. G. Horsfall for their editorial help in the preparation of this manuscript.

## References

Akai, S., Hiroyasu, T., and Noguchi, K. (1958). On the mechanism of starch accumulation in tissues surrounding lesions on rice leaves due to attack of *Cochliobolus miyabeanus*. I. Observation on starch accumulation in tissues surrounding spots. *Nippon Shokubutsu Byori Gakkaiho* 23, 111–116.

Akutsu, K., Imaseki, H., and Uritani, I. (1966). Induced respiration of sweet potato roots infected by the black rot fungus. *Shokubutsugaku Zasshi* 79, 644–653.

Allen, P. J. (1953). Toxins and tissue respiration. *Phytopathology* 43, 221–229.

Asahi, T., and Majima, R. (1969). Effect of antibiotics on biogenesis of mitochondria during aging of sliced sweet potato root tissue. *Plant Cell Physiol.* 10, 317–323.

Asahi, T., Honda, Y., and Uritani, I. (1966). Increase of mitochondrial fraction in sweet potato root tissue after wounding or infection with *Ceratocystis fimbriata*. *Plant Physiol.* 41, 1179–1184.

Barash, I., and Klein, L. (1969). The surface localization of polygalacturonase in spores of Geotrichum candidum. *Phytopathology* 59, 319–324.

Bateman, D. F., and Basham, H. G. (1976). Degradation of plant cell walls and membranes by microbial enzymes. *In* "Physiological Plant Pathology" (R. Heitefuss and P. H. Williams, eds.), Encyclopedia of Plant Physiology, Vol. 4, pp. 316–355. Springer-Verlag, Berlin and New York.

Bateman, D. F., and Beer, S. V. (1965). Simultaneous production and synergistic action of oxalic acid and polygalacturonase during pathogenesis by Sclerotium rolfsii. *Phytopatholgy* **55**, 204–211.

Biehn, W. L., and Dimond, A. E. (1971). Effect of galactose on polygalacturonase production and pathogenesis by Fusarium oxysporium f. sp. lycopersici. *Phytopathology* **61**, 242–243.

Bull, A. T., and Trinci, A. P. J. (1977). The physiology and metabolic control of fungal growth. *Adv. Microb. Physiol.* **15**, 1–84.

Chang, Y. F., and Feingold, D. S. (1970). D-glucaric acid and galactaric acid catabolism by Agrobacterium tumefaciens. *J. Bacteriol.* **102**, 85–96.

Chapman, A. G., and Atkinson, D. E. (1977). Adenic nucleotide concentrations and turnover rates. Their correlation with biological activity in bacteria and yeast. *Adv. Microb. Physiol.* **15**, 253–306.

Cowling, E. B. (1965). Microorganisms and microbial enzyme systems as selective tools in wood anatomy. In "Cellular Ultrastructure of Woody Plants" (W. A. Cote, ed.), pp. 341–368. Syracuse Univ. Press, Syracuse, New York.

Daly, J. M. (1976). Some aspects of host–pathogen interactions. In "Physiological Plant Pathology" (R. Heitefuss and P. H. Williams, eds.), Encyclopedia of Plant Physiology, Vol. 4, pp. 27–50. Springer-Verlag, Berlin and New York.

Dawes, E. F., and Senior, P. J. (1973). The role and regulation of energy reserve polymers in microorganisms. *Adv. Microb. Physiol.* **10**, 135–266.

Detroy, R. W., Lillehoj, E. B., and Ciegler, A. (1971). Aflatoxin and related compounds. In "Microbial Toxins" (A. Ciegler, S. Kadis, and S. J. Ajl, eds.), Vol. VI, pp. 3–178. Academic Press, New York.

Farkas, G. L., and Lovrekovich, L. (1965). Enzyme levels in tobacco leaf tissues affected by the wildfire toxin. *Phytopathology* **55**, 519–524.

Fry, W. E., and Evans, P. H. (1977). Association of formamide hydro-lyase with fungal pathogenicity to cyanogenic plants. *Phytopathology* **67**, 1001–1006.

Fry, W. E., and Millar, R. L. (1971). Development of cyanide tolerance in Stemphylium loti. *Phytopathology* **61**, 501–506.

Gander, J. E. (1974). Fungal cell wall glycoproteins and peptidopolysaccharides. *Annu. Rev. Biochem.* **28**, 103–119.

Gengenback, B. G., Miller, R. J., Koeppe, D. E., and Arntzen, C. J. (1973). The effect of toxin from *Helminthosporium maydis* (race T) on isolated corn mitochondria; swelling. *Can. J. Bot.* **51**, 2119–2125.

Greksak, M., Asahi, T., and Uritani, I. (1972). Increase in mitochondrial activity in diseased sweet potato root tissue. *Plant Cell Physiol.* **13**, 1117–1121.

Hackett, D. P., Haas, D. W., Griffiths, S. K., and Niederpruem, D. J. (1960). Studies on development of cyanide-resistant respiration in potato tuber slices. *Plant Physiol.* **35**, 8–19.

Heppel, L. A. (1971). The concept of periplasmic enzymes. In "Structure and Function of Biological Membranes" (L. I. Rothfield, ed.), pp. 223–247. Academic Press, New York.

Heppel, L. A., Rosen, B. P., Friedberg, I., Berger, E. A., and Weiner, J. H. (1972). Studies on binding proteins, periplasmic enzymes and active transport in *Escherichia coli.* In "The Molecular Basis of Biological Transport" (F. J. Woessner and F. Huising, eds.), Miami Winter Symposium, Vol. 3, pp. 133–156. Academic Press, New York.

Hirai, M., and Asahi, T. (1973). Membranes carrying acid hydrolases in pea seedling roots. *Plant Cell Physiol.* **14**, 1019–1029.

Holzer, H., Betz, H., and Ebner, E. (1975). Intracellular proteinases in microorganisms. *Curr. Top. Cell. Regul.* **9**, 103–156.

Ikuma, H. (1972). Electron transport in plant respiration. *Annu. Rev. Plant Physiol.* **23**, 419–436.

Imaseki, H., Asahi, T., and Uritani, I. (1968a). Investigations on the possible inducers of metabolic changes in injured plant tissues. *In* "Biochemical Regulation in Diseased Plants or Injury" (T. Hirai, ed.), pp. 189–201. Phytopathol. Soc. Jpn., Tokyo.

Imaseki, H., Teranishi, T., and Uritani, I. (1968b). Production of ethylene by sweet potato roots infected by black rot fungus. *Plant Cell Physiol.* **9**, 769–781.

Imaseki, H., Uritani, I., and Stahman, M. A. (1968c). Production of ethylene by injured sweet potato root tissue. *Plant Cell Physiol.* **9**, 757–768.

Inman, R. E. (1962). Relationships betwaen disease intensity and stage of disease development on carbohydrate levels of rust-affected bean leaves. *Phytopathology* **52**, 1207–1211.

Kahl, G. (1974). Metabolism in plant storage tissue slices. *Bot. Rev.* **40**, 263–314.

Kato, C., and Uritani, I. (1976). Changes in carbohydrate content of sweet potato in response to cutting and infection by black rot fungus. *Nippon Shokubutsu Byori Gakkaiho* **42**, 181–186.

Koch, A. L. (1971). The adaptive responses of Escherichia coli to a Feast and famine existence. *Adv. Micro. Physiol.* **6**, 147–217.

Koeppe, D. E., Malone, C. P., and Miller, R. J. (1973). An *in vivo* response of mitochondria in T cytoplasm corn to *Helminthosporium* maydis toxin. *Plant Physiol.* **51**, Suppl., 10.

Lee, S. G., and Chasson, R. M. (1966). Aging and mitochondrial development in potato tuber tissue. *Physiol. Plant.* **19**, 199–206.

Lehninger, A. L. (1973). "Bioenergetics," 2nd Ed. Benjamin, New York.

Levi, M. P., and Cowling, E. B. (1969). Role of nitrogen in wood deterioration. VI. Physiological adaptation of wood-destroying and other fungi to substrates deficient in nitrogen. *Phytopathology* **59**, 460–468.

Luckner, M. (1972). "Secondary Metabolism in Plants and Animals." Academic Press, New York.

Lunderstädt, J. (1964). Die Aktivität einiger Enzyme des Kohlenhydratstoffwechsels in Weizenkeimpflanzen nach Infektion mit *Puccinia graminis tritici. Phytopathol. Z.* **50**, 197–220.

MacDonald, P. W., and Strobel, G. A. (1970). Adenosine diphosphate-glucose pyrophosphorylase control of starch accumulation in rust-infected wheat leaves. *Plant Physiol.* **46**, 126–135.

Malca, I., and Zscheile, F. P., Jr. (1964). Dehydrogenase activity of the Helminthosporium leaf spot disease of maize. *Phytopathology* **54**, 1281–1282.

Matile, P. (1976). Vacuoles. *In* "Plant Biochemistry" (J. Bonner and J. E. Varner, eds.), 3rd Ed., pp. 189–224. Academic Press, New York.

Millar, R. L., and Higgins, V. J. (1970). Association of cyanide with infection of birdsfoot trefoil by Stemphylium loti. *Phytopathology* **60**, 104–110.

Miller, R. J., and Koeppe, D. E. (1971). Southern corn leaf blight: Susceptible and resistant mitochondria. *Science* **173**, 67–69.

Millerd, A., and Scott, K. (1956). Host–pathogen relations in powdery mildew of barley. II. Changes in respiratory pattern. *Aust. J. Biol. Sci.* **9**, 37–40.

Mirocha, C. J., and Zaki, A. I. (1966). Fluctuations in amount of starch in host plants invaded by rust and mildew fungi. *Phytopathology* **56**, 1220–1224.

Mirocha, C. J., DeVay, J. E., and Wilson, E. E. (1961). Role of fumaric acid in the hull rot disease of almond. *Phytopathology* **51**, 851–860.

Nakamura, K., and Asahi, T. (1976). Changes in properties of the inner mitochondrial membrane during mitochondrial biogenesis in aging sweet potato tissue slices in relation to the development of cyanide-insensitive respiration. *Arch. Biochem. Biophys.* **174**, 393–401.

Nakano, M., and Asahi, T. (1970). Biochemical studies on mitochondria formed during aging of sliced potato tuber tissue. *Plant Cell Physiol.* **11**, 499–502.

Ory, R. L., Yatsu, L. Y., and Kircher, H. W. (1968). Association of lipase activity with spherosomes of *Ricinus communis*. *Arch. Biochem. Biophys.* **123**, 255–264.

Rudolph, K. (1976). Non-specific toxins. *In* "Physiological Plant Pathology" (R. Heitefuss and P. H. Williams, eds.), Encyclopedia of Plant Physiology, Vol. 4, pp. 270–315. Springer-Verlag, Berlin and New York.

Sakano, K., and Asahi, T. (1971). Biochemical studies on biogenesis of mitochondria in wounded sweet potato root tissue. II. Active synthesis of membrane-bound protein of mitochondria. *Plant Cell Physiol.* **12**, 427–436.

Saier, M. H., Jr., and Moczydlowski, E. G. (1978). The regulation of carbohydrate transport in Escherichia coli and Salmonella tiphimurium. *In* "Bacterial Transport" (B. P. Rosen, ed.), pp. 103–125. Dekker, New York.

Samborski, D. J., and Shaw, M. (1956). The physiology of host–parasite relations. II. The effect of *Puccinia graminis tritici* Eriks. and Henn. on the respiration of resistant and susceptible species of wheat. *Can. J. Bot.* **34**, 601–619.

Schatz, G., and Mason, T. L. (1974). The biogenesis of mitochondrial proteins. *Annu. Rev. Biochem.* **43**, 51–87.

Scheffer, R. P. (1976). Host-specific toxins in relation to pathogenesis and disease resistance. *In* "Physiological Plant Pathology" (R. Heitefuss and P. H. Williams, eds.), Encyclopedia of Plant Physiology, Vol. 4, pp. 247–269. Springer-Verlag, Berlin and New York.

Schipper, A. L., Jr., and Mirocha, C. J. (1969). The mechanism of starch depletion of Phaseolus vulgaris infected with Uromyces phaseoli. *Phytopathology* **59**, 1722–1727.

Scott, K. J. (1976). Growth of biotropic parasites in axenic culture. *In* "Physiological Plant Pathology" (R. Heitefuss and P. H. Williams, eds.), Encyclopedia of Plant Physiology, Vol. 4, pp. 719–742. Springer-Verlag, Berlin and New York.

Scott, K. J., and Smillie, R. M. (1966). Metabolic regulation in diseased leaves. I. The respiratory rise in barley leaves infected with powdery mildew. *Plant Physiol.* **41**, 289–297.

Scott, K. J., Craigie, J. S., and Smillie, R. M. (1964). Pathway of respiration in plant tumors. *Plant Physiol.* **39**, 323–327.

Shaw, M. (1963). The physiology and host–parasite relations of the rust. *Annu. Rev. Phytopathol.* **1**, 259–294.

Singer, S. J., and Nicolson, G. L. (1972). The fluid mosaic model of the structure of cell membrane. *Science* **175**, 720–731.

Solomos, T. (1977). Cyanide-resistant respiration in higher plants. *Annu. Rev. Plant Physiol.* **28**, 279–297.

Strobel, G. A. (1977). Bacterial phytotoxins. *Annu. Rev. Microbiol.* **31**, 205–224.

Turner, J. F., and Turner, D. H. (1975). The regulation of carbohydrate metabolism. *Annu. Rev. Plant Physiol.* **26**, 159–186.

Uritani, I. (1976). Protein metabolism. *In* "Physiological Plant Pathology" (R.

Heitefuss and P. H. Williams, eds.), Encyclopedia of Plant Physiology, Vol. 4, pp. 509–525. Springer-Verlag, Berlin and New York.

Uritani, I., and Akazawa, T. (1959). Alteration of the respiratory pattern in infected plants. *In* "Plant Pathology" (J. G. Horsfall and A. E. Dimond, eds.), Vol. 1, pp. 349–390. Academic Press, New York.

Verleur, J. D., and Uritani, I. (1965). Respiratory activity of the mitochondrial fractions isolated from healthy potato tubers and tuber tissue incubated after cutting or infection with *Ceratocystis fimbriata*. *Plant Physiol.* **40**, 1008–1012.

*Chapter 4*

# Predisposition by the Environment

## JOHN COLHOUN

## I. DEVELOPMENT OF THE CONCEPT

It has long been realized that plant diseases are greatly influenced by the environment. For example, Theophrastus (370–286 B.C.) believed that the position and character of the land made a considerable difference with respect to attack of rusts on cereals. He believed that crops growing on elevated land exposed to the wind were not liable to rust, or were less so than those growing on lowlands not exposed to wind. It was accepted at that time that microorganisms associated with diseased plants arose spontaneously from the plants or possibly from the environment. This theory of spontaneous generation dominated the thoughts of those studying plant diseases for more than 2000 years. Indeed, when epidemics of potato blight occurred in Ireland in 1845 and 1846, cool weather was blamed for having upset the normal growth of the potato plants, so that internal breakdown occurred.

Some workers during the eighteenth century did not accept the almost universally held theory of spontaneous generation; one of these was Joseph Pitton de Tournefort (1705). He foresaw that fungi are autono-

PLANT DISEASE, VOL. IV

mous organisms rather than things that arise by spontaneous generation. He also recognized that they could induce diseases in plants but was unable to provide experimental proof that this was so. Almost another hundred years passed before Prevost (1807), in his classic study of wheat bunt, demonstrated experimentally that the cause of the disease was a fungus; he was able to show the indirect and secondary causal role of environment in the etiology of the disease.

Prior to Prevost, opinion as to the cause of wheat bunt varied. Tillet (1755) attributed wheat bunt to such factors as fog, wind, excessive soil moisture, improper plant nutrients, and soil texture. Tessier (1783) attributed the cause of cereal rust to mist blocking the transpiration of the plant. Unger (1833) believed that fungi associated with diseased plants were not parasites, that they had no independent existence, and that under certain environmental conditions every plant had the potentiality of producing them from the waste material of the plant. Fungi were still regarded as the product rather than the cause of plant diseases, which were still attributed to environmental factors. Acceptance of the idea that fungi can induce plant diseases, as shown by Prevost for wheat bunt, came only as a result of the contributions of Berkeley (1846, 1847) on potato blight and vine mildew and the proof by Speerschneider (1857) and de Bary (1861) that the fungus *Phytophthora infestans* is the actual cause of potato blight.

Once it was generally accepted that fungi were capable of causing diseases of plants, much attention and effort were directed to naming and describing those fungi associated with plant diseases. It should, however, be remembered that the effects of the environment were not entirely forgotten. Although Kühn (1858) accepted the nature of fungi as pathogens in plant diseases, he continued to believe that environmental conditions could cause some diseases. Hallier (1868) emphasized the influence of soil, and the predisposition of the plant. Sorauer (1874) urged consideration of all variables including (1) the direct cause of the disease, (2) the accessory conditions favoring the development of parasites, and (3) the disposition of the host organism. This clearly shows that Sorauer appreciated the role of environmental conditions in predisposing plants to disease. That there was considerable delay in general acceptance of these ideas can be clearly seen by inspection of the textbooks published immediately after 1874.

Following Sorauer, one of the first plant pathologists to appreciate the significance of the environment in relation to disease was Hartig (1882). He stated that a plant contracts disease only when subjected to definite preexisting conditions and that a predisposition or tendency to disease must exist. He believed that, to a certain extent, the origin of disease could be determined by the cooperation of two factors. One of these factors is the external cause of the disease, while the other has its inception in a peculiar condition of the organization of the plant. The latter may be present only at certain times, it may be peculiar to, and innate in, only certain plants, or it may have been acquired under the influence of definite external conditions. Among the factors that may predispose plants to disease, Hartig included the condition of vegetation in relation to season of the year and amount of water in plants as determined by the weather. He recognized that a plant suffering from one disease may be predisposed to another. Hartig, in the third edition of his book published in 1900, unreservedly acknowledged the existence of predisposition and differentiated between local, temporary, individual, acquired, and morbid predisposition (Sorauer, 1914). Not only did Hartig clearly appreciate the occurence of what we now call "disease escape" through the absence of inoculum at the time when a plant is susceptible to attacks by a particular fungus, but he understood fully that most of the phenomena that dispose plants to disease are not inherited.

It must be admitted that most of our current ideas of predisposition rest on the contributions of Marshall Ward. After his primary university education at Manchester and Cambridge, Ward studied coffee rust in Ceylon and as a result become a convinced predispositionist. In his return to England he continued his research at Manchester, the Royal Indian Engineering College, London, and at Cambridge. Ward (1890) pointed out that it is the reactions of the plant's constitution or internal disposition and its variations induced by changes in the physical environment that are so often and so persistently overlooked. He stated that unless these factors are taken into consideration an attempt to understand any disease is hopeless. He clearly recognized that a disease can become epidemic when the combined effects of the physical environment are unfavorable to the host but are not so unfavorable, or may even be favorable, to the parasite. Later, Ward (1901) propounded his ideas that when the roots of a plant are maintaining their greatest functional activity but the leaves, because of the lack of light, excess moisture, or too low air temperatures, are functionally depressed, abnormal conditions arise in the plant. These abnormalities predispose the plant to disease. He recognized that no disease can be efficiently caused by an organism alone, since its abilities as a parasite are affected by its nonliving environment as well as by the host plant.

In considering the question of predisposition and immunity in plants, Ward (1902) carried out a series of experiments involving brown rust of brome grasses. From the results obtained he concluded that the capacity for infection, or for resistance to infection, is independent of the anatomical structure of the leaf but depends on some other internal factor or factors in the plant. He stated that we have to consider the properties of the cell or the constitution of the plant in order to explain relative immunity or predisposition to disease.

From this examination of the development of the concept of predisposition to disease due to environmental factors, it is clear that there has been a change from the early theory that all plant diseases are caused by environmental conditions. We now appreciate that a few diseases, for example, those of a physiological nature, may be caused by environmental factors, while the rest are caused primarily by pathogens. In the case of the latter the environment may play a secondary but nonetheless very important role with respect to incidence of disease. Environmental factors may influence the occurrence and development of a disease by favoring (1) the production or longevity of inoculum, (2) spore germination, (3) the infection process, and (4) the later development of the pathogen within the plant. In addition, environmental factors may produce conditions within the plant that predispose it to disease, and such effects can be achieved prior to infection by the pathogen.

A survey of the literature on predisposition due to all factors including the environment has been provided by Yarwood (1959). In this chapter attention will be directed only to predisposition as it is influenced by the environment. The phenomena to be considered are therefore more restricted than those previously discussed by Colhoun (1973) when dealing with the effects of environmental factors on plant disease, but of necessity there must be some overlap.

## II. DEFINITION OF THE TERM "PREDISPOSITION BY THE ENVIRONMENT"

For the purpose of this discussion predisposition is defined as the effects of nongenetic conditions on susceptibility of plants to pathogens when these conditions act solely on the plant before infection occurs. Predisposition by the environment introduces a further limitation on the type of conditions with which we are concerned, and these can be regarded as narrowed down to include environmental conditions of a physical or biotic nature. This definition should be regarded as involving either an increase or a decrease in plant susceptibility.

An example of predisposition as it is defined here was elegantly provided by Bawden and Roberts (1948) when studying photosynthesis in relation to predisposition of plants to infection by certain viruses. They showed that when plants raised under standard conditions were exposed to shading or darkness for relatively short periods immediately before inoculation and were thereafter all maintained under uniform conditions, their susceptibility to viruses usually increased. These effects on the predisposition of plants to disease were compared with those that resulted when plants raised under similar conditions prior to inoculation were exposed to different conditions after inoculation. If this preinoculation treatment was prolonged, however, a decrease in susceptibility could result.

It is important to emphasize that, when one claims that plants exhibit predisposition to disease, it must be clear that the preinoculation treatment has affected the host plant and that the changes in susceptibility that result cannot be attributed to effects on the pathogen. When increased susceptibility occurs as a result of predisposition to disease, this should not be capable of being attributed, for example, to effects on plant vigor if this influences the microclimate around the plants and so could aid germination of the spores of a pathogen and their ability to penetrate the host tissue. It is also possible that changes in the microclimate could affect the ability of a pathogen to sporulate on the host and by increasing the level of inoculum could result in a greater intensity of attack by the pathogen. Effects such as these cannot be attributed to predisposition, which may perhaps be involved together with effects of the treatment on the pathogen. When such combined effects occur, it is difficult to determine the extent to which the change in susceptibility can be attributed to effects on the pathogen. When it is claimed that predisposition occurs, the experiments providing the data require careful design, so that the claim can be fully justified. Very few modern plant pathologists have deliberately designed their experiments to show that predisposition is involved when changes in susceptibility occur as a result of alterations in environmental factors.

It is surprising and even disturbing that the term "predisposition" is absent from the index of most modern textbooks of plant pathology and that it does not occur in the glossaries of phytopathological terms issued by the American Phytopathological Society (1940), the British Mycological Society (1950), and the Federation of British Plant Pathologists (1973) or in the "Dictionary of the Fungi" (Ainsworth, 1971).

## III. TERMS CONSIDERED COMPARABLE WITH PREDISPOSITION

Yarwood (1959) considered such terms as "disease proness," "disease potential," "acquired susceptibility," "physiological susceptibility," "preconditioning," "induced susceptibility," and "acquired disposition" as being comparable with "predisposition."

Disease proneness was considered in detail by Gäumann (1950). He recognized that if invasion of a plant by a fungus succeeds, we describe the plant as susceptible to infection; if, in addition, disease results we describe the plant as susceptible to disease. He regarded these two forms of susceptibility as forming the basis of disease proneness, which he considered to be comprised of the inherited, genotypic, constitutional conditions of the behavior of the host and the disposition of the host, including its developmental changes, with modifications due to environmental factors. It therefore follows that the term "disease proneness," as employed by Gäuman, embraces a much wider range of phenomena than does predisposition as defined here. It is interesting that Gäumann was convinced that completely susceptible plants become diseased regardless of environmental conditions, whereas completely resistant plants never become diseased, so that neither group shows any "disposition." He stated that environmental conditions effect modifications in types with intermediate disease proneness.

Grainger (1968) proposed that the major ecological suitability of a host for attack by a pathogen can be measured by assessing the relative amount of carbohydrate the pathogen has available for its own development. He considered that this could be done by determining the $C_p/R_s$ ratio within a plant, where $C_p$ is weight of total carbohydrate in the whole plant and $R_s$ is the residual (carbohydrate-free) dry weight of the shoot. Grainger's hypothesis rests on the statement that "if the host has an adequately high exploitable level of carbohydrate, a fungal or bacterial pathogen can attack but it cannot even start to do so if the $C_p/R_s$ ratio value is sufficiently low." This idea is well illustrated by considering the changing degree of proneness to disease of potato plants at different stages of development. Young plants and those approaching maturity were regarded as subject to severe attacks by *Phytophtora infestans* but, according to Grainger, there is a period in midgrowth when the fungus does not infect, or attacks only very slightly, even when inoculum is present and external conditions suitable for infection are provided. Grainger was convinced that such disease reactions must be caused by physiological factors operating within the host and changing during the period of growth of the individual plant. He therefore proposed that in relation to these changes such terms as "disease potential," "recepitivity," or "disease proneness" should be adopted. It is clear that these terms cannot be regarded as synonymous with "predisposition by environment," since the changes can occur as a result of plant development or aging and not necessarily directly as a result of environmental changes prior to inoculation.

Undoubtedly the terms "preconditioning," "induced susceptibility," and "acquired disposition" may be used in the same sense as "predisposition by environment," but when they are used it is desirable that the experimental evidence produced should be examined with care, before it is concluded that they are synonyms of "predisposition by environment" as defined here.

## IV. ENVIRONMENTAL FACTORS BRINGING ABOUT PREDISPOSITION

In this consideration of the effects of individual factors on predisposition to disease no attempt has been made to provide a comprehensive review of the literature on the subject. Publications have been selected to illustrate developments.

### A. Temperature

Temperature has been regarded as one of the most important variables affecting the development of biological systems, and its effects on the occurrence and development of many diseases have already been discussed by Colhoun (1973). Many of its influences on the incidence and development of diseases are due to effects on the pathogen or on the interaction between host and pathogen, but some have been clearly established to result from effects on the host.

Heat treatments prior to inoculation do not uniformly affect susceptibility of plants to infection by fungi. However, when viruses are concerned such pretreatment usually predisposes the plant to make it more susceptible.

A detailed study of the effects of exposing plants to relatively high temperatures prior to inoculation with various viruses was made by Kassanis (1952). He established that when various plants were kept at 36°C for periods as short as 6 hours prior to inoculation, the number of local lesions was increased as compared with the number in plants held at 18°–23°C. In most instances the greatest effects were achieved after exposure for 2 days, but the time taken for the maximal response was related to the age and general physiological state of the plants. In general, it was found that the older the plants, the longer the period of treatment required to obtain the greatest effects. Although Kassanis reported a reduction in the content of total carbohydrates in tobacco plants after being held at 36°C for 24 hours, he did not consider that the increased suceptibility associated with the treatment was due solely to these changes.

In agreement with the findings of Kassanis (1952), Yarwood (1956) reported that when bean leaves were immersed in hot water for a few seconds before inoculation, their susceptibility to various viruses increased. When leaves were immersed in water at 40°C for 100–600 seconds, the result was an increase in infection by tobacco mosaic virus, the greatest increase occurring at about 360 seconds. As the temperature was raised the period of immersion necessary to obtain maximal increases was reduced until, at 55°C, 1 second was sufficient. When leaves were immersed at 45°C for 60 seconds, the increased susceptibility to tobacco mosaic virus that resulted lasted for at least 3 days. When susceptibility to *Uromyces appendiculatus* ( = *U. phaseoli*), *Erysiphe polygoni*, and *Colletotrichum lindemuthianum* was considered, the response to heat treatment was similar to that for the viruses.

A fairly early example of predisposition by temperature treatment was demonstrated by Blodgett (1936). Plants of currant or gooseberry that had been growing at 19°C for 30 days were then placed in chambers at 12°, 19°, or 28°C for about 4 weeks. The leaves were detached, inoculated with *Pseudopeziza ribis*, and incubated at 20°C. Leaves from plants held at the highest temperature were more susceptible than those held at the lower temperatures. Observations were made with both old and young leaves. It was considered unlikely that the differences observed as a result of the preinoculation treatment could be explained by morphological changes. I believe that it is desirable whenever possible to avoid using detached leaves when testing reactions to fungi. Warren *et al.* (1971), for example, proved that, when potato leaves were inoculated with *Phythophthora infestans*, hypersensitivity or absence of symptoms was more pronounced in attached than in detached leaves. Also the lesions on detached leaves expanded more rapidly than those on attached leaves. It should also be emphasized that, in view of results obtained by Warren *et al.* (1971) and also those obtained by Hyde (1976, 1977) with wheat leaves inoculated with *Erysiphe graminis* f. sp. *tritici*, comparisons of leaf reaction should be made using leaves of similar physiological age. Not only does tissue aging affect the susceptibility of various hosts in different ways but, according to Hyde (1977), specific factors, including environmental conditions before, during, and after inoculation, may be important.

Tapke (1951) placed young barley plants outdoors in autumn for 6½ weeks and then brought them into the glasshouse and inoculated them with *Erysiphe graminis*. They were much more resistant than plants that had been grown throughout in the glasshouse at a higher temperature. The plants that had been growing out-of-doors showed considerable morphological differences from glasshouse-grown plants.

The plants grown out-of-doors lost their resistance if held in a glasshouse for 1 month before inoculation. The changes in the plant associated with the change in reaction were therefore reversible. Tapke noted that barley seedlings reacted differently from older plants to the preinoculation environment.

Recently, Khamees (1977) studied predisposition of ryegrass plants to attacks by *Fusarium nivale*. He found that plants exposed to a warm pretreatment (15°–18°C for 7–15 days for seedling and 12–36 days for tillering plants) suffered considerably more winter damage than those acclimatized to cold pretreatment in the open. He regarded plant size as less important than acclimatization to cold conditions in determining mortality.

From the above discussion it is clear that plants can be rendered more susceptible to infection by relatively high temperatures before inoculation. There are several records which do not conform with these findings. For example, Foster and Walker (1947) found that tomatoes could be rendered more susceptible to *Fusarium oxysporum* f. sp. *lycopersici* by exposure for 50 days to soil or air temperatures near the optimum for growth of the host as compared with higher or lower temperatures. Almost diametrically opposite results were reported by Kendrick and Walker (1948) regarding the same host for the bacterial disease induced by *Corynebacterium michiganense*. In this instance, the plants treated at a soil temperature of 24°C succumbed to the disease less rapidly than plants given a preinoculation treatment at higher or lower soil temperatures. It may be noted here that *C. michiganense* is a phloem invader as opposed to *Fusarium oxysporum*, which invades the xylem. I can find no data to explain this fascinating phenomenon. In experiments with rice and *Pyricularia oryzae*, Ramakrishnan (1966) showed that larger lesions developed on seedlings grown for 15 days at a night temperature of 20°C as compared with night temperatures of 25° or 30°C. After inoculation all the experimental plants were incubated for 24 hours at room temperature and then maintained under uniform conditions of 30°C during the day and 20°C during the night. In one rice variety it was observed that penetration of the leaves occurred only in plants raised at the lowest night temperature, and so it is possible that morphological changes had occurred, but these may well have also been accompanied by physiological changes.

An interesting series of experiments was made by Schulz and Bateman (1969) in which seeds of bean, pea, cucumber, and corn were pretreated for 24 hours at 5°C or 25°C while being allowed to absorb moisture when supported by moist filter paper, glass microbeads, or steamed or nonsteamed soil. After this pretreatment the seeds were planted and maintained in a glasshouse at 22°–30°C, or in controlled environmental chambers at 24°C, in the presence or absence of *Rhizoctonia solani*. When examined 16 days later, seedlings arising from seeds pretreated at 5°C for 24 hours exhibited increased susceptibility to *R. solani* over those pretreated at 25°C. Schulz and Bateman pointed out that although pea and corn seeds are liable to be injured by the low temperature during the initial 24 hours of germination, they had no evidence that bean or cucumber seeds are subject to such damage. When low-temperature injury occurs, the increased susceptibility to *R. solani* may be related to increased exudations from the seeds of organic constituents that favor the fungus, and so predisposition is not really involved.

It may be mentioned that not all plants show predisposition to disease as a result of temperature treatments. For example, Hill and Green (1968) found no significant response from three different preinoculation temperature treatments of tobacco plants prior to inoculation with *Peronspora tabacina*.

Numerous instances have been recorded in which susceptibility to disease has been altered by the temperature to which storage organs like fruits and potato tubers have been subjected before inoculation. Vasudeva (1930) established that by heating apples at 30°C for 17 days they became susceptible to attack by *Botrytis allii*, whereas similar fruits held at the normal laboratory temperature prior to inoculation remained immune to attack. It was demonstrated that the water lost from the fruit during the heat treatment was not responsible for the resulting susceptibility. The difference in susceptibility between the treated and untreated apples could not be explained in terms of alterations in acidity, total sugar content, or nitrogen content. It has since been suggested that the increased susceptibility caused by heat treatment may be associated with changes in the phenolic constituents resulting from suppression of an enzyme system (Epton, personal communication). Similarly, Chona (1932) showed that maintaining apples at 35°C for 1 week prior to inoculation rendered them liable to slight attack by *Fusarium solani* var. *coeruleum* (= *F. caeruleum*), to which they are not susceptible under normal conditions. Moreover, apples receiving this heat treatment were more susceptible when inoculated with *F. lateritium* (= *F. fructigenum*) than were those kept at 15°C prior to inoculation.

Gregg (1952) maintained potato tubers at 35°C and also at laboratory temperatures prior to inoculation with *Erwinia aroideae* (= *Bacterium aroideae*) or *Erwinia carotovora* (= *Bacterium carotovorum*). Pretreatment at 35°C brought about an increase in susceptibility to *E. aroideae* after 5–21 days and to *E. carotovora* after 14–21 days. When the heated tubers were held at laboratory temperatures for 6 days before inoculation, the increased susceptibility no longer persisted. Gregg suggested that the increased susceptibility brought about by the heat treatment could result from changes in the composition of the middle lamella, which rendered it more readily hydrolyzed, or by lessening the effect of some factor which slowed down enzyme action. She was unable, however, to detect by microscopic examination any changes in the middle lamella as a result of the heat treatment. In a similar experiment made with three cultivars of potato, Mishra (1953) found that preheating for 7 days at 35°C did not alter the susceptibility of the tubers to *Pythium debaryanum*. Boyd (1952) recorded that the higher the temperature of storage of potatoes (up to 14°C, range 13°– 26°C) prior to inoculation with *Fusarium solani* var. *coeruleum* the higher the susceptibility of the tubers when the period of treatment was 6 weeks or longer. This result was obtained in spite of the time lag in the initiation of wound periderm caused by previous storage at a temperature of 2°C (Steward, 1943). Boyd stated that the differences in susceptibility he observed between tubers preheated at 4° and 14°C could be explained by changes in the composition of the tuber rather than in terms of water loss occurring at the higher temperature.

## B. Moisture

Weather conditions may have substantial effects on both soil moisture and atmospheric humidity, but they may also predispose plants to disease in other ways. For example, it has been noted that during storms tobacco leaves often develop water-soaked areas due to flooding of the intercellular spaces. If infection by *Pseudomonas tabaci* ( = *Bacterium tabaccum*) follows, very large lesions can develop within 48 hours, although without water soaking about a week is required for the development of large halo lesions (Clayton, 1936). Later work by Johnson (1947) showed that water congestion of tobacco seedlings, resulting from placing glasshouse-grown seedlings under outdoor conditions for a few days prior to inoculation, greatly increased their susceptibility to *Pseudomonas tabaci* ( = *Phytomonas tabaci*). Predisposition to infection resulting from outdoor exposure could be significantly reduced by transference to glasshouse conditions for 2 days or longer prior to inoculation. Similar results have been obtained with other hosts and a range of bacterial and fungal pathogens. Plants normally immune to certain bacteria may become susceptible when the tissues are water soaked (Johnson, 1937).

It has been suggested (Colhoun, 1962) that high atmospheric humidity during the growing season induces greater resistance of lenticels on apple fruits to penetration by *Penicillium expansum* during their later storage life.

Exposure of tomato plants to low soil moisture for 30 days prior to inoculation with *Fusarium oxysporum* f. sp. *lycopersici* made plants more susceptible (Foster and Walker, 1947). Kendrick and Walker (1948) found that tomato plants grown at the optimum soil moisture prior to inoculation with *Corynebacterium michiganense* succumbed more rapidly than when pretreated at higher or lower moisture levels in the same type of soil. The predisposing effect of soil moisture in relation to these two pathogens of tomato is therefore similar to that reported for temperature and, unfortunately, no explanation is available.

Some workers have implicated water stress and wilting in predisposition. Thus, Ghaffar and Erwin (1969) concluded that water stress is more important than temperature in predisposing cotton plants to severe attacks of *Macrophomina phaseolina* (= *Macrophomina phaseoli*). Similarly, low soil moisture prior to inoculation of tobacco plants with *Peronospora tabacina* predisposed them toward greater susceptibility (Rotem *et al.*, 1968), and tomato seedlings showed increased susceptibility to *Alternaria solani* as the period of wilting of seedlings increased from 0–6 hours to 12–24 hours (Moore and Thomas, 1943).

The effect of wilting or turgidity of plant tissues on the susceptibility of cereal plants to *Erysiphe graminis* is not clear. Tapke (1951), in his studies of the influence of the preinoculation environment, decided that severe mildew occurred on turgid plants kept continually in that state, although it had previously been reported (Rivera, 1924) that green organs at the height of their maximal turgidity proved to be resistant.

Using a number of mechanically transmitted viruses, Tinsley (1953) showed that increasing the amount of water supplied to tobacco plants during the 4 weeks prior to inoculation increased the number of local lesions per leaf and per unit area of leaf. It was suggested that this effect could be explained in terms of differences in cell structure. The thicker cuticle and the greater compactness of cells in leaves of plants grown in dry soil before inoculation would probably increase resistance to injury when rubbed, and virus particles might penetrate less readily.

## C. Light

Light, darkness, or shading may be predisposing to attacks by fungi or viruses. Samuel and Bald (1933) established that shading tobacco plants for 3 days made them more susceptible than unshaded plants to the spotted wilt virus. Furthermore, if half a leaf was covered with black paper for 4 days, the shaded part became more susceptible. As already stated, Bawden and Roberts (1947, 1948) proved very clearly that a number of hosts became more susceptible to viruses if kept in the dark for short periods before inoculation. The increased susceptibility persisted for some time after plants were returned to normal glasshouse conditions. The increase in susceptibility resulting from dark treatment appeared to be roughly proportional to the amount by which photosynthesis was reduced. This suggested that some of the labile products of photosynthesis were able to combine with the virus particles and render them noninfective, but no positive evidence was produced in support of this theory. Further experiments by Wiltshire (1956a,b) demonstrated that plants maintained for a time in the light but without carbon dioxide prior to inoculation became as susceptible as plants maintained in the dark. On the other hand, darkened plants were equally susceptible in the presence or absence of carbon dioxide. More recently Kimmins and Litz (1967) established a correlation between the greater susceptibility of French bean leaves to infection with tobacco necrosis virus and an increase in the osmotic pressure of the epidermal cells. According to Helms and McIntyre (1967a,b), bean plants exposed to light for 2–3 minutes after a dark period were more susceptible to tobacco mosaic virus than those without the light period. Results obtained by Coast and Chant (1970) showed that plants exposed to blue or red light before inoculation with tobacco necrosis virus were less susceptible than those exposed to green light. These results support the hypothesis of Bawden and Roberts (1948) that susceptibility is related to photosynthetic activity because chlorophylls a and b absorb more light energy of the red and blue wavelengths than of the green wavelengths.

Changes in the susceptibility of plants to attacks by fungi can also be caused by light treatments prior to inoculation. Brooks (1908) found that spores of *Botrytis cinerea* in drops of water did not attack leaves of normal green lettuce plants, but if the leaves were kept in the dark for 5 days they were readily invaded. Tomato plants can be rendered more susceptible to *Fusarium oxysporum* f. sp. *lycopersici* if exposed to a low light intensity for 30 days prior to inoculation or by being exposed to short day lengths as compared with long day lengths for the same period (Foster and Walker, 1947).

Horsfall and Dimond (1957) attempted to explain the effects of dark treatments on the susceptibility of plants to fungi in terms of sugar being used up by respiration in the dark so that such plants become more susceptible to low-sugar diseases such as those caused by *Fusarium* sp. Darkness would induce increased resistance to high-sugar diseases such as those caused by the rusts and powdery mildew. This hypothesis is not supported by the findings of Jones (1975), in that oats can be rendered more resistant to powdery mildew by being exposed to long days or higher light intensity prior to inoculation.

## D. Nutrition

It has been assumed that when manurial treatments of plants are associated with changes in susceptibility, these are due to changes in the host. Thus, Moore (1944) considered that deficiency of potassium or available phosphate in soil predisposes beans to attack by *Botrytis fabae*.

Few experiments have been made in which the susceptibility of plants has been clearly shown to be altered by the nutrition of plants prior to inoculation. However, Foster and Walker (1947) found that the susceptibility of tomatoes to *Fusarium oxysporum* f. sp. *lycopersici* is increased if the nutrient they receive for 30 days prior to inoculation is low in nitrogen or phosphorus or high in potassium. In experiments on the root rot of chrysanthemums caused by *Phoma chrysanthemicola* it has been established that heavy applications of fertilizers, particularly nitrogen and phosphate, gives good control (Peerally and Colhoun, 1969; Menzies and Colhoun, 1976a,b). Menzies and Colhoun (1976b) showed that high nutrient levels provided prior to inoculation in most instances rendered the roots more susceptible than when no nutrients were supplied, but plants that were supplied with nutrients before inoculation and not afterward were usually less susceptible than those that received nutrients throughout the experiments. In these experiments commercial control of the disease is probably determined by the ability of the plant to regenerate roots, and this determines growth of the aerial parts.

Increased nitrogen may increase the susceptibility of apple fruits of one variety to invasion by *Cytosporina ludibunda* but decrease the susceptibility of another variety (Horne, 1939) depending, as Colhoun (1948) has shown, on the acidity of the varieties involved. Evidence has also been obtained (Colhoun, 1962) that in an orchard of low fertility when the supply of nitrogen, phosphate, and potash in combination is increased the susceptibility of lenticels on the fruit to penetration by *Penicillium expansum* is greatly increased.

Particularly during the last half-century many nonprofessional gardeners have advocated providing plants with organically based nutrients on the basis that the plants receiving these are more resistant to disease.

However, it is difficult to support such claims by experimental evidence insofar as predisposition is concerned.

## E. Soil pH

While soil reaction has been shown to affect the occurrence of such diseases as club root of crucifers (Colhoun, 1958), its relationship with predisposition has not been so clearly established. In most instances in which effects of soil reaction on disease incidence have been reported these can be explained by effects on the pathogen rather than on the host. The statement by Foster and Walker (1947) that nutrients of low pH, by their action prior to inoculation with *Fusarium oxysporum* f. sp. *lycopersici*, rendered tomato plants more susceptible clearly involves predisposition.

## F. Radiation and Chemicals

When whole tomato plants (cv. Bonny Best) were given X-ray irradiation before inoculation, resistance to *F. oxysporum* f. sp. *lycopersici* was increased (Waggoner and Dimond, 1956). Similar results were obtained when the roots were irradiated, but shoot irradiation resulted in decreased resistance, which was not significant, however.

When root injury accompanies the application of several chemicals to soil, susceptibility to fungal attack may be altered. Formalin, hydrogen peroxide, potassium bromide, and several other chemicals used in this way were found by Keyworth and Dimond (1952) to reduce susceptibility of tomatoes to *Fusarium oxysporum* f. sp. *lycopersici*. They considered that this resulted from a change in host resistance and not merely from inhibition of the pathogen at the injured root surfaces. Plants whose roots had been chemically injured had a higher soluble carbohydrate content and less potassium, calcium, and phosphates than the uninjured controls.

## G. Presence of Other Organisms

Some plant pathologists undoubtedly do not accept that the concept of predisposition by the environment embraces a change in the reaction of a host to one pathogen because of a prior attack by another pathogen. Nevertheless, it is now clear that certain pathogens can alter the physiological environment within their hosts, and this has an effect on the hosts' reactions to other pathogens. For this reason some consideration is now given to biotic factors, although these must to some extent enter into subjects dealt with in the two chapters that follow and in Volume V.

Research workers have for a long time been interested in the mechanisms of resistance that prevent a pathogen from attacking one host species when it readily infects another. In this connection much attention has been devoted to powdery mildews of cereals. Salmon (1905) reported that when conidia of *Erysiphe graminis* from wheat were placed on barley leaves nearly all of the incipient haustoria formed soon stopped growing and became disorganized. However, Moseman *et al.* (1965) showed that *E. graminis* f. sp. *hordei* can infect resistant barley varieties and wheat when these plants have already been attacked by a virulent culture of *E. graminis*. They considered that physiological and metabolic changes in the host, either induced directly by the virulent fungus or resulting from the interaction of the compatible combination, may explain the growth and propagation of avirulent cultures on previously resistant hosts.

Further developments have taken place in Japan (Ouchi *et al.,* 1974a,b). It has been demonstrated that preliminary inoculation of barley leaves with an incompatible or nonpathogenic race of *E. graminis* induced resistance to a normally compatible race, at least in the infected site. On the contrary, previous infection with a compatible race rendered the leaves accessible to a normally incompatible or non-pathogenic race. Moreover, *E. graminis* and *Sphaerotheca fuliginea* can infect leaves of melons and barley, respectively, if the leaves have been predisposed by being previously attacked by the primarily compatible pathogen. It is suggested that the induction of accessibility may be a kind of nonspecific permission by the host cells against invading organisms so that the cells in which accessibility is induced can no longer recognize nonpathogens as foreign entities. When cells are inoculated with a nonpathogen and a competent rejection reaction is induced, they also recognize as a nonassociable entity what should normally be a compatible race or pathogen.

Kuć *et al.* (1975) found that when a leaf of cucumber was inoculated with a race of *Colletotrichum lagenarium* to which it is susceptible the leaf above was protected against attacks by this pathogen. The factors that confer protection are therefore able to operate further away from the site originally infected than was shown by Ouchi *et al.* (1974a).

Damage to roots by nematodes may provide suitable points of entry for pathogenic fungi, but there is also evidence that nematodes may alter the physiology of roots so that they become more suitable for invasion by fungi (Powell and Nusbaum, 1960). It has been suggested that infestation of roots of celery by nematodes partially or entirely inhibits attacks on the aerial parts by *Septoria apicola* (= *S. api*) because the foliage is starved by disruption of the vascular system and by withdrawal of food material in the production of galls (Thomas, 1921).

There are instances in which infection of a plant by a virus may protect it against later infection by another virus, and this is of particular interest when it introduces the possibility of a mild strain of a virus providing protection against a more virulent strain (Posnette and Todd, 1955). These resistance reactions are discussed much further in Volume V.

## V. INTERACTION OF FACTORS

It has been pointed out (Colhoun, 1973) that the study of the effects of individual factors can be regarded as the first step in understanding the influence of environment on the incidence of crop diseases. Yet if an analysis of such factors is to be meaningful in relation to field conditions, it is necessary to design experiments that will involve various levels of a number of factors. Such studies have been undertaken with respect to a limited number of diseases, and better understanding of their epidemiology has resulted. Few workers, however, in studies of predisposition by the environment have attempted to study the interaction of the individual factors involved. It is suggested that this may be a profitable field for further work, especially systems analysis.

## VI. SIGNIFICANCE OF PREDISPOSITION
## IN PLANT PATHOLOGY

An understanding of predisposition by environment assists in disease diagnosis and epidemiological studies. For example, attacks by *Pseudomonas tabaci* on inoculated tobacco plants were less severe than in commercial seed beds until it was realized that water congestion of tissues predisposed plants to infection (Clayton, 1936; Johnson, 1947). According to Noon and Colhoun (1979), some of the symptoms observed in naturally infected yam tubers could not be reproduced by inoculation methods. It was suggested that some microorganisms colonized the chilled yams but could not invade healthy tubers.

The common practice of exposing plants to darkness or shading prior to inoculation so that their susceptibility to viruses is increased indicates the contribution made to disease transmission studies by a better knowledge of predisposition. Such knowledge also provides better means for testing plants for resistance to pathogens.

If, through studies of predisposition, we can acquire means of increasing the resistance of genetically susceptible plants without interfering seriously with their satisfactory growth, a contribution can be made to disease control. Such an impact can be made earliest under glasshouse conditions, where environmental conditions can be more readily controlled than in the field. As we come to understand better the protection phenomena it may be possible that early attacks by a mild disease-producing agent can reduce losses from later infection by a more virulent pathogen. Recent studies of the effects of the consecutive infection of a host by two separate pathogens suggest that when interactions do occur greater susceptibility is involved but, according to Hyde (1978), such interactions may not occur consistently.

The challenge that remains is whether we can reduce such synergistic effects.

## References

Ainsworth, G. C. (1971). "Ainsworth and Bisby's Dictionary of the Fungi," 6th Ed. Commonw. Mycol. Inst., Kew.

American Phytopathological Society (1940). Report of the Committee on Technical Words. *Phytopathology* **30**, 361–368.

Bawden, F. C., and Roberts, F. M. (1947). The influence of light intensity on the susceptibility of plants to certain viruses. *Ann. Appl. Biol.* **34**, 286–296.

Bawden, F. C., and Roberts, F. M. (1948). Photosynthesis and predisposition of plants to infection with certain viruses. *Ann. Appl. Biol.* **35**, 418–428.

Berkeley, M. J. (1846). Observations, botanical and physiological, on the potato murrain. *J. Hort. Soc. (London)* **1**, 9–34. (Reprinted, Phytopathological Classics, No. 8, 1948, Amer. Phytopathol. Soc., Ithaca.)

Berkeley, M. J. (1847). *Gard. Chron.* 1847, p. 779.

Blodgett, E. C. (1936). The anthracnose of currant and gooseberry caused by *Pseudopeziza ribis*. *Phytopathology* **26**, 115–152.

Boyd, A. E. W. (1952). Dry rot disease of the potato. VII. The effect of storage temperature upon subsequent susceptibility of tubers. *Ann. Appl. Biol.* **39**, 351–357.

British Mycological Society (1950). Definitions of some terms used in plant pathology. *Trans. Br. Mycol. Soc.* **33**, 154–160.

Brooks, F. T. (1908). Observations on the biology of *Botrytis cinerea*. *Ann. Bot. (London)* **22**, 479–487.

Chona, B. L. (1932). Studies on the physiology of parasitism. XIII. An analysis of the factors underlying specialization of parasitism with special reference to certain fungi parasitic on apple and potato. *Ann. Bot. (London)* **36**, 1033–1050.

Clayton, E. E. (1936). Water soaking of leaves in relation to development of the wildfire disease of tobacco. *J. Agric. Res.* **52**, 239–269.

Coast, E. M., and Chant, S. R. (1970). The effect of light wavelength on the susceptibility of plants to virus infection. *Ann. Appl. Biol.* **65**, 403–409.

Colhoun, J. (1948). Nitrogen content in relation to fungal growth in apples. *Ann. Appl. Biol.* **35**, 638–647.

Colhoun, J. (1958). "Club Root Disease of Crucifers caused by *Plasmodiophora brassicae* Woron." Commonw. Mycol. Inst., Kew.

Colhoun, J. (1962). Some factors influencing the resistance of apple fruits to fungal invasion. *Trans. Br. Mycol. Soc.* **45**, 429–430.

Colhoun, J. (1973). Effects of environmental factors on plant disease. *Annu. Rev. Phytopathol.* **11**, 343–364.

de Bary, A. (1861). "Die Gegenwärtig Herrschende Kartoffelkrankheit, ihre Ursache und ihre Verhütung." A. Felix, Leipzig.

de Tournefort, J. P. (1705). Observations sur les maladies des plantes. *Mem. Acad. R. Sci., Paris* 1705, pp. 332–345.

Federation of British Plant Pathologists (1973). "A Guide to the Use of Terms in Plant Pathology," Phytopathol. Pap. No. 17. Commonw. Mycol. Inst., Kew.

Foster, R. E., and Walker, J. C. (1947). Predisposition of tomato to *Fusarium* wilt. *J. Agric. Res.* **74**, 165–185.

Gäumann, E. (1950). "Principles of Plant Protection" (Engl. transl. by W. B. Brierley). Crosby Lockwood, London.

Ghaffar, A., and Erwin, D. C. (1969). Effect of soil water stress on root rot of cotton caused by *Macrophomina phaseoli*. *Phytopathology* **59**, 795–797.

Grainger, J. (1968). $C_p/R_s$ and the disease potential of plants. *Hortic. Res.* **8**, 1–40.

Gregg, M. (1952). Studies in the physiology of parasitism. XVII. Enzyme secretion by strains of *Bacterium carotovorum* and other pathogens in relation to parasitic vigour. *Ann. Bot. (London)* **16**, 235–250.

Hallier, E. (1868). "Die Krankheiten der Culturgewächse für Land- und Forstwirthe, Gärtner und Botaniker."

Hartig, R. (1882). "Lehrbuch der Baumkrankheiten." Springer-Verlag, Berlin. (English transl. by W. Somerville and H. Marshall Ward. Country Life, London, 1894.)

Helms, K., and McIntyre, G. A. (1967a). Light-induced susceptibility of *Phaseolus vulgaris* L. to tobacco mosaic virus infection. I. Effects of light intensity, temperature, and the length of the preinoculation dark periods. *Virology* **31**, 191–196.

Helms, K., and McIntyre, G. A. (1967b). Light-induced susceptibility of *Phaseolus vulgaris* L. to tobacco mosaic virus infection. II. Daily variation in susceptibility. *Virology* **32**, 482–488.

Hill, A. V., and Green, S. (1968). The role of temperature in the development of blue mould (*Peronospora tabacina* Adam) disease in tobacco seedlings. III. The effect of pre-inoculation temperature and of shade on plant growth and disease development. *Aust. J. Agric. Res.* **19**, 759–766.

Horne, A. S. (1939). "The Resistance of the Apple to Fungal Invasion," Rep. Food Invest. Board, London, 1938, p. 173. HM Stationery Off., London.

Horsfall, J. G., and Dimond, A. E. (1957). Interactions of tissue sugar, growth substances, and disease susceptibility. *Z. Pflanzenkr. (Pflanzenpathol.) Pflanzenschutz* **64**, 415–419.

Hyde, P. M. (1976). Comparative studies of the infection of flag leaves and seedling leaves of wheat by *Erysiphe graminis*. *Phytopathol. Z.* **85**, 289–297.

Hyde, P. M. (1977). The effect of leaf age on infection of wheat seedlings by

*Erysiphe graminis* and on subsequent colony development. *Phytopathol. Z.* **88**, 299–305.

Hyde, P. M. (1978). A study of the effects on wheat of inoculation with *Puccinia recondita* and *Leptosphaeria nodorum* with respect to possible interactions. *Phytopathol. Z.* **92**, 12–24.

Johnson, J. (1937). Relation of water-soaked tissues to infection by *Bacterium angulatum* and *Bacterium tabacum* and other organisms. *J. Agric. Res.* **55**, 599–618.

Johnson, J. (1947). Water-congestion in plants in relation to disease. *Univ. Wis., Res. Bull.* No. 160.

Jones, I. T. (1975). The preconditioning effect of day length and light intensity on adult plant resistance to powdery mildew in oats. *Ann. Appl. Biol.* **80**, 301–309.

Kassanis, B. (1952). Some effects of high temperature on the susceptibility of plants to infection with virus. *Ann. Appl. Biol.* **39**, 358–369.

Kendrick, J. B., and Walker, J. C. (1948). Predisposition of tomato to bacterial canker. *J. Agric. Res.* **77**, 169–186.

Keyworth, W. G., and Dimond, A. E. (1952). Root injury as a factor in the assessment of chemotheraputants. *Phytopathology* **42**, 311–315.

Khamees, M. A. F. (1977). Studies of *Fusarium nivale* (Fr.) Ces. and winterkill on *Lolium* spp. M.S. Thesis, Aberdeen Univ., Aberdeen.

Kimmins, W. C., and Litz, R. E. (1967). The effect of leaf water balance on the susceptibility of French bean to tobacco necrosis virus. *Can. J. Bot.* **45**, 2115–2118.

Kuć, J., Shockley, G., and Kearney, K. (1975). Protection of cucumber against *Colletotrichum lagenarium* by *Colletotrichum lagenarium*. *Physiol. Plant Pathol.* **7**, 195–199.

Kühn, J. (1858). "Die Krankheiten der Kulturgewächse, ihre Ursachen und ihre Verhutung," Berlin.

Menzies, S. A., and Colhoun, J. (1976a). Control of *Phoma* root rot of chrysanthemums by the use of fertilizers. *Trans. Br. Mycol. Soc.* **67**, 455–462.

Menzies, S. A., and Colhoun, J. (1976b). Mechanisms of control of *Phoma* root rot of chrysanthemums by the use of fertilizers. *Trans. Br. Mycol. Soc.* **67**, 463–467.

Mishra, J. N. (1953). Resistance of potato tubers to certain parasitic fungi. *Phytopathology* **43**, 338–340.

Moore, W. C. (1944). Chocolate spot of beans. *Agriculture* (*London*) **51**, 266–269.

Moore, W. D., and Thomas, H. R. (1943). Some cultural practices that influence the development of *Alternaria solani* on tomato seedlings. *Phytopathology* **33**, 1176–1184.

Moseman, J. G., Scharen, A. L., and Greeley, L. W. (1965). Propagation of *Erysiphe graminis* f. sp. *hordei* on wheat. *Phytopathology* **55**, 92–96.

Noon, R. A., and Colhoun, J. (1979). Market and storage diseases of yams imported into the United Kingdom. *Phytopathol. Z.* **94**, 289–302.

Ouchi, S., Oku, H., Hibino, C., and Akiyama, I. (1974a). Induction of accessibility and resistance in leaves of barley by some races of *Erysiphe graminis*. *Phytopathol. Z.* **79**, 24–34.

Ouchi, S., Oku, H., Hibino, C., and Akiyama, I. (1974b). Induction of accessibility to a nonpathogen by preliminary inoculation with a pathogen. *Phytopathol. Z.* **79**, 142–154.

Peerally, M. A., and Colhoun, J. (1969). The epidemiology of root rot of chrysanthemums caused by *Phoma* sp. *Trans. Br. Mycol. Soc.* **52**, 115–125.

Posnette, A. F., and Todd, J. McA. (1955). Virus diseases of cacao in West Africa. IX. Strain variation and interference in virus 1A. *Ann. Appl. Biol.* **43**, 433–453.

Powell, N. T., and Nusbaum, C. J. (1960). The black shankroot-knot complex in flue-cured tobacco. *Phytopathology* **50**, 899–906.

Prevost, B. (1807). "Mémoire sur la Cause Immediate de la Carie ou Charbon des Blés, et de Plusieurs Autres Maladies des Plantes, et sur les Préservatifs de la Carie." Bernard, Paris. (English transl. by G. W. Keitt, "Phytopathological Classics," No. 6. 1939, Amer. Phytopathol. Soc. Ithaca.)

Ramakrishnan, L. (1966). Studies in the host–parasite relations of blast disease of rice. III. The effect of night temperature on the infection phase of blast disease. *Phytopathol. Z.* **57**, 17–23.

Rivera, V. (1924). Cryptogamic epidemics and the environmental factors that determine them. *Int. Rev. Sci. Pract. Agric.* **2**, 604–609.

Rotem, J., Cohen, Y., and Spiegel, S. (1968). Effect of soil moisture on the predisposition of tobacco to *Peronspora tabacina*. *Plant Disease Rep.* **52**, 310–313.

Salmon, E. S. (1905). On the stages of development reached by certain biologic forms of *Erysiphe* in cases of non-infection. *New Phytol.* **4**, 217–222.

Samuel, G., and Bald, J. G. (1933). On the use of the primary lesions in quantitative work with two plant viruses. *Ann. Appl. Biol.* **20**, 70–99.

Schulz, F. A., and Bateman, D. F. (1969). Temperature response of seeds during the early stages of germination and its relation to injury by *Rhizoctonia solani*. *Phytopathology* **59**, 352–355.

Sorauer, P. (1874). "Handbuch der Pflanzenkrankheiten." Weigandt, Hempel & Parey, Berlin.

Sorauer, P. (1914). "Manual of Plant Diseases. Vol. 1: Non-Parasitic Diseases" (transl. by F. Dorrance). Record Press, Wilkes-Barre, Pennsylvania.

Speerschneider, J. (1857). Die Urasche der Erkrankung der Kartoffelknolle durch eine Reihe Experimente bewiesen. *Z. Bot.* **15**, 122–124.

Steward, F. C. (1943). The effect of low temperature storage on meristematic activity of the cells of potato tuber. *Ann. Bot. (London)* **7**, 242–244.

Tapke, V. F. (1951). Influence of preinoculation environment on the infection of barley and wheat by powdery mildew. *Phytopathology* **41**, 622–632.

Tessier, M. (1783). "Traité des Maladies des Grains." La Veuve Herissant, Paris.

Thomas, H. E. (1921). The relation of the health of the host and other factors to infection of *Apium graveolens* by *Septoria apii*. *Bull. Torrey Bot. Club* **48**, 1–29.

Tillet, M. (1755). "Dissertation sur la Cause qui Corrompt et Noircit les Grains de Bled dans les Épis; et sur les Moyens de Préventir ces Accidens. Brun, Bordeaux. (Engl. transl. by H. B. Humphrey, "Phytopathology Classics, No. 5, 1937), Amer. Phytopathol. Soc. Ithaca.)

Tinsley, T. W. (1953). The effects of varying the water supply of plants on their susceptibility to infection with viruses. *Ann. Appl. Biol.* **40**, 750–760.

Unger, F. (1833). "Die Exantheme der Pflanzen." C. Gerold, Vienna.

Vasudeva, R. S. (1930). Studies in the physiology of parasitism. XI. An analysis of the factors underlying specialization of parasitism, with special reference to the fungi *Botrytis allii* Munn, and *Monilia fructigena* Pers. *Ann. Bot. (London)* **44**, 469–493.

Waggoner, P. E., and Dimond, A. E. (1956). Altering disease resistance with ionizing radiation. *Phytopathology* **46**, 125–127.

Ward, H. M. (1890). On some relation between host and parasite in certain epidemic diseases of plants. *Proc. R. Soc. London* **47**, 393–443.

Ward, H. M. (1901). "Disease in Plants." Macmillan, London.
Ward, H. M. (1902). On the question of "predisposition" and "immunity" in plants. *Proc. Cambridge Philos. Soc.* **11**, 307–328.
Warren, R. C., King, J. E., and Colhoun, J. (1971). Reaction of potato leaves to infection by *Phytophthora infestans* in relation to position on the plant. *Trans. Br. Mycol. Soc.* **57**, 501–514.
Wiltshire, G. H. (1956a). The effect of darkening on the susceptibility of plants to infection with virus. I. Relation to changes in some organic acids in the French bean. *Ann. Appl. Biol.* **44**, 233–248.
Wiltshire, G. H. (1956b). The effect of darkening on the susceptibility of plants to infection with virus. II. Relation to changes in ascorbic acid content of French bean and tobacco. *Ann. Appl. Biol.* **44**, 249–255.
Yarwood, C. E. (1956). Heat-induced susceptibility of beans to some viruses and fungi. *Phytopathology* **46**, 523–525.
Yarwood, C. E. (1959). Predisposition. *In* "Plant Pathology" (J. G. Horsfall and A. E. Dimond, eds.), Vol. 1, pp. 521–562. Academic Press, New York.

Chapter 5

# External Synergisms among Organisms Inducing Disease

COLIN H. DICKINSON

## I. INTRODUCTION

A broad survey of almost any group of pathogens reveals the astonishing extent to which they have adapted their life styles to fit in with those of their green plant hosts. From the first signs of activity in long-dormant propagules to the final act of sporulation or renewed dormancy the pathogen's every move is geared to its host's phenology, morphology, anatomy, physiology, and biochemistry. Studies in greenhouses and growth rooms have, for example, shown that epiphytic hyphal growth can be orientated by the submicroscopic structure of leaf cuticle and that reproduction of pathogens, in many instances, is synchronized with the development of flowers or newly expanded leaves.

From a consideration of such facts it is, however, easy to gain the impression that pathogen–host interactions always proceed along inevitable routes. This analysis is not correct. Superimposed on these neat and efficient bipartite arrangements is the influence of a third party, the environment. This adds an additional dimension of considerable complexity. In pathological terms, the environment includes meteorological

97

PLANT DISEASE, VOL. IV
Copyright © 1979 by Academic Press, Inc.

and soil factors together with a biological component whose significance is still not fully appreciated.

It may be argued that progress in understanding plant pathogens has been retarded by our habit of elevating them to pedestals above other microorganisms. Such emphasis is understandable considering the economic importance of many pathogens. It has, however, resulted in many pathologists viewing their charges in splendid isolation. This has led to numerous detailed studies on pathogens being carried out under regimes that are remote from conditions pertaining to the field.

There are, of course, exceptions to this generalization. Some involve plant viruses. It has become abundantly clear that mixed infections involving distinct strains of the same virus or two or more unrelated viruses commonly occur in nature (Kassanis, 1963). Other exceptions concern pathogens whose natural milieu is on or in subterranean plant tissues. Here the greater emphasis on associated nonpathogenic organisms probably follows from the relatively advanced state of soil microbiology, as compared with studies of microbial populations in other habitats. Soil appears to contain many excellent microhabitats for microbial colonization, and hence it has been studied more intensively than any other ecosystem. As a result a number of schools of research have been developed in which pathogens are treated as a part of the soil ecosystem. Their interrelationships with other microorganisms have been explored, and the biological factors in the environment have thus been emphasized in a more balanced and realistic approach.

In considering such ecological investigations of plant pathogens, there would appear, a priori, to be no particular reason to expect their relationships with nonpathogens to be biased toward either antagonism or cooperation. However, in practice there is a general emphasis throughout pathology on control, and hence any indications of antagonism between pathogenic and saprobic microorganisms have been avidly explored. The obvious goal of creating or enhancing biological systems for the management of pathogens has spurred many pathologists to devise experimental systems in which maximal antagonistic pressure is applied at all vulnerable phases of pathogen development.

Relatively recently these studies have attained a considerable degree of success. Details of many of the relationships between soilborne pathogens and the associated nonpathogenic microorganisms have been described by Garrett (1970) in an account that chronicles his pioneering work in this area of microbial ecology. More recently, Baker and Cook (1974) have formulated a number of principles that will facilitate our understanding of the nature of the interrelationships involved in biological control. These authors also emphasize the paucity of information concerning the biological component of the aboveground environment.

Set in this context it is hardly surprising that situations in which the activities of pathogens are enhanced by the biological component of the environment have received relatively short shift. The majority of ecologically orientated pathologists have simply neglected this arena in favor of one that appears to promise more academic (and financial) rewards.

## II. SYNERGISM

The term "synergism" appears to have been used initially by bacteriologists to describe situations in mixed cultures in which two organisms act in a complementary manner (Holman and Meekison, 1926). Fawcett (1931) extended the use of the term to cover plant pathogens. He described a number of investigations in which mixed inocula resulted in more extensive damage than either organism could produce when acting independently. In the present context it is of interest, however, that in at least two of the instances he described one of the partners involved in the synergism was not a recognized pathogen, and the majority of the damage was ultimately caused in each case by the known pathogen. Hence, it is possible that one organism may have simply increased the efficiency of the other in the initial stages of infection.

This chapter is concerned with interactions between two or more pathogens or between pathogens and other nonpathogenic organisms that contribute to the success of one or more of the pathogens. Some of these interactions are clearly examples of synergism, which was defined in "A Guide to the Use of Terms in Plant Pathology" (Federation of British Plant Pathologists, 1973) as an association of two or more organisms acting at one time and effecting a change that one only is not able to make. The limitation imposed in this definition by the inclusion of the phrase "at one time" seems to be unnecessarily restrictive. Many examples have been described in which sequential development is normal, and indeed there are few proved instances of simultaneous mixed infections. In other respects the definition appears to cover the wide range of interactions that have been included under this term.

Outside the host, biological factors affecting the pathogen may range from those directly stimulating its propagules or vegetative organs to others that subtly modify the host's surface defenses, making it easier for the pathogen to penetrate and become established. Further discussion of this topic will be from two different, but complementary, perspectives. The first part will be concerned with the main problems facing a pathogen whose life in a particular host is about to be terminated. At that time, it must reorganize its activities so as to spread to other hosts. This spread frequently involves a more or less extended period of dormancy when survival becomes imperative. Following successful dispersal and/or survival, the pathogen must become reactivated at an appropriate time, and then it must overcome the defenses of potential hosts. The final part of the chapter will be concerned with qualitative aspects of the relationships between pathogens and other organisms that contribute to the success of the former.

Although this chapter deals mainly with synergisms among parasitic organisms, similar effects have also been reported involving abiotic agents of disease. For example, marginally toxic concentrations of various air pollutants have been shown to increase the amount of disease induced by various fungal pathogens, including *Botrytis cinerea, Uromyces phaseoli, Erysiphe graminis,* and tobacco mosaic virus (Heagle, 1973). In addition, direct synergisms between various air pollutants, including sulfur dioxide and ozone, sulfur dioxide and nitrogen dioxide, and sulfur dioxide and hydrogen fluoride, have been described by Reinert *et al.* (1975).

## III. PROBLEMS FACED BY PATHOGENS IN SPREADING FROM PLANT TO PLANT

### A. Dispersal

It is axiomatic that pathogens require efficient dispersal systems that are related both to the distribution patterns of their hosts and to the conditions prevailing in the environment. Some pathogens simply grow from one individual host plant to another. Others have developed dispersal systems involving independent deciduous propagules that are shot off, blown about, or splashed from plant to plant. A number of pathogens avoid the need for elaborate propagule dispersal systems by becoming dependent to a greater or lesser extent upon vectors.

Clearly the viruses provide the best demonstrations of the importance of vectors, and in the most complex situations transmission occurs only when there is a precise combination of two viruses, one termed the dependent and the other the helper, and a specific vector (Rochow, 1972). Vectors are also employed by many plant pathogenic fungi and bacteria, although in several instances the relationships involved seem more a matter of accident than design. Many vector–pathogen relationships are highly specific, with the habits of the vector being closely geared to the life style of the pathogen. In these instances the pathogen is never exposed to the external environment as it passes from its host plant to the vector and back to the host plant. The vector takes over the responsibility for ensuring that infection is successful, and this is accomplished with special efficiency when the vector itself is parasitic on the same host plant. The latter type of situation clearly opens possibilities for the vector to increase the damage caused by the pathogen and vice versa, although it might be argued that such synergistic effects would be more likely to occur when the vector is itself a pathogen rather than merely a parasite. Hence, such fungal pathogens as *Olpidium, Spongospora,* and *Synchytrium,* which transmit viruses (Teakle, 1969), might be more dramatically affected by the subsequent virus infections than those aphid, nematodes, and leaf hoppers that are also virus vectors. However, not a great deal is known about this type of interaction, although one report indicates that *Olpidium* is in fact less successful in virus-infected roots than in healthy tissues (Fry and Campbell, 1966).

## B. Survival

As this phase is one involving minimal pathogen activity, there are few opportunities for the development of synergistic relationships. Nevertheless, possibilities do exist whereby refuge from adverse environmental conditions is provided by other organisms. In some instances this again involves vectors, which may or may not themselves become inactive.

## C. Infection

Dormant pathogen propagules often require specific stimuli before they recommence active growth. Their subsequent vegetative development on the surface of a potential host is then more or less extensive depending on their biology and on the availability of nutrients and suitable infection sites. Efficient penetration may follow the development of appressoria or infection cushions and depends on the ability of infection structures to force their way through the host's physical barriers and withstand its chemical defenses.

All these processes are unnecessary if the pathogen is conveniently deposited directly into the host tissues by an obliging vector. There are also fewer problems if the pathogen normally enters through wounds, although it is arguable that in this case substantial limitations are placed on the pathogen by the need to discover and exploit such wounds before they have healed.

Nonpathogenic microorganisms associated with the rhizosphere, rhizoplane, and phylloplane have been regarded as playing an integral part in the plant's defenses, by virtue of their antagonistic behavior toward pathogens. However, while it cannot be denied that this is an important role for some members of the epiphytic populations, there are numerous other organisms on and around roots, leaves, flowers, and fruits for which no such antagonistic properties have been demonstrated. These organisms are clearly able to moderate the environment within which pathogens germinate, explore the host's surface, and commence penetration. It is therefore conceivable that many of these epiphytic organisms actually assist in the establishment of pathogens. Such assistance may be rendered directly, by their acting as a source of allochthonous nutrients (that is, food originating from outside the habitat, which in this case is the plant's surface), by moderation of the surface pH, or by detoxification of chemicals produced by the host. Alternatively, these epiphytic populations may have indirect effects, as, for instance, when activity of saprobic organisms results in biodeterioration of the host's surface layers.

A number of possible scenarios may be envisaged covering a range of interactions pertaining to infection: (a) parasite A infects a plant that has already been infected by parasite B; (b) parasite A and parasite B infect the same plant simultaneously; and (c) parasite A infects via infection courts created or modified by one or more nonparasites.

### 1. Infection of Plants Previously Colonized by Another Pathogen

The simplest type of example involves pathogens whose entry into a host is made possible by the activities of another parasitic organism. A spectacular example of this type of opportunistic behavior occurs in cotton bolls, where *Aspergillus* and *Rhizopus* enter through holes made by insects (Ashworth *et al.*, 1971). Another example is provided by the nematode *Pratylenchus,* which interacts synergistically with the wilt fungus *Verticillium* on eggplant (McKeen and Mountain, 1960) and peppermint (Faulkner *et al.*, 1970). In the latter instance the nematode has been shown to open infection courts for the fungus by its feeding activities in the developing root tips. In these examples the plant's surface defences are breached and openings are created, but there is also damage to the adjacent cells, and the relative importance of these two factors is as yet unknown.

More subtle interactions included in this category involve changes in the external environment that result from pathogenic invasion of tissues. Infection by viruses may change the chemical composition of the tissues and their cell membrane structures, and hence they alter the quality and quantity of exudates that leak out onto the host's external surfaces. Beute and Lockwood (1968) showed that the infections by bean yellow mosaic virus increased the concentration of amino acids in root exudates such that *Fusarium solani* was stimulated to greater levels of activity than occurred around the roots of healthy plants. A similar situation was described by Tu and Ford (1971) for maize dwarf mosaic virus. Virus infection encouraged subsequent root infections by *Gibberella* and *Helminthosporium*. Nematode infections may also alter the exudates from roots with consequent effects on soilborne pathogens (Bergeson *et al.*, 1970). Whitney (1971) attributed synergism between *Heterodera* and *Pythium* to an increase in the growth of the fungus in the soil around the roots when the nematode was present. This growth presumably led to an increased number of infections, which would finally result in a more devastating fungal attack.

A number of other examples of successive infections have been described for both subterranean and foliar tissues, but in many cases it is not yet known what change or changes are responsible for the success of the second pathogen (e.g., Pieczarka and Abawi, 1978). In some instances changes induced by an initial local infection can influence the success of a subsequent pathogenic attack elsewhere on the plant. Sidhu and Webster (1977) were able to demonstrate that the resistance of tomato cultivars to *Fusarium* wilt was diminished following invasion by root-knot nematodes. The factor produced or induced by these nematodes was translocated over a substantial distance from the roots, but no evidence was given as to its nature. Exudates leak onto the surface of leaves and other aerial plant parts (Tukey, 1971; Godfrey, 1976), but less is known of their importance with regard to the superficial development of pathogens than has been discovered for root exudates in the rhizosphere. For example, changes in exudate quality or quantity may be responsible for the secondary attacks of *Alternaria* and *Erysiphe betae* on sugar beet leaves suffering from beet mild yellowing virus (Russell, 1970). These fungal pathogens can cause considerable crop losses, as judged by the efficiency of fungicidal spray treatments in limiting damage due to this disease complex.

Exudates can also originate from the primary pathogen or pest itself.

Nematodes attacking roots in soil can themselves produce enzymes, which may stimulate the development of dormant propagules. Aphids feeding on green tissues produce honeydew, which clearly fuels the growth of many phylloplane saprophytes. It would seem likely that this substance is also available for pathogen spores. Even individual spores can alter the nutrient composition of the water drops within which they germinate and this nutrient supplement may well encourage the concurrent growth of other organisms, such as the bacteria (Blakeman and Brodie, 1976).

Another obvious area in which such interactions may be important concerns bacteria entering lesions after an initial attack by fungal pathogens or animal pests. Frequent reference is made in the literature to secondary bacteria that are thought to colonize lesions induced by pathogens, such as *Phytophthora infestans* and *Phoma exigua* var. *foveata*. However, progress in understanding these interactions is likely to be slow until reliable and rapid methods are developed for identifying bacteria in natural environments.

Numerical taxonomy (Goodfellow *et al.*, 1976) offers some hope of progress in this field. The method can be used to provide detailed comparisons of the bacterial populations in disease lesions with those on healthy tissues. This enables those bacteria that are positively associated with different stages in the pathogen's development to be identified. Subsequently, representative isolates can be tested to determine the nature of their relationship with the pathogen. We have carried out such a study using cabbage leaves attacked by *Alternaria brassicicola*. At an early stage of infection there were clear quantitative and qualitative differences between the bacterial populations on diseased leaves and those on healthy leaves (C. H. Dickinson and H. Al Hadithi, unpublished observations). Our data suggest that at least two of the groups of bacteria whose activity is enhanced by the growth of the pathogen are able to exploit the lesions created by the fungus and thus increase the severity of this disease.

### 2. Simultaneous Infection by Two or More Pathogens

At first sight it might appear unlikely that the propagules of two or more pathogens would be in the same place and commencing activity at the same time. However, considering the similarities of the dispersal processes adopted by many airborne pathogens, there is a strong possibility that high concentrations of propagules accumulate on leaves and other aerial tissues. These concentrations would be further increased by the fluctuating, stop/go nature of the aerial environment, which operates so as to alternatively inhibit or actively promote germination.

Hence, dormant spores collect on the plant's surfaces until a favorable combination of environmental conditions triggers mass germination and growth.

Given such a situation, it would seem appropriate to consider whether synergism is likely to occur between the germ tubes or penetration hyphae of adjacent spores. Garrett (1970) discussed this situation in some detail and advanced several reasons to explain why large numbers of spores are often required for successful experimental infections by plant pathogens. He did not, however, rule out the possibility of synergism occurring in certain instances, especially where spores remain naturally clumped together and germinate simultaneously. Such conditions occur with *Tilletia caries*, causing bunt of wheat. Synergism is also likely to be important for those tree pathogens that produce a number of infection hyphae from mycelial strands or rhizomorphs.

Little work appears to have been done on this aspect of synergism with regard to mixed inoculations in field conditions. Clearly any such study would need to be closely monitored to follow the progress of each of the organisms. It would be unsatisfactory to merely record the amount of disease after a period of incubation, as this would compound interactions at infection with those during the subsequent colonization processes. Judging by the large number of successful infections that regularly occur on the foliar parts of many crop plants, it would seem unlikely that the results of any such investigation would be wholly negative.

### 3. The Influence of Nonpathogens on Infection Processes

The populations of nonpathogenic microorganisms surrounding both subterranean and aerial plant tissues may influence pathogens directly or indirectly via an effect on the host. Direct effects include the provision of nutrients that enable the pathogen to become established. Such nutrients could be released during the decomposition of macromolecules derived from the organic matter in the soil, root cell debris, and allochthonous material landing on leaves.

Some of the most fascinating possibilities involving indirect effects concern the influence of the epiphytic microflora on the surfaces of roots and leaves. Bacteria and fungi have been shown to be capable of active growth on the surfaces of living roots of numerous plants, but the role played by these rhizoplane organisms is not yet fully appreciated. Some may be involved in destructive activities that modify the root surface defenses. Developing root apices are covered with a layer of acidic polysaccharide slime (Barlow, 1975), and this material has been shown to be extensively colonized by bacteria (Campbell and

Rovira, 1973). These bacteria may account for the eventual disappearance of the slime from older parts of the root. In addition to effects on the slime, scanning microscopy has shown that the outer cells of *Ammophila* roots are visibly damaged beyond about 5 mm from their tips (Marchant, 1970). This deterioration could result from the activity of the numerous fungi and bacteria seen growing amongst the cell debris. If such surface damage were found to be a widespread phenomenon, then it would have important implications for pathogens that invade via the root system.

The root thus comprises several zones ranging from the structurally intact, newly formed apex, the cells of which are continuously replaced throughout the growing season, to the most mature proximal portions, which will have been exposed to biodeterioration for long periods. In contrast, most leaves offer a more uniform habitat for pathogens, as their development is relatively rapid and finite. The leaf surface is protected by a complex cuticle (Jeffree *et al.*, 1976), which covers the entire surface and is apparently more durable than the outer defenses of the more distal, fibrous parts of the root system.

The discovery that the aerial surfaces of most plants are inhabited by extensive populations of microorganisms has, however, led to speculation that the cuticle might be degraded by these organisms. A large number of bacteria and yeasts inhabiting the phylloplane are lipolytic (Austin *et al.*, 1978), but it is not yet known how much wax is actually broken down by these organisms in the field. It has been shown that epicuticular wax is lost from leaves as they age, but this may be due to erosion by wind and rain. The significance of this process was demonstrated by Shriner (1977) using simulated acid rain which removed wax from oak and bean leaves. Cutin, the other main component of the cuticle, is eventually broken down by saprobic organisms in the litter layer (Martin and Juniper, 1970) but there is at present little information as to the possibility of such breakdown occurring while the leaf is still alive. We may, however, be optimistic in expecting any significant degradation of cutin to occur on living leaves. Recent experiments have shown that detached, wax-free cuticles were not visibly degraded by any phylloplane organisms and clear signs of decay were only seen after they had been buried in soil for three months (O. MacNamara, unpublished observations).

Should either or both of these components of the leaf's protective layer be substantially degraded, pathogens might be expected to benefit from an increased outflow of exudates and by an easier entry route into the epidermal cells.

Another profitable avenue of study of interactions involving nonpathogenic organisms in the phylloplane concerns their influence on leaf surface nutrient regimes. Evidence has been presented which shows that some epiphytic microorganisms release various indole auxins and B vitamins during their growth and these compounds may well stimulate the development of associated pathogens (Klincāre et al., 1971). The growth of these epiphytic organisms will, however, be at the expense of the carbohydrates and nitrogenous compounds which accumulate on the aerial plant surfaces. Hence plant pathogens will be deprived of exogenous supplies of such nutrients. The resulting starvation will drastically curtail the activity of some pathogens but for others shortages of nutrients may trigger a switch in development from hyphal growth to appressorium formation (Blakeman and Parbery, 1977). Hence production and consumption of leaf surface nutrients by microbial epiphytes may affect all processes from the onset of germination of a pathogen's spore until its entry into the leaf.

## IV. QUALITATIVE ASPECTS OF THE EXTERNAL RELATIONSHIPS BETWEEN PATHOGENS AND OTHER ORGANISMS

### A. Highly Specific Relationships

It goes without saying that highly specific relationships are best exemplified by many of the virus/insect combinations. These relationships are often long-term, and they cover all aspects of the pathogen's activities outside the host. Some fungal and bacterial pathogens are also closely associated with particular vectors.

### B. Moderately Specific Relationships

Examples here are drawn mainly from relationships in which only one or two of the pathogen's problems are solved, albeit in a very efficient manner, by a relationship with another organism. Entry into a host may be facilitated by the increased availability of wounds, superficial growth may be encouraged by the increased amounts of exudates, and transport between hosts may be provided without any further assistance with either survival or reentry. Entry through wounds is commonplace amongst numerous postharvest pathogens, and several are able to capitalize on damage caused by insects and other pests. Examples have been cited earlier concerning the enhanced development of *Verticillium* and *Fusarium* around already diseased or damaged roots, which are leaking larger amounts of exudates. A number of bacteria and fungi are transported between hosts by other organisms whose activities are also centered on the relevant plant. *Erwinia amylovora* is spread by pollinating insects, and *Verticillium* conidia are carried through the soil by animals that feed on root debris.

## C. Casual Relationships

The success of many pathogens, especially those with necrotrophic tendencies, is frequently affected by numerous factors that operate in a very general manner to either encourage or deter the pathogen. It is likely that pathogens on roots and leaves are frequently stimulated as a result of changes in the environment and/or the host brought about by the growth of nonpathogenic organisms. Exudates from other microbes, nutrients released during decomposition processes, and the destruction of the host's cuticular membrane are examples of factors that may influence the success or failure of many pathogens, such as *Botrytis cinerea, Pythium ultimum,* and the sooty molds, *Alternaria alternata* and *Cladosporium cladosporioides.* Indeed, we have probably only begun to understand the interrelationships among the teeming populations of the rhizoplane and phylloplane.

## V. CONCLUSIONS

Pathogens do not function in a microbiological vacuum. This is especially true for those parts of their life cycles when they are not actually engaged in parasitism. Interactions with other organisms may benefit or harm the pathogen's development. Beneficial interactions range from specific one-to-one relationships, which solve most of the problems involved in spread and reinfection, to nonspecific systems, in which a single aspect of the pathogen's development is enhanced by the general activities of the nonpathogenic microbial population. Some of these interactions may be regarded as examples of synergism, but others falls outside most definitions of this term.

There have, to date, been relatively few detailed studies of synergism in plant pathology. Furthermore, many experiments have not been carried out in such a manner that the precise timing and nature of the relationship can be ascertained. Sometimes it is possible to make useful deductions, as, for example, when data indicated that prior infection of barley by *Puccinia* influenced the subsequent infection of the leaves by *Erysiphe* (Simkin and Wheeler, 1974). More often one can only guess at the probable course of events.

We are continually refining our knowledge of the factors that influence the development of pathogens. Precise control of temperature, humidity, and light have yielded fascinating information concerning the influence of these factors on various pathogens. Perhaps before we go too far along this road we should reevaluate our attitude toward the biological component of the environment. Fawcett attempted to inject a note of caution in 1931, and the fragmentary nature of the evidence that has accumulated since then suggests that his warning needs repeating. We are indeed still a long way from the first volume on synergistic interactions, which would compare with that on the opposite topic by Baker and Cook (1974).

### References

Ashworth, L. J., Jr., Rice, R. E., McMeans, J. L., and Brown, C. M. (1971). The relationship of insects to infection of cotton bolls by *Aspergillus flavus*. *Phytopathology* **61**, 488–493.

Austin, B., Goodfellow, M., and Dickinson, C. H. (1978). Numerical taxonomy of phylloplane bacteria isolated from *Lolium perenne*. *J. Gen. Microbiol.* **104**, 139–155.

Baker, K. F., and Cook, R. J. (1974). "Biological Control of Plant Pathogens." Freeman, San Francisco, California.

Barlow, P. W. (1975). The root cap. In "The Development and Function of Roots" (J. G. Torrey and D. G. Clarkson, eds.), pp. 21–54. Academic Press, New York.

Bergeson, G. B., Van Gundy, S. D., and Thomason, I. J. (1970). Effect of *Meloidogyne javanica* on rhizosphere microflora and *Fusarium* wilt of tomato. *Phytopathology* **60**, 1245–1249.

Beute, M. K., and Lockwood, J. L. (1968). Mechanism of increased root rot in virus-infected peas. *Phytopathology* **58**, 1643–1651.

Blakeman, J. P., and Brodie, I. D. S. (1976). Inhibition of pathogens by epiphytic bacteria on aerial plant surfaces. In "Microbiology of Aerial Plant Surfaces" (C. H. Dickinson and T. F. Preece, eds.), pp. 529–557. Academic Press, New York.

Blakeman, J. P., and Parbery, D. G. (1977). Stimulation of appressorium formation in *Colletotrichum acutatum* by phylloplane bacteria. *Physiol. Pl. Path.* **11**, 313–325.

Campbell, R., and Rovira, A. D. (1973). The study of the rhizosphere by scanning electron microscopy. *Soil Biol. Biochem.* **5**, 747–752.

Faulkner, L. R., Bolander, W. J., and Skotland, C. B. (1970). Interaction of *Verticillium dahliae* and *Pratylenchus minyus* in *Verticillium* wilt of peppermint: Influence of the nematode as determined by a double root technique. *Phytopathology* **60**, 100–103.

Fawcett, H. S. (1931). The importance of investigations on the effects of known mixtures of micro-organisms. *Phytopathology* **21**, 545–550.

Federation of British Plant Pathologists (1973). "A Guide to the Use of Terms in Plant Pathology," Phytopathol. Pap. No. 17. Commonw. Mycol. Inst., Kew.

Fry, P. R., and Campbell, R. N. (1966). Transmission of tobacco necrosis virus by *Olpidium brassicae*. *Virology* **30**, 517–527.

Garrett, S. D. (1970). "Pathogenic Root-Infecting Fungi." Cambridge Univ. Press, London and New York.

Godfrey, B. E. S. (1976). Leachates from aerial parts of plants and their relation to plant surface microbial populations. *In* "Microbiology of Aerial Plant Surfaces" (C. H. Dickinson and T. F. Preece, eds.), pp. 433–439. Academic Press, New York.

Goodfellow, M., Austin, B., and Dawson, D. (1976). Classification and identification of phylloplane bacteria using numerical taxonomy. *In* "Microbiology of Aerial Plant Surfaces" (C. H. Dickinson and T. F. Preece, eds.), pp. 275–292. Academic Press, New York.

Heagle, A. S. (1973). Interaction between air pollutants and plant disease. *Annu. Rev. Phytopathol.* **11**, 365–388.

Holman, W. L., and Meekison, D. M. (1926). Gas production by bacterial synergism. *J. Infect. Dis.* **39**, 145–172.

Jeffree, C. E., Baker, E. A., and Holloway, P. J. (1976). Origins of the fine structure of plant epicuticular waxes. *In* "Microbiology of Aerial Plant Surfaces" (C. H. Dickinson and T. F. Preece, eds.), pp. 119–158. Academic Press, New York.

Kassanis, B. (1963). Interactions of viruses in plants. *Adv. Virus Res.* **10**, 219–255.

Klincāre, A. A., Krēslina, D. J., and Mishke, I. V. (1971). Composition and activity of the eiphytic microflora of some agricultural plants. *In* "Ecology of Leaf Surface Micro-Organisms" (T. F. Preece and C. H. Dickinson, eds.), pp. 191–201. Academic Press, New York.

McKeen, C. D., and Mountain, W. B. (1960). Synergism between *Pratylenchus penetrans* (Cobb) and *Verticillium albo-atrum* R. & B. in eggplant wilt. *Can. J. Bot.* **38**, 789–794.

Marchant, R. (1970). The root surface of *Ammophila arenaria* as a substrate for micro-organisms. *Trans. Br. Mycol. Soc.* **54**, 479–483.

Martin, J. T., and Juniper, B. E. (1970). "The Cuticles of Plants." Arnold, London.

Pieczarka, D. J., and Abawi, G. S. (1978). Effect of interaction between *Fusarium, Pythium* and *Rhizoctonia* on the severity of bean root rot. *Phytopathology* **68**, 403–408.

Reinert, R. A., Heagle, A. S., and Heck, W. W. (1975). Plant responses to pollutant combinations. *In* "Responses to Pollution" (A. Stern, ed.), pp. 159–177. Academic Press, New York.

Rochow, W. F. (1972). The role of mixed infections in the transmission of plant viruses by aphids. *Annu. Rev. Phytopathol.* **10**, 101–124.

Russell, G. E. (1970). Interactions between diseases of sugar beet leaves. *NAAS Q. Rev.* **87**, 132–138.

Shriner, D. S. (1977). Effects of simulated rain acidified with sulphuric acid on host parasite interactions. *Water Air Soil Pollut.* **8**, 9–14.

Sidhu, G., and Webster, J. M. (1977). Predisposition of tomato to the wilt fungus (*Fusarium oxysporum lycopersici*) by the root-knot nematode (*Meloidogyne incognita*). *Nematologica* **23**, 436–442.

Simkin, M. B., and Wheeler, B. E. J. (1974). Effects of dual inoculations of *Puccinia hordei* and *Erysiphe graminis* on barley, cv. Zephyr. *Ann. Appl. Biol.* **78**, 237–250.

Teakle, D. S. (1969). Fungi as vectors and hosts of viruses. *In* "Viruses, Vectors and Vegetation" (K. Maramorosch, ed.), pp. 23–54. Wiley (Interscience), New York.

Tu, J. C., and Ford, R. E. (1971). Maize dwarf mosaic virus predisposes corn to root rot infection. *Phytopathology* **61**, 800–803.

Tukey, H. B., Jr. (1971). Leaching of substances from plants. *In* "Ecology of Leaf Surface Micro-Organisms" (T. F. Preece and C. H. Dickinson, eds.), pp. 67–80. Academic Press, New York.

Whitney, E. D. (1971). Synergistic effect between *Heterodera schachtii* and *Pythium ultimum* on damping off of sugar beet vs. additive effect of *H. schachtii* and *P. aphanidermatum*. *Phytopathology* **61**, 917.

*Chapter 6*

# Internal Synergisms among Organisms Inducing Disease

## N. T. POWELL

## I. INTRODUCTION

The germ theory of disease is by far the most significant concept ever developed in pathology. This principle (the word "principle" can now replace "theory," I believe) holds that the influence of a primary inducing factor (pathogen, parasite, "germ"), along with other necessary conditions such as favorable environment, is responsible for disease.

Today it is difficult to imagine the thoughts and attitudes of those who were concerned with plant disease before the germ principle became accepted and understood. A "pathological" world without such an understanding is completely foreign to us because this principle provides the foundation on which our entire science rests. It permits us to understand disease causality, infection processes, disease develop-

PLANT DISEASE, VOL. IV

ment, symptom expression, development of epidemics, and, most importantly, the principles of disease management and control.

Many great scientific discoveries lead to simple generalizations that illuminate complicated phenomena. The germ principle is a good example—it made plant disease comprehensible for the first time. However, this powerful principle also led to some oversimplification in our perceptions of disease causality.

Koch's postulates provided part of the experimental rationale on which the germ principle was founded. They also provided the means by which the specific etiological agents of hundreds, even thousands of diseases of plants, animals, and human beings were identified. The germ principle and Koch's postulates became coupled in our minds. The doctrine of specific etiology came to dominate the way plant pathologists thought, talked, and wrote in our scientific papers. For further discussion of these matters see also Volume I, Chapter 2, and this volume, Chapter 1.

It is a natural goal of scientists to seek the simplest explanation possible. In plant pathology, for example, we have attempted to ascribe disease resistance to one or a few "factors," to explain the physiology of diseased plants on the basis of a single toxin, one growth hormone, or a few enzymes, to minimize the complexities of predisposition, and to overlook the possibilities of multiple-pathogen etiology. The simple explanation may be entirely appropriate in many instances—but not in all. Nature rarely takes things "one at a time."

Plants are prisoners of their environment. They generally lack the ability to move from place to place and to escape from a less favorable environment to one that would be more favorable. Also, they lack the ability to modify their surroundings substantially, to obtain materials that they need, or to avoid influences that may be detrimental. Thus, plants are subjected to whatever the environment offers, or fails to offer.

From the plant's perspective, biotic pathogens and other stress factors such as drought and air pollutants constitute threatening components of their environment. Moreover, these threatening components may include many members of a given group of pathogens, such as the fungi, or there could be diverse types of pathogens present at the same time, including fungi, bacteria, nematodes, viruses, and others. An individual plant may well be influenced by one, several, or all of these potential pathogens. If a plant can be attacked by any one of these organisms under suitable environmental conditions, why should it not be attacked by more than one? After a plant is infected by one entity, does it then become "off limits" to subsequent attack by others? It seems naive even to think of pathogens "standing in line" waiting for the effect of one to be complete before it begins its own infective processes, especially if our ultimate interest is practical disease control. To do so is to disregard the way things really are in nature. As Fawcett (1931) put it many years ago, "Nature does not work with pure cultures." In many cases, diseases of complex etiology may be more common than diseases of specific etiology. Yet the preponderance of phytopathological literature deals with monopathogen disease.

If, or when, a plant is jointly infected by more than one pathogen, it would also seem reasonable to expect the activities and influences of one pathogen to influence the activities and influences of other pathogen(s). A healthy plant once infected is never the same again; it is altered in some way, at first, perhaps in only a single way, but later many different physiological functions are affected. This is inferred in definitions of disease that refer to "injurious physiological activity" (Whetzel, 1935), or "injurious alterations of . . . ordered processes" (Bateman, Chapter 3, Volume III). In other words, the physiology of the host plant is changed in such a way that its susceptibility to invasion by other pathogens is also changed. Such interactions may result in diminished susceptibility, as in the case of cross-protection among viruses (see Volume V), or it may result in synergism, i.e., the concurrent or sequential pathogenesis of a host plant by two (or more) pathogens in which the combined effects of the pathogens are greater than the sum of the effects of each pathogen acting alone (adapted from Agrios, 1969).

In this chapter, we will discuss the concept of internal synergism and certain aspects of multiple pathogen etiology. The objective is not to review the literature dealing with these subjects, but rather to discuss the theory of synergistic relationships and the importance of these phenomena and to point out selected examples. Nematode interactions with other pathogens will be emphasized because our understanding of synergism involving nematodes is more fully developed than our understanding of that involving only other types of pathogens.

This discussion will be restricted to the influence that one pathogen exerts within the host; factors external to the host, such as root exudates and vector relationships are treated in Chapter 5.

## II. COMPONENTS OF A SYNERGISTIC RELATIONSHIP

In its simplest form, the doctrine of specific etiology provides a rationale for analysis of pathogenesis as a two-component system—the host and the pathogen, which interact over time. In the case of synergistic diseases complexes, at least three components are involved—the host and two or more pathogens, which become engaged with the host either sequentially or concomitantly. We shall discuss each system in turn.

## A. Sequential Etiology

Sequential disease complexes are those in which one component of the pathogen complex (the "primary" pathogen) infects and alters the host in advance of subsequent invasion by other ("secondary") pathogen component(s). The terms "primary" and "secondary pathogens" are appropriate for two major reasons: (1) Primary pathogens invade the host prior to invasion by secondary pathogens; and (2) primary pathogens govern the development of the synergistic association by altering the host in such a way that its tissues become suitable for colonization by secondary pathogens.

By definition, all primary pathogens possess the *inherent, specific,* and *independent* capability of causing disease. Thus, they can induce disease whether or not secondary pathogens become involved. Of course, secondary pathogens can drastically alter the pattern of disease development and may eventually dominate the symptom syndrome of the disease complex.

It is interesting, and perhaps significant, that many primary pathogens are obligate parasites that induce various disturbances (galls, syncytia, etc., but rarely necrosis) of host tissues. Nematodes and viruses are the most commonly recognized primary pathogens.

Secondary pathogens, unlike primary ones, *may or may not* have the inherent and specific capability of inducing disease; thus, they may be dependent or independent with respect to the primary pathogen.

Fungi and bacteria are the most commonly recognized secondary pathogens. Most of them are facultative pathogens; obligately parasitic organisms rarely develop as secondary pathogens.

A distinction must be made between "secondary pathogens" and "secondary invaders." Secondary pathogens actively participate in and alter the course of pathogenesis, whereas secondary invaders merely colonize dead cells that remain in lesions induced by primary pathogens.

The time interval between invasion by the primary and secondary pathogens can vary enormously—from a day or so to several weeks. In most sequential disease complexes, the secondary pathogen eventually dominates the clinical picture. In some cases, this domination can be so complete that the primary pathogen virtually disappears from the infected host. This should not be interpreted as a basis for argument about which pathogen is primary and secondary. Primary pathogens enter first and critically alter the host so as to enhance its suitability for colonization by the secondary pathogen(s). Thus, the role of the primary pathogen is so important that control of sequential disease complexes is dependent primarily on the efficacy of strategies and tactics aimed at suppression of populations of the primary pathogen.

"Biopredisposition" is another term that is applicable to diseases of sequential etiology. The nature of the predisposition is determined by the primary pathogen as it goes about its pathogenic activities. The question of whether predisposition leads to synergism is determined also by the size and invasive force of the populations of secondary pathogens in the host environment. An extreme case of biopredisposition is the disease complex involving infection of tobacco roots by root-knot nematodes followed by extensive decay of the infected roots by such organisms as *Trichoderma* and *Curvularia*. Without the primary nematode pathogen, these fungi are incapable of penetrating, let alone inducing disease in, tobacco roots (Powell *et al.*, 1971).

## B. Concomitant Etiology

Concomitant disease complexes occur when two (or perhaps more) pathogens infect the host simultaneously. Usually, the resulting disease is entirely different from, or at least more severe than, that which is induced by either pathogen acting alone. In these associations, it is not appropriate to consider primary or secondary pathogens; both (or all) pathogens are equally necessary to produce the particular disease syndrome.

The rugose mosaic disease of potato is a particularly good illustration of concomitant synergism. Symptoms of rugose mosaic appear only in plants infected jointly by the potato virus X and potato virus Y; infection by either virus alone results in very different and much less severe symptoms.

The term "biopredisposition" is difficult to apply to diseases of concomitant etiology. These diseases appear to involve interactions *between the pathogens* rather than between the host and each of its pathogens. Relatively few concomitant synergistic associations have been elucidated, but nature may harbor many more that are yet to be discovered.

## III. POSSIBLE MECHANISMS OF INTERNAL SYNERGISM

The mechanisms by which primary pathogens increase susceptibility to secondary pathogens are not well understood, but then neither are the mechanisms of susceptibility to the primary pathogens themselves. Investigators of susceptibility in general might well consider using secondary invaders as test organisms. It would give them an additional reference line in their efforts to discover the mechanisms involved.

In working with single pathogens, only two points are available:

a healthy host and a diseased host. If a secondary pathogen is added, one can look at a healthy host, at a host modified by a primary pathogen, and at the same host modified by two pathogens. What does the primary pathogen do to the host? Elucidating this would help to understand susceptibility to the secondary pathogen. Potato leaf roll, for example, inhibits movement of photosynthate out of the leaf. As a result the leaf is engorged with a surfeit of photosynthate, while the normal carbohydrate sinks in the stem, root, and tuber are starved. What do these easily established effects have on possible secondary invaders? We shall examine this further in the following paragraphs.

For convenience of discussion, one can consider three theoretical mechanisms of biopredisposition involving two pathogens: (1) The primary pathogen may make the host tissue more susceptible to the secondary pathogen; (2) the primary pathogen may enhance the activity of the secondary pathogen; and (3) the secondary pathogen may even enhance the activity of the primary pathogen. These three categories, although arbitrary and artificial, may be helpful in thinking about these relationships.

## A. Increased Host Susceptibility

### 1. Increased Penetration

For many years it was assumed that wounding by nematodes increased the susceptibility of plants to invasion by other pathogens. This assumption seemed reasonable because of the manner of nematode ingress into plants. It still appears logical and unquestionably applies in some cases such as diseases involving migratory nematodes. However, in other instances (in fact, in most cases) much more complicated mechanisms appear to be involved.

If wounding were of paramount importance, enhanced susceptibility would be at its maximum at the time recently hatched larvae penetrate host roots. In fact, the susceptibility induced by root-knot nematodes often peaks 3 or 4 weeks later. Also wounding could not account for the infection by secondary pathogens far away from the nematode infection. Clearly, we must look further for an explanation for many cases of increased susceptibility.

### 2. Transmission of a Message by the Host

Faulkner et al. (1970) devised experiments in which detached peppermint stems were rooted at two places, thereby providing two connecting plants, each with its own distinct root system. Lesion nematodes increased the incidence and severity of verticillium wilt, even when the two pathogens were inoculated onto the separate root systems of connected plants. Thus, a translocatable "messenger" seems to be involved, although its nature is not known. Sidhu and Webster (1977) demonstrated a similar phenomenon with root-knot nematodes and *Fusarium oxysporium lycopersicae*. Perhaps the messenger is a growth-regulating substance produced in the nematode gall, since Sequeira (1963) states that "hyperauxiny occurs in wilt diseases of both fungal and bacterial origin."

If the phloem tissue in the petiole or stem of a potato plant is infected by the potato leaf roll virus (Richardson and Doling, 1957), the leaves above the virus infection become more susceptible to infection by *Phytophothora infestans*. When the petoile or stem is damaged, carbohydrates cannot be translocated from the leaf blades to the roots and other metabolic sinks below. Then they back up into the leaf and increase its susceptibility to *P. infestans*. A similar effect is found when wheat grains are killed by *Tilletia tritici*. When the metabolic sink is gone, the leaves below become engorged with sugars (Gaümann, 1950) and leaf rust (a high-sugar disease) builds up (Dillon-Weston, 1927).

The stalk-rot and root-rot diseases of maize are low-sugar diseases caused by *Gibberella* and *Fusarium*. *Helminthosporium* infection on the leaves, partial defoliation, and smut galls on the stalk all sharply decrease the amount of sugar in the stalks and their susceptibility to stalk rot (Michaelson, 1957). The reverse happens if the ears of the corn plant are removed; under these circumstances the sugars move downward, and this reduces susceptibility of the stalks to these low-sugar diseases.

A more detailed explanation of the effect is shown by the early blight of tomato, a low-sugar disease induced by *Alternaria solani*. The fungus can invade and thrive on the sugars in young leaves without damaging them. Sands and Lukens (1974) have shown that sugars inhibit the action of the destructive enzymes of *A. solani*. When the tomato fruits are formed, sugar is translocated from the old leaves into the fruits. When the surplus of sugar disappears from the leaves, the fungus activates its pectinolytic and cellulolytic enzymes to provide its sugar supply and, in the process, causes damage to the leaves. In passing, we might refer again to the increase in leaf sugar associated with stem and petiole infections with the potato leaf roll virus. In such plants, *A. solani* has all the sugar it needs, so it lives on the leaves without damaging them, and no disease is visible on the leaf roll-affected plants.

### 3. Shortened Incubation Period

Bergeson (1963) states that *Pratylenchus penetrans* shortens the incubation period of *Verticillium* in peppermint. Also, Faulkner *et al.* (1970) showed that the incubation period is shortened for *Verticillium* even on that portion of peppermint plants emanating from the non-infected part of a divided root system. Clearly, the secondary pathogen enjoys a better life in the nematode-invaded plant than in nematode-free plants; apparently it finds more easily the energy needed to speed up its growth and development.

### 4. Shift in Optimal Temperature

Faulkner and Bolander (1969) observed a curious phenomenon with *Pratylenchus minus* and *Verticillium*. With *Verticillium* alone, the wilt developed optimally at 24°C, but this shifted upward to 27°C when the nematode was present. The population of nematodes in the soil was greatest at 30°C. This suggests that the nematode shifts susceptibility of the plant to wilt disease towards its own optimum temperature.

### 5. Inherent Host Resistance Overcome

We now come to a difficult twilight zone between susceptibility of the host and virulence of secondary pathogens. This is the phenomenon by which primary pathogens overcome the resistance of resistant culti-vars. I believe that this is more likely to be due to exaggerated host susceptibility than to increased pathogen virulence.

Root-knot nematodes worsen the attack of fusarial wilt on the resistant Alaska pea, of fusarial wilt on resistant tomato cultivars, and of *Phytophthora parasitica* on resistant tobacco (reviewed in Powell, 1971). As mentioned above, perhaps auxins are involved.

In a related case, root-knot nematodes can markedly increase the attack on tobacco by certain soil fungi such as *Pythium ultimum* and others that are normally harmless on this host (Powell *et al.*, 1971).

### 6. Saprophytes Converted to Pathogens

Results such as those outlined due to root-knot nematodes in combination with nonagressive pathogens on tobacco have led to work using *Meliodogyne incognita* with a variety of soilborne fungi that are not regarded as pathogens of this crop (Powell *et al.*, 1971). *Curvularia trifolii, Botrytis cinerea, Aspergillus ochraceus, Penicillium martensii,* and *Trichoderma harzianum* are common inhabitants of soils on which tobacco is grown, but none of these fungi have been reported to be pathogenic on tobacco. Some of them have been incriminated as disease causal agents for certain other plants. When any of these organisms are added to tobacco roots in the greenhouse, no necrosis follows; there is no evidence of invasion. On the other hand, if any of these fungi are added to roots that have been exposed to *M. incognita* for 3–4 weeks, fungal invasion occurs, followed by extensive decay. In fact, root destruction by any one of these fungi in the presence of

*M. incognita* is as great as that due to either *Pythium* or *Rhizoctonia* under similar conditions.

This work underscores the outstanding abilities of root-knot nematodes to induce susceptibility in plants to attack by other organisms. These abilities seem to apply for whatever fungus happens to be present in the rhizosphere—whether or not they are recognized pathogens. This, in turn, causes one to wonder about the nature of the physiological changes induced in plants by *Meloidogyne* spp. that produce the effect.

## B. Enhancing the Capabilities of Secondary Pathogens

It appears that root-knot nematodes frequently improve the nutritional status of invading necrogenic fungi and perhaps other invaders as well.

Histological studies of roots infected jointly with certain sedentary nematodes and various fungus pathogens have given some insight into possible mechanisms of synergism. Powell and Nusbaum (1970) have shown that galled roots are more susceptible to fungus invasion than nongalled roots. Giant cells are particularly susceptible to invasion and degenerate rapidly following colonization by several fungi. Areas surrounding the developing nematode, as well as hyperplastic tissues, are also readily colonized. These histopathological findings suggest changes in nematode-infected roots that render them very susceptible to subsequent fungus invasion (Melendéz and Powell, 1967).

*Meloidogyne* galls are very rich in amino acids in comparison to healthy tissues, and most of these increases are found within giant cells. According to Owens and Specht (1966), the magnitude of amino acid increase depends on the age and metabolic state of the giant cells. Viglierchio and Yu (1968) found several different auxins in nematode-induced galls and stated that tissues infected with different nematode species may contain different forms of auxin. Some unknown growth-promoting and growth-inhibiting substances have been reported by Bird (1962). Sandstedt and Schuster (1966) suggested that root-knot nematodes, in addition to producing unknown growth regulators, may also enable infected tissues to retain and utilize auxins that would otherwise be transported to other plant parts.

Nucleotides in nematode galls increased as compared to healthy tissues (Owens and Specht, 1966). Also, galls contained more RNA, DNA, and both inorganic and organic phosphorus. This was correlated with bursts of DNA and RNA synthesis accompanying giant cell formation.

Owens and Bottino (1966) studied changes in the composition of host cell walls induced by root-knot nematodes and found that total wall components decreased from 70% of the total cell solids in normal roots to 60% in galls. Major quantitative differences were an increase in polyhexuronic acid and a decrease in cellulose. In healthy roots, xylans, arabans, and mannans constituted most of the noncellulosic polysaccharides, and xylans and gluocosans were the major polymers in galls. Except for sucrose, which decreased by about 20%, free sugar content of galled tissue was reduced to slightly more than one-half that of healthy root tissues.

Obviously, the changes mentioned above represent tremendous departures from the normal but constitute only examples of what might occur in nematode-infected plants. Such important effects must play a part in predisposing infected plants to infection by other ogranisms. It is to these changes that one must look to define the physiological basis for many complexes.

## C. Increasing the Capabilities of Primary Pathogens

It is difficult to illustrate cases of internal synergism in which the activity of the primary pathogen is enhanced as a result of subsequent activity by a secondary pathogen. However, it must be acknowledged in principle as a possible factor in such relationships. With migratory plant parasitic nematodes, for example, populations increase when any one of several fungi are present in the plant roots. Furthermore, larvae aggregate around such fungal-infected roots and penetrate in greater numbers than is the case with fungal-free roots. It is obvious, therefore, that circumstances exist in which the primary pathogen is enhanced in some fashion by the secondary pathogen after the relationship has been established (reviewed in Powell, 1971).

## IV. SELECTED EXAMPLES OF INTERNAL SYNERGISM

The following paragraphs contain examples of internal synergism among various types of pathogens. The cases included here are by no means exhaustive. They are included mainly to show both the extent and limitations of our present knowledge of multiple-pathogen complexes.

## A. Nematodes

### 1. Root-Knot Nematodes (Meloidognye sp.)

Root-knot nematodes have long been known for their ability to predispose plants to infection by secondary pathogens. The first published account of biopredisposition is by Atkinson (1892), who noted a very strict relationship between *Meloidogyne* sp. and fusarium wilt in cotton. The two pathogens were virtually inseparable in plants exhibiting fusarium wilt symptoms, and the combination of pathogens always contributed to more severe losses from wilt. This still holds true today, not only with cotton, but with many other crops such as tomato, tobacco, and soybean (Powell, 1971). In fact, the root knot–*Fusarium* interaction has become the best known nematode–fungus complex in plant pathology.

Probably the first example of careful analysis of a nematode–fungus complex dates back to the classic description of root-knot symptoms by Göldi (de Souza, 1978). He stated:

> Among thousands of samples from roots of different diameters from diseased and dying coffee plants, very seldom indeed have I found one sample without the mycelium of a fungus in and around the vascular vessels . . . this fungus is inseparable from this coffee disease. . . . So we have two causal agents, one of animal nature and the other of plant nature, each one working by different methods, but both having the same end which is the destruction of the plant host.

This fungus is currently thought to have been *Rhizoctonia* and/or *Rosellinia* sp.

*Meloidogyne* sp. are involved with other fungi. Reports from all over the world show that root-knot nematodes can predispose plants to invasion by different species of *Phytophthora, Rhizoctonia, Verticillium,* and *Pythium.* All of these fungi are recognized as pathogens of certain hosts in their own right; e.g., they do not *require* biological predisposition. Nevertheless, root-knot nematodes substantially increase the susceptibility of plants to their damaging effects.

Stronger evidence of predisposing effects of *Meloidogyne* sp. is provided in the reports of Powell *et al.* (1971) regarding the capacity of fungi in the genera *Curvularia, Botrytis, Penicillium,* and *Aspergillis* to colonize tobacco roots previously infected by root-knot nematodes. Since none of these fungi normally infect tobacco, they are absolutely dependent on the primary pathogen. The fact that *Trichoderma harzianum,* normally an innocuous or even beneficial soil saprophyte, causes extensive decay of *Meloidogyne*-infected roots attests further to the "power" of this predisposition (Powell *et al.* 1971). These results show that root-knot nematodes are capable of effectively extending host ranges of these secondary pathogens by some form of biopredisposition.

Bacterial infections also are promoted by either previous or simultaneous infection by *Meloidogyne* sp. Examples may be found in the cases of *Pseudomonas solanacearum* (which induces bacterial wilt) on several different hosts and *Corynebacterium michiganense,* the inducer of bacterial canker on tomatoes (Johnson and Powell, 1969; de Moura *et al.*, 1975). In the latter case, predisposing effects are noted in stem tissue considerably above the root infection site of the nematodes. Plants that harbor root-knot nematodes frequently show very different and much more severe symptoms of virus infection (Osores-Duran, 1967). In fact, the entire disease syndrome may be altered in such mixed infections, even though infection sites, infection processes, and other important differences exist in the nature of the two pathogens. All of these pathogens are appropriately regarded as secondary but independent pathogens.

Although the influence of root exudates from plants infected by one pathogen or infection by other pathogens was mentioned in Chapter 5, proper perspective mandates that they be considered also in relation to biopredisposition by *Meloidogyne* spp. The outstanding work by Van Gundy *et al.* (1977) has clearly shown the role of root exudates in certain root knot–fungus interactions.

## 2. Cyst Nematodes (Heterodera sp.)

The cyst-forming nematodes, *Heterodera* sp., constitute important plant pathogens of many crops in various parts of the world. They may be limiting factors to successful production of plants such as potato, sugar beet, and soybean. As are *Meloidogyne* sp., cyst nematodes are sedentary, that is, they are stationary in relation to the host. Syncytia are formed, although such morphological changes in the host are less extensive than those induced by root-knot nematodes. The two groups are closely related from evolutionary viewpoints.

Disease complexes in which *Heterodera* sp. are primary pathogen components are well-documented and also represent cases of internal synergism. The secondary pathogens involved include members of several fungus genera such as *Fusarium*, *Phytophthora*, and *Rhizoctonia*, among others (Powell, 1971). Disease development often becomes so severe that control of the disease complex is very difficult and uncertain.

Interactions of cyst nematodes with other organisms have not been subjected to as much or detailed analysis as those relating to root-knot nematodes. Nevertheless, research has generally indicated that the two nematode pathogens are quite similar in their participation in complexes. This is to be expected when one realizes the similarities between them and their host-pathogen relations. At any rate, evidence of the complicity of *Heterodera* in internal synergistic association is convincing.

## 3. Migratory Nematodes

Although sedentary nematode plant pathogens are considered by many plant pathologists and nematologists to be the more important forms, certain genera of migratory nematodes are of great significance in many areas of the world. *Pratylenchus* sp., often commonly referred to as "lesion nematodes," are important migratory plant pathogens. All stages of the life cycle are mobile, but they are endoparasitic in that they feed and move entirely within the plant root. Gross host morphological changes typically do not take place, but pockets of necrosis result from nematode feeding, and some tissue disruption occurs due to nematode migration from one feeding site to another.

*Pratylenchus* sp. do not alter the host as obviously as do sedentary nematode parasites, but they function as primary pathogens in several synergistic associations with other organisms. Verticillium wilt is often much more serious when the *Verticillium* fungus (independent secondary pathogen) acts together with lesion nematodes. This multiple-pathogen complex is frequently observed in crops such as cotton, potato, eggplant, and peppermint. This strongly suggests that synergism occurs here and is manifested in effects on disease development.

Other fungi. such as *Phytophthora*, may also serve as independent secondary pathogens with *Pratylenchus* sp. These interactions have not been subjected to the critical analysis afforded verticillium wilt on peppermint (Faulkner *et al.*, 1970), but development of the complex indicates that wounding and tissue disruption is a strong predisposing factor with these migratory endoparasites.

Most migratory nematode plant pathogens are ectoparasites; that is, they remain outside the host and feed by inserting only a part of their head into host tissue. With some forms, only the stylet is actually inserted into the host. Such nematode pathogens predispose plants to infection by certain fungi and bacteria.

Mechanical injury by the nematode seems to be the main predisposing effect. The nematode wounds the plant, and this probably paves the way for invasion by secondary pathogens that would be inconsequential without such injury. Thus, the secondary pathogen components are usually those that are enhanced by such factors.

This does not deny the possible influence of other more intrinsic predisposing factors. The work of Faulkner *et al.* (1970), for example, implys that, at least with *Pratylenchus*, as yet unknown internal predisposing factors are translocated from the site of nematode activity to other areas within the plant.

The sting nematodes, *Belonolaimus* sp. are notable primary pathogens; they exert powerful predisposing effects on certain plants, rendering the latter increasingly susceptible to fusarium wilt and possibly other diseases. This relationship has been especially noted in cotton (Yang *et al.*, 1976). Stunt nematodes, *Trichodorum* sp., are also reportedly involved in complexes with pathogens of other types (Liu and Ayola, 1970).

Studies of the synergistic relationships between migratory nematodes and various secondary pathogens, although few in number, indicate that the mechanisms of synergism involved are quite different from those of sedentary nematodes. This is to be expected, since the diseases induced by the primary pathogens are so obviously different. With migratory nematode forms, the profound anatomical changes and accompanying changes in root physiology are lacking; rather, necrosis and other manifestations of destruction develop. There is not nearly the balanced or compatible situation between host and pathogen. This clearly is related to the migratory versus sedentary style of life.

## B. Fungi

Thousands of plant diseases are induced by fungi, but few involve synergism with other pathogens.

There are instances in which one fungus pathogen *impedes* development of another pathogen (Caruso and Kuć, 1977a,b; Hammerschmidt *et al.*, 1976), but this is just the opposite of synergism and will be discussed in Volume V of this treatise.

In addition to the examples of synergism cited earlier in the discussion on possible mechanisms, certain other cases are worthy of mention. For example, *Tilletia tritici* increases susceptibility of wheat foliage to infection by *Puccinia glumarum* (Dillon-Weston, 1927). Similarily, powdery mildew increases susceptibility of wheat to leaf rust (Simkin and Wheeler, 1974).

## C. Viruses

Like many of the sedentary nematodes, viruses are discrete pathogens; they depend on a living host for their own perpetuation and rarely incite necrosis. Virus infections are studies in compatibility between host and pathogen. The pathogen—the virus particle itself—*becomes a part of the host*. Truly, then, a plant so diseased can never be the same again—there will forever be, as a part of its makeup, something foreign that induces it to appear and respond in ways different from the "normal." Thus, viruses serve as primary pathogens in several synergistic disease complexes.

Potato plants infected by potato virus Y show increased susceptibility to infection by *Alternaria solani* (Hooker and Fronek, 1960). A similar synergistic association has been noted between the leaf-roll virus and *Phytophthora infestans* (Richardson and Doling, 1957; de Cubillos and Thurston, 1975). A classic example of a virus serving as a primary pathogen involves the barley yellow dwarf pathogen along with a fusarial root rot in barley. Here the effects of the secondary pathogen (the fungus) eventually dominate the disease syndrome.

Quite commonly, two or perhaps more viruses relate synergistically within a given host plant to produce increased disease losses, or even completely different symptom expressions, when compared to plants harboring only one virus pathogen. Examples exist in certain soybean diseases and with the potato mosaic viruses (Ross, 1969; Rochow and Ross, 1955). Undoubtedly, many situations of this kind exist in nature that have not as yet been resolved experimentally but appear to be concomitant associations.

### D. Bacteria

Although the number of genera containing phytopathogenic bacteria is small, these "biochemical factories" seem virtually ubiquitous where plants are grown and may serve as "secondary invaders," often leading to decay and substantial economic losses. As such, they become secondary pathogens and are usually dependent on the primary one.

The secondary invader's role often played by phytopathogenic bacteria is consistent with their typical mode of pathogenesis and relationships with hosts. Bacteria rarely, if ever, gain entry into host cells, at least while the latter are physiologically active and normal. Host alterations induced by other pathogens thus, understandably, influence subsequent bacterial activity, especially if alterations result in death or significant cell debilitation.

Synergistic associations involving bacteria, which may not be classified as secondary invaders, do exist, however. For example, basal soft rot of carnation is measurably increased when both *Pseudomonas caryophylli* and *Corynebacterium* sp. are present together as compared to cases wherein *P. caryophylli* occurs alone. *Corynebacterium* alone causes no visible symptoms. Media studies have confirmed this synergistic effect, as have studies with detached carnation leaves (Brathwaite and Dickey, 1970, 1977). Also, disease in bean due to *Pseudomonas phaseolicola* is increased in the presence of an *Achromobacter* sp. regardless of various inoculation techniques (Maino *et al.*, 1974). The fact that *Achromobacter* may be isolated from noninoculated, symptomless plant parts suggests that this organism is involved in disease only when present along with another pathogen. These situations are obviously different from those involving secondary invaders, wherein the bacterial synergistic components are merely "opportunists" (see Kelman, Chapter 10). They serve either as secondary or primary pathogens or are involved in concomitant relationships.

## V. SOME THOUGHTS ABOUT EXPERIMENTAL APPROACHES

Precise suggestions regarding concepts and methodology appropriate for experimentation on synergistic complexes have been given by Wallace (1978) in a recent review. Only a few thoughts will be included here.

Synergistic disease complexes are not easy to work with and interpret experimentally. By nature, they are somewhat more complicated than those involving a single pathogen; perhaps this is one reason for the relative paucity of reports on experiments dealing with synergism. At any rate, various approaches that have proved useful for experimentation on these diseases are mentioned briefly below.

Experimental analysis of synergistic disease complexes usually is aimed at answering one or more of the following questions: (1) Is the disease of complex etiology? (2) What pathogens are involved? (3) Do the pathogens interact synergistically? (4) If so, is the interaction sequential or concomitant? (5) If sequential, which pathogen(s) are primary and which are secondary? (6) What is the time sequence (epidemiology) of the interaction? (7) What anatomical changes are induced by each pathogen? (8) What physiological mechanisms of synergism are involved? (9) What practices are useful in minimizing losses due to the disease complex involved?

At least indications relative to questions (1) and (2) above are based initially on observation rather than experimentation. Are multiple pathogens usually associated with the disease syndrome? Pathogen isolation, using selective media if appropriate, is usually necessary to identify complex components. This step provides the foundation for a working hypothesis on which to base further investigation. After the hypothesis has been developed, one or more of several approaches can be used to gain additional information.

### A. Greenhouse and Growth Chamber Experiments

Greenhouses and other types of plant growth facilities are helpful and even vital in determining the roles of suspected or potential partners in synergistic complexes, and their use follows logically the observations mentioned above. Experiments in greenhouses yield additional information on questions (1) and (2) and may also be designed to give data on questions (3), through (6).

These facilities permit control over many potential influences that are not possible in the field. It is also possible to follow precisely the development of disease as well as changes in the host and in component populations. Care should be exercised, however, especially with inoculation techniques and dosages, when sequential relationships are suspected. For example, with soilborne pathogens, special systems such as inoculation tubes may be needed to permit treatment applications without disturbing host plants (for description, see Powell *et al.*, 1971). The type of data taken and the time intervals for gathering such data depend on the disease syndrome. Often, disease indices must be devised.

Phytotrons, or precisely operating growth chambers, afford opportunities for careful experimentation. These facilities permit more complete control of extraneous factors that may influence results. Ultimately, however, the field is the only real proving ground.

## B. Field Experiments

Field (or "microplot") investigations are absolutely necessary for a full appreciation of natural synergistic associations and may be designed to answer questions (1) through (6) and finally (9). Two major approaches to field research are feasible. The first is to select and treat the field site to eliminate as much microflora and microfauna as possible and selectively add various complex components as desired, based on observations and preliminary tests. Treatment with a broad-spectrum biocide such as methyl bromide, applied uniformly perhaps more than once, is satisfactory when nonleguminous hosts are used. This permits overall control in that treatments can be imposed by the investigator in various dosages, at certain time intervals, and with other considerations in mind. Selective control measures, such as differentialy resistant varieties, can also be superimposed in this system.

Although this technique permits control, it is still quite unnatural. Normally present entities have been virtually negated, and the investigator could be missing contributions made by them. Nevertheless, I have often used this system in research dealing with soil-borne microorganisms.

The second approach involves taking field areas as they exist naturally and attempting to unravel the presence, absence, magnitude, and nature of the associations present in relation to a given host population. This approach usually involves the application of control treatments that have a selective effect on one or another suspected primary or secondary pathogen. For example, if a nematode is suspected as a primary pathogen, a nematicide may be used on certain experimental plots. If disease fails to develop or develops differently on nontreated plots when compared to treated ones, the assumption that nematodes are so involved is strengthened. Selective fungicides or varieties differentially resistant to certain pathogen components permit the same type of assessment, perhaps with even more precision. In fact, this technique is responsible for some of the earliest indications of occurrences of synergistic relationships.

However, this method has one obvious disadvantage—the investigator has a very incomplete picture of the composition and complexity of the experimental system. Thus, although it may be possible to reach conclusions regarding the influence of certain components of a complex, the influences of others remain unknown and could be responsible for major effects.

## C. Laboratory Experiments

The histopathology and physiology laboratories afford opportunities to gain understanding of synergistic complexes, especially relating to questions (7) and (8) mentioned previously. Histopathological observations of the pathogen components as they reside in the common host, along with host responses, can reveal intimacies of their lives together. Affected host tissues can be seen and studied, preferences of pathogens for various types of tissues can be measured, and the relationships between components can be studied. In sequential associations, vital time relationships can be measured and understood. Conventional histological techniques are usually appropriate.

Physiological analysis of jointly infected tissues compared with non-infected tissue or that infected by one pathogen should reveal much about the true basis for synergism. Although work in this area is meager, it is accelerating and information is gradually accumulating. The types of analysis—the changes looked for—should be mandated by a knowledge of the physiology of the host under normal conditions along with that of each pathogen component.

## VI. REFLECTIONS

Evidence and observations are slowly but surely pointing to the frequency with which synergistic complexes occur and the significance that they may have in agricultural and forest production. Even so, this aspect of our discipline has been barely touched as far as knowledge and understanding are concerned. At times this seems paradoxical. After all, plant pathologists are applied biologists and, as such, have an appreciation for living things, their nature, and interrelationships. Surely, we know that pathogens do not exist as "isolationists;" we must determine the magnitude of their interactions in nature, even if initially only in a descriptive fashion. One of the most important uses of our present knowledge of synergistic complexes may be to inform plant pathologists generally that there are many, many more such instances of which nothing is known.

Overall, very little is understood as to the reasons why certain types of pathogens are involved in internal synergism in certain hosts. Many factors may be involved as pathogen components, hosts, and other conditions vary. For example, nutritional changes in host populations appear to be important in root-knot nematode influences on other pathogens. Subsequent colonization by bacteria may well be affected by host cell wall changes and/or changes in host cell physiology brought about by another pathogen. With some fungi and viruses, inherent host resistance could be lessened or overcome by production of an essential metabolite that would be absent under other conditions. The resolution of these questions awaits research, but one can discern commonality based on knowledge of how the various synergistic components relate individually to the host, along with known requirements of each component.

It is obvious, however, that synergistic components differ sufficiently to merit investigation as individual complexes. The nature and behavior of all the varied organisms that may participate in these relationships are too diverse to permit complete generalization, although there are consistent principles that apply in many different cases.

Although a knowledge of the basis of synergism is interesting and important, *effective disease control is not necessarily predicated upon such knowledge.* For example, over 50 years ago fusarium wilt in certain crops was first controlled in part at least by controlling root-knot nematodes, long before there was a good understanding of the nature of the interaction.

There is ample opportunity in the area of synergism or disease complexes for strong research (and teaching) emphasis. To me, the priority lies in determining the existence and magnitude of these interactions in crop production. Studies of the nature of association, conditions under which they occur, and control should follow or accompany this priority. Until synergistic complexes receive appropriate attention, the discipline of plant pathology will remain fragmented, incomplete, and irrelevant to nature.

### References

Agrios, G. N. (1969). "Plant Pathology." Academic Press, New York.

Atkinson, G. F. (1892). Some diseases of cotton. *Ala. Polytech. Inst., Agric. Exp. Stn., Bull.* No. 41, 64–65.

Bergeson, G. B. (1963). Influence of *Pratylenchus penetrans* alone and in combination with *Verticillium albo-atrum* on growth of peppermint. *Phytopathology* **53**, 1164–1166.

Bird, A. F. (1962). The inducement of giant cells by *Meloidogyne javanica*. *Nematologica* **8**, 1–10.

Brathwaite, C. D., and Dickey, R. S. (1970). Synergism between *Pseudomonas caryophyllii* and a species of *Corynebacterium*. *Phytopathology* **60**, 1046–1051.

Brathwaite, C. D. ,and Dickey, R. S. (1971). Role of cellular permeability alteration and pectic and cellulolytic enzymes in the maceration of carnation tissue by *Pseudomonas caryophyllii* and *Corynebacterium* sp. *Phytopathology* **61**, 476–483.

Caruso, F. L., and Kuć, J. (1977a). Protection of watermelon and muskmelon against *Colletotrichum lagernarium* by *Colletotrichum lagernarium*. *Phytopathology* **67**, 1285–1289.

Caruso, F. L., and Kuć, J. (1977b). Field protection of cucumber, watermelon and muskmelon against *Colletotrichum lagernarium* by *Colletotrichum lagernarium*. *Phytopathology* **67**, 1290–1292.

de Cubillos, F. C., and Thurston, H. D. (1975). The effect of viruses on infection by *Phytophthora infestans* (Mont.) de Bary in potatoes. *Am. Potato J.* **52**, 121–126.

de Moura, R. N., Echandi, E., and Powell, N. T. (1975). Interaction of *Corynebacterium michiganense* and *Meloidogyne incognita* on tomato. *Phytopathology* **65**, 1332–1335.

de Souza, P. (1978). A disease complex of coffee involving *Meloidogyne exigua* and *Rhizoctonia solani*. Ph.D. Thesis, North Carolina State Univ., Raleigh.

Dillon-Weston, W. A. R. (1927). Incidence and intensity of *Puccinia glumarum* Eriks. & Henn. on wheat-infected and non-infected with *Tilletia tritici* Wint. showing the apparent relationships between susceptibility of wheat plants to yellow rust and to bunt. *Ann. Appl. Biol.* **14**, 105–112.

Faulkner, L. R., and Bolander, W. J. (1969). Interaction of *Verticillium dahliae* and *Pratylenchus minyus* in Verticillium wilt of peppermint: Effect of soil temperature. *Phytopathology* **59**, 868–870.

Faulkner, L. R., Bolander, W. J., and Skotland, C. B. (1970). Interaction of *Verticillium dahliae* and *Pratylenchus minyus* in Verticillium wilt of perppermint; Influence of the nematode as determined by a double root technique. *Phytopathology* **60**, 100–103.

Fawcett, H. S. (1931). The importance of investigations on the effects of known mixtures of organisms. *Phytopathology* **21**, 545–550.

Gäumann, E. (1950). "Principles of Plant Infection." Lockwood, London.

Hammerschmidt, R., Acres, S., and Kuć, J. (1976). Protection of cucumber against *Colletotrichum lagernarium* and *Cladosporium cucumerinum*. *Phytopathology* **66**, 790–793.

Hooker, W. J., and Fronek, F. R. (1960). The influence of virus Y infection on early blight susceptibility in potato. *Conf. Potato Virus Dis. Proc.* **4**, 76–81.

Johnson, H. A., and Powell, N. T. (1969). Influence of root-knot nematodes on bacterial wilt development in flue-cured tobacco. *Phytopathology* **59**, 486–491.

Liu, L. J., and Ayola, A. (1970). Pathogenicity of *Fusarium moniliforme* and *F. roseum* and their interaction with *Trichodorum christiei* on sugar cane in Puerto Rico. *Phytopathology* **60**, 1540. (Abstr.)

Maino, A. L., Schroth, M. N., and Vitanza, V. B. (1974). Synergy between *Achromobacter* sp. and *Pseudomonas phaseolicola* resulting in increased disease. *Phytopathology* **64**, 277–283.

Melendez, P. L., and Powell, N. T. (1967). Histological aspects of the Fusarium wilt–root-knot complex in flue-cured tobacco. *Phytopathology* **57**, 286–292.

Michaelson, M. E. (1957). Factors affecting development of stalk rot of corn caused by *Diplodia zeae* and *Giberella zeae*. *Phytopathology* **47**, 499–503.

Osores-Duran, A. (1967). Interactions of root-knot nematodes and systemic viruses in tobacco. M.S. Thesis, North Carolina State Univ., Raleigh.

Owens, R. G., and Bottino, B. F. (1966). Changes in host cell wall composition induced by root-knot nematodes. *Contrib. Boyce Thompson Inst. Plant Res.* **23**, 171–180.

Owens, R. G., and Specht, H. N. (1966). Biochemical alterations induced in host tissues by root-knot nematodes. *Contrib. Boyce Thompson Inst. Plant Res.* **23**, 181–198.

Powell, N. T. (1971). Interactions between nematodes and fungi in disease complexes. *Annu. Rev. Phytopathol.* **9**, 253–274.

Powell, N. T., and Nusbaum, C. J. (1960). The black shank-root-knot complex in flue-cured tobacco. *Phytopathology* **50**, 899–906.

Powell, N. T., Melendéz, P. L., and Batten, C. K. (1971). Disease complexes in tobacco involving *Meloidogyne incognita* and certain coil-borne fungi. *Phytopathology* **61**, 1332–1337.

Richardson, D. E., and Doling, D. A. (1957). Potato blight and leaf-roll virus. *Nature (London)* **180**, 866–867.

Rochaw, W. F., and Ross, A. F. (1955). Virus multiplication in plants doubly infected by potato viruses X and Y. *Virology* **1**, 10–27.

Ross, J. P. (1969). Effect of time and sequence of inoculation of soybean with soybean mosaic and bean pod mattle virus on yield and seed characters. *Phytopathology* **59**, 1404–1408.

Sands, D. L., and Lukens, R. L. (1974). Effect of glucose and adenosine phosphates on production of extracellular carbohydrates of *Alternaria solani*. *Plant Physiol.* **54**, 666–669.

Sandstedt, R., and Schuster, M. L. (1966). The role of auxins in root-knot nematode-induced growth on excised tobacco stem segments. *Physiol. Plant.* **19**, 960–967.

Sequeira, L. (1963). Growth regulators in plant disease. *Annu. Rev. Phytopathol.* **1**, 5–30.

Sidhu, G., and Webster, J. M. (1977). Predisposition of tomato to the wilt fungus (*Fusarium oxysporum lycopersicae*) by the root-knot nematode (*Meloidogyne incognita*). *Nematologica* **23**, 436–442.

Simkin, M. B., and Wheeler, B. E. J. (1974). Effects of dual infection of *Puccinia hordei* and *Erysiphe graminis* on barley, cv. Zephyr. *Ann. Appl. Biol.* **78**, 237–250.

Van Gundy, S. D., Kirkpatrick, J. D., and Golden, J. (1977). The nature and role of metabolic leakage from root-knot nematode galls and infection by *Rhizoctonia solani*. *J. Nematol.* **9**, 113–121.

Viglierchio, D. R., and Yu, P. K. (1968). Plant growth substances and plant parasitic nematodes. II. Host influence on auxin content. *Exp. Parasitol.* **23**, 88–95.

Wallace, H. R. (1978). The diagnosis of plant diseases of complex etiology. *Annu. Rev. Phytopathol.* **16**, 379–402.

Whetzel, H. H. (1935). The nature of disease in plants. *In* "Elementary Plant Pathology," Lecture Notes. Cornell Univ., Ithaca, New York.

Yang, H., Powell, N. T., and Barker, K. R. (1976). Interaction of concomitant species of nematodes and *Fusarium oxysporum* f. sp. *vasinfectum* on cotton. *J. Nematol.* **8**, 81–86.

Chapter 7

# How the Defenses Are Breached

### R. L. DODMAN

## I. INTRODUCTION

The means by which pathogens enter their host plants is an important part of the complex interaction leading to disease. During the 60 years since Brown and his colleagues began their work on this subject (see review by Brown, 1936), significant advances have been made in some areas. Very detailed observations of penetration structures are now possible with transmission and scanning electron microscopy. Although these instruments allow us to see much more than could be observed with the ordinary light microscope, interpretation of fixed and sectioned material still presents problems. Thus, gaps in our knowledge still remain, especially in the dynamic aspects of penetration.

Some of these aspects have been long-standing subjects of argument and in fact remain so today. For example, what factors control fungal growth and initiation of specialized infection structures on plant surfaces? What is the contribution of chemical action to the penetration of a surface covered with cuticular waxes? Studies of these and other questions involving the great diversity of interactions that occur at the plant surface during penetration will be aided by a better understanding of the physiological processes occurring within the pathogen. Although the technical difficulties associated with these aspects of fungal physiology are great, increased attention to the problems will be very worthwhile.

Examining how the defenses of plants are breached has always been fascinating, but it can be very useful as well. As pathologists we study

135

PLANT DISEASE, VOL. IV

the sequence of events associated with disease development, always seeking points of weakness in the pathogen and exploiting them in ways most likely to reduce disease losses.

The process of penetration commonly consists of a sequence of events, with each event being an important prerequisite for successful penetration. Some pathogens have evolved specialized associations with other organisms that enable them to bypass one or more of these stages (see this volume, Chapter 5). Successful penetration involves a complexity of interactions between pathogen, plant, and the numerous abiotic and biotic environmental influences. The effects of predisposing environmental factors are covered in Chapter 4 of this volume. The processes of colonization that follow after penetration are discussed by Durbin in Chapter 8 of this volume.

## II. EVENTS BEFORE THE PENETRATION PROCESS

The establishment of an interaction between plant and pathogen begins with the arrival of the pathogen on the plant surface or with the pathogen coming within the zone of influence of plant exudates in the soil. Usually, this is followed by the initiation of new growth from the resting propagule and subsequent growth and orientation on the plant surface prior to the formation of infection structures and penetration. With some pathogens, growth or motile stages may be directed toward suitable penetration sites. The importance of these activities in successful penetration is considered in the next two sections.

### A. Initiation of Growth

#### 1. Growth on Aboveground Plant Parts

Propagules of plant pathogens may arrive at the plant surface in a number of ways, including dry deposition of spores and hyphal fragments from the air, wet deposition in rain drops or rain splash, and transfer from different parts of insects, pruning and harvesting equipment, and other vectors. Once deposited, these propagules usually have little opportunity for effective redistribution. Thus, survival depends on growth and reproduction either on the plant surface or within the plant after the establishment of a parasitic (and sometimes pathogenic) relationship.

In general, fungal propagules are too large to enter through any of the natural openings on plant surfaces. Thus, penetration is dependent on initiation of growth. For growth to begin, the propagule must have or be able to obtain the necessary nutrients, and the environment must be favorable. The spores of many pathogens that attack aboveground parts of plants are able to germinate readily in distilled water, indicating that they have no exogenous requirements for germination. However, this does not mean that nutrients on the plant surface are unimportant. Even when the percentage of spores that germinate is high in the absence of added nutrients, both the time taken for germination to begin and also the rate of germination may be altered by nutrients from the host, with significant implications for disease development. This is illustrated by work on *Botrytis cinerea* in which materials from grape berries stimulated both an earlier and more rapid germination of conidia than occurred in water (Kosuge and Hewitt, 1964). In this disease situation, the period of wetness of berries is often a limiting factor in infection; thus, both early and rapid germination increase the likelihood of successful penetration.

Various factors may alter the ability of propagules to germinate without exogenous nutrients (Blakeman, 1971; Godfrey, 1976). Materials that inhibit germination may be released from plant tissues into water present on the plant surface, while substances within the spores themselves can also inhibit germination, especially when there is a high concentration of spores in the infection drop. Inhibitory materials also can be produced by the microflora on the plant surface. Apart from inhibitors, the loss of endogenous reserves from spores due to leaching and respiration may lead to dependence on external nutrients. For the pathogen to obtain the factors needed to overcome the effects of inhibitors and supplement endogenous reserves it must compete with other organisms on the plant surface. The outcome of these complex interactions will influence the percentage of spores that germinate and thus the proportion of propagules that are able to continue growth and attempt penetration.

## 2. Growth on and around Belowground Plant Parts

In soil there are many different types of propagules of pathogenic bacteria, fungi, and nematodes. These propagules include spores, eggs, cysts, sclerotia, and hyphal fragments and aggregates, all of which may occur in the soil or in organic debris associated with previous pathogenic or saprophytic activity. In many cases these propagules exist in soil in an inactive state and require some external stimulus for the initiation of renewed activity. Without such a stimulus the population of the pathogen will tend to decline, with the rate of decline being influenced by physical, chemical, and biotic components of the soil environment.

The introduction of the plant into this relatively stable situation brings about some marked changes. Large amounts of readily usable sources of carbon and nitrogen become available, first from the seed as it germinates and then from the roots, particularly the root tip as it moves through the soil leaving behind sloughed-off cells and exudates associated with its intense metabolic activity. Other factors, such as soil moisture, concentration of oxygen and carbon dioxide, and pH (particularly near the root surface), are also altered both by the activities of the root itself and by the growth of microorganisms within the rhizosphere. In many cases, a pathogen must respond quickly to the sources of nutrients and energy released by the plant and use these materials to penetrate the plant before competition becomes too severe.

It is now well known that exudates from seeds and roots can stimulate the germination of spores in soil (Rovira, 1965). The interactions between plant and pathogen are well illustrated by work on fusarium root rot of beans. Chlamydospores of *Fusarium solani* f. sp. *phaseoli* germinate close to bean seeds and the tips of primary, lateral, and adventitious roots, but not in soil away from roots or around older roots (Schroth and Snyder, 1961). It was subsequently shown (Schroth *et al.*, 1963) that exudates from bean seeds and roots contain sugars and amino acids and that these materials stimulate germination of chlamydospores when added to soil. Quantity of exudate was also found to be important. Seeds, which exude large amounts of sugars and amino acids, stimulate a much higher percentage of germination than hypocotyls, which exude smaller amounts of sugars and only a trace of amino acids (Cook and Snyder, 1965).

Many other factors, including plant age, plant nutrition, temperature, soil moisture, light intensity, infection by other pathogens, the application of foliar sprays, and damage caused either by soil fauna or by toxins from decaying plant debris, may influence exudation and thus affect the initiation of growth by propagules of soilborne pathogens.

In some cases, the germination of pathogenic propagules may be stimulated by specific materials released only from plants that are susceptible to that particular pathogen. For instance, the germination of sclerotia of *Sclerotium cepivorum* is highly host specific; it occurs on and around the roots of hosts such as onion and other members of the genus *Allium* but not around the roots of plants in other genera from the family Liliaceae or plants from many other families (Coley-Smith and Holt, 1966). Such a high degree of specificity provides an excellent means for maintaining the population of the pathogen during periods when there is either no crop present or when nonhost crops are grown. Although specific relationships may exist between other host–pathogen combinations, there is often a low percentage of prop-agules that germinate in the presence of nonhosts. This may lead to a gradual decline in the population of the pathogen unless some growth and production of new propagules can be achieved by saprophytic growth.

In addition to materials inducing germination, plant exudates may contain substances that inhibit initiation of growth. The role of these factors in resistance will be considered in Volume V, but an illustration of the nature of possible interactions between stimulation and inhibition emphasizes the complexity of the soil situation. In studies of fusarium root rot of peas, Kraft (1977) has shown that the pigment delphinidin is produced in the testa of certain types of peas. Delphinidin inhibits germination of conidia of the pathogen and could help to prevent the initial development of root rot. However, the fungistatic effect of delphinidin can be overcome by glucose, and thus the susceptibility of varieties that produce the pigment depends on the balance between the stimulatory and inhibitory factors released from the seed.

In studies of very complex interactions, it is often necessary to examine different aspects of growth initiation under carefully controlled and simplified conditions. However, to test the validity of the impor-tance of any interaction discovered *in vitro,* it is necessary to consider this interaction in relation to the real situation as it occurs in the field.

## B. Growth and Movement on and around the Plant

Before the process of penetration begins, there is often a period of growth, either on or around the plant. This may involve the develop-ment of germ tubes and mycelium, the multiplication of bacteria, the production and release of motile spores from an encysted sporangium, or the continued growth and molting of nematode larvae. In some cases these activities may be essential for successful penetration, while in others they may have an important effect on penetration either by altering the number of units of inoculum or by altering the nutritional status of the inoculum. During this phase, the location of infection sites may be accomplished through growth of hyphae or movement of motile stages of the pathogen. In many cases the pathogen is particularly vulnerable to both the rigors of the physical environment and competi-tion from the biotic environment during this stage of the infection process. Thus, it is important for the pathogen to make only the growth that is essential for penetration so that it can escape from the hostile environment outside the host as quickly as possible.

In the complex situation on and around the plant, there are many factors that may have both a direct effect on the pathogen and an indirect effect by influencing interactions in which the pathogen is involved. Direct effects of physical environmental factors such as temperature, moisture, light, etc., on the behavior of the pathogen prior to penetration have been described for many pathogens but are particularly well illustrated by the detailed studies carried out on the powdery mildews (Mount and Slesinski, 1971). By carefully controlling environmental conditions, it has been shown that there are optimal environmental conditions for different phases of the infection process and that any deviation from these optima leads either to less infection or to an increase in the time required to establish successful infection.

Direct effects on the growth of the pathogen may also be exerted by biotic components of the environment, including the plant and also the microflora on and around the plant surface. The importance of plant exudates in stimulating growth and increasing the nutritional status of inoculum is shown by the interaction between cotton and *Rhizoctonia solani*. Exudate from cottonseed contains large amounts of sugars and amino acids; these materials stimulate the growth of *R. solani*, with the amount of growth being related to the quantity of nutrients exuded (Hayman, 1970). Related studies by Kamal and Weinhold (1967) and Weinhold *et al.* (1972) have suggested that growth of the pathogen from organic debris in soil onto cotton hypocotyls may occur, but successful infection is dependent on the utilization of nitrogenous materials present in seed exudate.

In addition to direct effects on pathogens, physical environmental factors can influence plant exudation and thus alter behavior of the pathogen. In a disease of strawberries caused by *Rhizoctonia fragariae*, Husain and McKeen (1963) found that there was a good correlation between disease severity at different temperatures and the amount of growth of the pathogen around strawberry roots enclosed in cellophane bags. A similar correlation was found with exudates from strawberry roots, and it was shown that the quantities of amino acids were much greater in exudates collected from roots grown at lower temperatures than those grown at higher temperatures. It was suggested that these materials played an important part in the stimulation of growth by the pathogen with the result that at low temperatures the disease was more severe than at high temperatures.

Besides the effects of the environment on exudation, there are several other factors that may influence the release of materials from plants and thus affect growth of the pathogen on and around the plant surface. Robinson and Hodges (1977) found that various types of nitrogenous fertilizers applied to soil increased the length and number of branches in germ tubes of *Drechslera sorokiniana* growing on the leaves of *Poa pratensis*. It was suggested that effects on the pathogen by the fertilizers were due to changes in the amounts of certain sugars and amino acids in the leaf tissue and associated changes in the composition of leaf exudates. Another factor that may influence release of materials from plant cells, but has received little attention, is the ability of the pathogen while growing on the plant to produce materials leading to increased exudation. Such exudates might be used by the pathogen to increase its capacity to penetrate.

The microflora on and around the plant surface is another component of the complex interaction in which the pathogen is involved prior to penetration. All of the factors that directly affect the pathogen may also affect the microflora, which in turn may influence the response of the pathogen. Such a situation was described by Cook and Papendick (1970) for foot rot of wheat caused by *Fusarium roseum* f. sp. *cerealis* 'Culmorum.' Although the disease is favored by dry soils, it was suggested that this is not due to the direct effect of water potential on germination and growth of the pathogen; rather, the relation was achieved indirectly via the interaction between pathogen, antagonists, and soil water potential. In wet soils, germ tubes arising from chlamydospores are rapidly lysed by soil bacteria. However, in drier soils (below −10 bars) bacterial activity is markedly reduced and little or no lysis occurs, thus facilitating continued growth and infection by the pathogen. Interactions of this nature involving the pathogen, plant exudates, and other chemicals on the plant surface, physical environmental factors, and the microflora on the plant are probably commonplace. It is thus important that we attempt to integrate information obtained from studies of the different components of a system under artificial conditions, so that we gain a better understanding of the whole system as it operates in the field.

Another aspect of pathogen behavior before penetration involves the direction of growth of germ tubes and hyphae and the direction of movement of motile organisms. This growth and movement is often controlled to facilitate rapid location of suitable infection sites, thereby reducing both the time taken and the amount of energy used prior to penetration. The factors associated with this behavior and the physiology of the responses are generally not well understood.

On the plant surface, several different types of germ tube and hyphal growth have been observed, each indicating that growth does not occur at random. Where stomata serve as important penetration sites, growth may be directed toward them in two ways. First, the tips of hyphae and germ tubes may grow toward stomata under the influence of some stimulus associated with these openings in the plant surface. This type of behavior by fungal pathogens has not been reported frequently, and evidence concerning the controlling factors is generally lacking. Rathaiah (1977) has shown that germ tubes from spores of *Cercospora beticola* grow toward stomata on sugar beet leaves provided the leaf surface is not continuously wet. Because of the increased frequency of penetration when leaves are intermittently dried and wetted, Rathaiah has suggested that hydrotropism is the factor controlling germ tube growth of *C. beticola* under these conditions. However, further studies of this and other interactions in which hyphae are attracted to stomata are needed to determine the nature of the tropic stimuli and the physiology of the fungal response.

The second way in which growth is directed toward stomata involves the orientation of a germ tube in such a way as to maximize its chances of locating a stoma as it grows. This behavior is common among rust pathogens on cereal leaves but also occurs with rusts on other hosts. On cereals, it was observed that germ tubes become oriented at right angles to the longitudinal axis of the leaf; because of the arrangement of stomata in staggered rows parallel to this axis, a germ tube growing across the leaf would have the maximal chance of contacting a stoma. It was originally suggested (Johnson, 1934) that growth was at right angles to the junctions of cell walls that ran longitudinally along the leaf and that this was related to differences in the leaf surface along these junctions. Detailed examination of the wheat leaf surface with the scanning electron microscope (Lewis and Day, 1972) showed that there is a lattice of wax crystals covering the cuticle; it was suggested that the germ tube, growing in contact with this lattice, shows a thigmotropic response to the regular arrangement of wax crystals in the lattice.

With *Uromyces phaseoli* on bean leaves, Wynn (1976) showed that germ tube growth is also related to the topography of the leaf surface. In this case, cuticular ridges encircle stomata, and, by growing at right angles to these ridges, germ tubes are directed to the stomata. Although the wax crystals found on wheat leaves are not present on bean leaves or replicas of bean and wheat leaves, growth on all three surfaces was at right angles to the cuticular ridges or their replicas. However, it was also reported that germ tubes of *U. phaseoli* failed to grow at right angles to the cuticular ridges on wheat leaves, indicating that various characteristics of the plant surface can play a part in the location of infection sites. Similar conclusions were made by Heath (1977) in studies of three unrelated rust fungi on several hosts and nonhosts.

Hyphal growth on the plant surface frequently follows the lines of junction of underlying epidermal cell walls. Although this location is a preferred site for the formation of appressoria by a few pathogens (Preece *et al.*, 1967; Clark and Lorbeer, 1976), there is little evidence that this pattern of growth is essential for successful penetration by the many pathogens that behave in this way. Occasional reports (Flentje, 1957; Borges *et al.*, 1976) indicate that growth along cell wall junctions occurs on susceptible hosts but not on resistant hosts. However, further investigations should be carried out to determine the significance of this behavior and also the factors controlling it. Suggestions that growth is directed along cell wall junctions because of greater amounts of exudate in that area or because of physical characteristics of the plant surface have not been substantiated experimentally.

In addition to showing directional growth on the plant surface, germ tubes and hyphae may be attracted toward belowground parts of plants in soil. A positive response to roots was shown by germ tubes from encysted zoospores of *Pythium aphanidermatum* (Royle and Hickman, 1964), while hyphae of *Gaeumannomyces graminis* var. *tritici* were also attracted toward wheat roots (Pope and Jackson, 1973). In both cases, similar responses were obtained with root exudates. The materials producing this attraction have not been identified, but it seems likely that they are nonspecific carbon and nitrogen sources, since *P. aphanidermatum* responded to a wide range of plants. It is possible that this type of directional response occurs commonly in soil but often is not detected because of difficulties associated with observation of fungal growth in soil.

Some fungi and other plant pathogens produce motile stages that are capable of movement either in liquid on the plant surface or in soil, provided there is sufficient water in the soil pores. A few pathogens have been studied intensively so that there is considerable information concerning the behavior of *Phytophthora* zoospores in soil, but much less is known about other pathogens and especially those that move about over aboveground plant surfaces.

In reviewing the behavior of zoospores of Phycomycetes, Hickman and Ho (1966) cited examples of zoospores accumulating around stomata and along the lines of junction of underlying epidermal cell walls. It was suggested that various materials, including $K^+$, $H^+$, and $OH^-$ ions and phosphatides, induced these responses. More recently, Royle and Thomas (1973) reported that zoospores of *Pseudoperonospora humuli* are attracted to open stomata on hop leaves but settle at random on the leaf surface when stomata are closed. They showed that there are two independent processes in the response. The first directs zoospores toward open stomata and is conditioned both by a chemical stimulus associated with photosynthesis and by a physical stimulus related to the structure of open stomata. The second part of the response involves a shortening of the motile period and is controlled only by the chemical stimulus. Zoospores of *Phytophthora citrophthora* are also attracted to open stomata on the leaves of *Pieris japonica*, and it has been suggested that this may be related to an electrochemical stimulus (Gerlach *et al.*, 1976).

The behavior of zoospores around roots has been investigated extensively in the last 20 years, with considerable emphasis on various species of *Phythophthora* and *Pythium*. It is now well known that zoospores of soilborne pathogens are attracted toward roots, where they usually accumulate in the area just behind the root tip. Most of this work has been done under controlled conditions with roots being dipped into zoospore suspensions. However, there are reports that accumulation does occur around roots in soil and that this response is important in diseases caused by some pathogens (Hickman, 1970).

There is much evidence that root exudates are associated with the attraction of zoospores to roots. However, attempts to purify and identify individual components of exudates and reproduce the attraction *in vitro* with known chemicals or mixtures of chemicals have been largely unsuccessful. The work has shown, however, that the response of zoospores consists of several stages (attraction, trapping, encystment, and germination) and that different substances may control different stages. Combinations of different factors may also be important, with $H^+$ ions (Ho and Hickman, 1967), electrical forces and surface charges on roots (Khew and Zentmyer, 1974), and chemotaxis all playing a part in the various stages of zoospore behavior around roots.

Zoospore attraction is usually nonspecific, with similar responses occurring around roots of both hosts and nonhosts (Hickman, 1970) and on roots of hosts differing in susceptibility (Irwin, 1976; Malajczuk and McComb, 1977). An early report that *Phytophthora cinnamomi* was attracted to avocado roots but not to roots of nonhosts (Zentmyer, 1961) was not confirmed in recent work (Ho and Zentmyer, 1977). Instead, it was suggested that specificity is probably related to events after penetration has occurred.

Roots may also influence the behavior of other pathogens in soil. Evidence reviewed by Endo (1975) indicates that nematodes are attracted to roots and that this attraction is related to chemical factors associated with root growth, particularly carbon dioxide and some amino acids. The zone of influence of these factors may extend for several centimeters away from the root. In general, nematodes are attracted to both hosts and nonhosts, although some examples are known in which specificity may be at least partly related to the extent of attraction (Rohde, 1972; Endo, 1975).

There is still a great deal to be done to develop our understanding of the numerous simple and complex interactions that take place during this phase of growth and development on and around the plant surface. Progress has been slow in part because individual facets of the complex interactions involved cannot be readily interpreted in isolation from the whole. However, these investigations play an important part in locating and exploiting weaknesses in the pathogen's offensive capabilities and thereby provide opportunities for management of disease.

## III. MODE OF PENETRATION

Pathogens penetrate plants in many ways. Some take advantage of natural openings in the plant surface, some rely on damaged plant tissue to provide entry, others have developed a close association with vectors, and a great many are able to penetrate intact plant surfaces. These different types of penetration are described, and some of the factors that influence penetration are discussed.

Injured plant tissue may become the site for invasion by pathogens that lack the ability to penetrate the intact plant surface. In such cases, disease development is controlled by the factor(s) causing the injury. However, some pathogens that are able to penetrate uninjured tissue also use areas of damage as alternative penetration sites. There are many ways in which a plant may be injured, with the resulting damage varying greatly in extent and severity. Microscopic cracks and abrasions may provide suitable avenues of entry for bacteria and viruses, while at the other extreme the large surface of a tree stump becomes an ideal site for invasion by *Fomes annosus*.

Propagules of a pathogen may arrive at an injured area of tissue in a number of ways. Spores that are dispersed by air movement may be deposited on wounds made by man during crop husbandry, by the activities of other animals, or perhaps by wind, hail, and other violent storms (see Chapter 14, Volume III). During harvesting and thinning operations in conifer forests, freshly cut stumps provide a suitable location for infection by basidiospores of *Fomes annosus* (Rishbeth, 1951) and thus enable the establishment of new centers of infection in a previously healthy stand. Another method of pathogen movement involves transfer by vectors of various types. In some diseases the wounding agent and vector are different, as shown by Moller and DeVay

(1968) for *Ceratocystis* canker of stone fruits. In this disease, wounds to tree trunks are caused by mechanical harvesting equipment; insects carrying the pathogen are attracted to these wounds, which then become infected and develop into cankers. In other diseases, the wounding agent itself may act as the vector, as can happen with the transfer of pathogens from diseased to healthy tissue by cultivating, harvesting, or pruning equipment. The transmission of nonpersistent viruses by aphids could be considered a very specialized example of transfer by the wounding agent, although there is still considerable debate on the actual transfer process (Pirone and Harris, 1977).

Wounds also serve as important points for penetration by soilborne pathogens. Hyphae of *Fusarium solani* f. *phaseoli* can grow over the surface of bean hypocotyls and invade cortical cells in damaged areas of the hypocotyl (Christou and Snyder, 1962). Rupturing of the hypocotyl surface by the emergence of adventitious roots is one way in which damaged areas are produced. With other pathogens that produce motile zoospores it has been shown that these often accumulate around wounds or the cut ends of roots. Irwin (1976) found that zoospores of *Phytophthora megasperma* accumulate at the point of emergence of lateral roots from the tap root of lucerne plants. Penetration of the damaged tissue then leads to severe infection of the tap root.

Since the external barriers to penetration are destroyed when the plant surface is wounded, the development of disease depends on several factors, including (1) the presence of inoculum when the wound is made or during the period when the wound is susceptible, (2) the ability of the pathogen to germinate and grow in the damaged tissue, (3) the rate and types of processes that occur during wound healing, and (4) the ability of the pathogen to compete with other microorganisms present at the wound site. An understanding of the way in which these factors control penetration provides an opportunity for a range of disease management tactics.

Natural openings in the plant surface such as lenticels, stomata, and hydathodes provide an avenue of entry for many pathogens. Although lenticels occur in the periderm of most plants, they are used as penetration sites by only a small number of pathogens (Royle, 1976). For some pathogens, lenticels represent the only pathway by which the periderm can be breached. However, detailed descriptions of the process of penetration through lenticels are lacking; also there appears to be little information on the physical and chemical processes used by the pathogen to effect entry through these structures.

In contrast, stomata and, to a lesser extent, hydathodes are commonly used as sites of penetration by many bacterial and fungal pathogens.

With nonmotile bacterial cells, the withdrawal of liquid from the plant surface through both stomata and hydathodes can cause the passive movement of the pathogen through these openings and into contact with underlying tissues. Motile bacteria are also able to migrate through these openings with the aid of their flagellae, provided moisture is present. In general, it appears that the size of the stomatal aperture prevents the entry of fungal propagules. There is also some evidence that the structure and condition of stomata can restrict or prevent entry of bacterial pathogens ( Royle, 1976).

Fungal penetration through stomata is the sole means of entry for some pathogens, while others may use stomata as an alternative to direct penetration of the intact plant surface. There are two ways in which entry through stomata can be effected. First, as occurs with many Phycomycetes and Fungi Imperfecti, germ tubes or hyphal tips may grow through the stomatal opening without forming any morphologically recognizable structure to assist penetration. It appears that in many cases the size and shape of the aperture does not influence the ability of the pathogen to pass through the opening, although it is not uncommon for a constriction in hyphal diameter to occur during penetration ( Dodman *et al.*, 1968; Patton and Johnson, 1970). In general, there is little information on the ability of pathogens to penetrate directly through closed stomata, although Royle ( 1976) stated that germ tubes from encysted zoospores of two downy mildew pathogens are able to penetrate both open and closed stomata. It appears that the hyphal tip is able to exert the forces needed to push through the closed lips of the guard cells, but how this force is applied and whether penetration is assisted by any chemical action is unknown.

The second method of entry through stomata involves the formation of an appressorium over the stoma and subsequent penetration by a fine infection peg from the base of the appressorium. This mode of infection is common among rust pathogens but also occurs with a range of other fungi. Appressoria are also involved in the penetration of the intact plant surface by a great many pathogens. Since many aspects of the infection process are similar in these two situations, it is possible to consider them together. I will point out any special features and differences in each type of penetration where appropriate.

The term "appressorium" was originally associated with the swelling formed at the end of a germ tube. Such easily recognizable structures are very commonly associated with stomatal penetration, but, where the intact plant surface is involved, the size, shape, complexity, and origin of infection structures can vary greatly. Emmett and Parbery ( 1975) suggested that the scope of the term "appressorium" should include all structures that have the capacity to adhere to the plant surface and the ability to effect penetration. They also considered, in some detail, the factors that contribute to the considerable variations in appressorium morphology; these include the fungal genotype, various physical environmental factors, and other influences associated with the host surface.

The function of the appressorium and the part that it plays in the process of penetration have been the subject of much research and debate. Evidence that the appressorium is well suited for the breaching of barriers by mechanical action was put forward for many fungi. The attachment of the appressorium to the plant surface by a deposit of mucilaginous material and the formation of a fine infection peg capable of concentrating the applied force in a restricted area were two important features of this evidence. Recent studies, particularly with the electron microscope, provide further illustrations of the presence of these binding substances around appressoria (Emmett and Parbery, 1975; Clark and Lorbeer, 1976; Harvey, 1977) and thus support the view that these materials assist in mechanical penetration. However, there is now little doubt that the passage of the infection peg through the intact plant surface is at least assisted by chemical activity (Emmett and Parbery, 1975; Ingram et al., 1976). Evidence favoring this interpretation has been obtained from detailed investigations with a range of microscopy procedures and microchemical techniques. These have shown that the tip of the infection peg may be rounded and protected only by the plasmalemma and that the host cuticle shows little sign of the application of force either by indentation beneath the infection peg or by tearing of the cuticle (McKeen, 1974; Harvey, 1977). Enzymes capable of degrading the cuticle are known to be produced by a number of fungi (Van den Ende and Linskens, 1974), and there is now evidence of their activity at the point of penetration (McKeen, 1974; Ingram et al., 1976; Mayama and Pappelis, 1977; Shaykh et al., 1977). With the renewed interest in this aspect of penetration and the development of better techniques for its study, it seems likely that there will soon be some resolution of the long-standing conflicts in this area and a better understanding of how penetration is effected.

Although the mechanism of penetration of the intact plant surface has been studied extensively, much less attention has been given to the mechanism of stomatal penetration. Royle (1976) discussed some aspects of entry through stomata, including the effect of stomatal function on penetration. It appears that the appressorium provides the means for forcing the infection peg through closed stomata and that in some situations stomatal closure is actually induced by appressorial attachment. However, where entry occurs through open stomata, it would seem that the need for the application of force would be either reduced or eliminated. Such an interpretation could explain the different modes of penetration of sugar beet leaves by *Cercospora beticola* (Rathaiah, 1976). It is also possible that with some pathogens, appressoria may not only be involved in the process of penetration but also serve as survival structures, making it possible for penetration to be temporarily interrupted during unfavorable environmental conditions (Emmett and Parbery, 1975).

Factors that influence the formation of appressoria have been studied for many years, with major emphasis on the specific stimuli that induce appressorium initiation. However, in recent years there has been an increasing awareness that the initiation and maturation of appressoria and subsequent penetration constitute a complex sequence of events, which are influenced by a wide range of factors. These include the pathogen genotype, physical environmental factors such as temperature, light, moisture, etc., and many factors associated directly or indirectly with the plant such as the physical and chemical characteristics of the plant surface, the release of exudates from plant cells, and the activity of the microflora on the plant surface. The response of the pathogen to specific factors may be difficult to determine because of interactions between components of the system.

The relation of these and other factors to appressorium development has been discussed in detail in a review by Emmett and Parbery (1975). They have proposed that appressorium formation is primarily controlled by the pathogen genotype and that the expression of this genotype may require a specific conducive environment. Although the importance of genotype has been stressed by these authors, at present there is only limited evidence to support this hypothesis. Within the conducive environment some factors may have a greater influence than others; however, the concept of a single controlling factor may apply only rarely.

Among the many factors that may affect appressorium formation, the physical characteristics of the plant surface and chemicals present on that surface are of major importance. The fine structure of the plant surface and particularly the waxy materials in the cuticle play a significant part in the formation of appressoria and their location at specific penetration sites such as stomata and junctions of anticlinal walls of epidermal cells (Lewis and Day, 1972; Wynn, 1976). The influence of chemicals on appressorium development is well known; such materials may be associated with the cuticular layer itself, released from the underlying cells, or produced by the surface microflora, or they may arrive from external sources. Interactions between the numerous components that affect the nature and quantity of chemicals on the plant surface may be extremely complex, making interpretation of the causes of observed responses very difficult.

Changes in disease severity with time, with different environmental conditions, or with different parts of the plant may often be associated with the effect of many of these factors on appressorium formation (Preece and Dickinson, 1971; Dickinson and Preece, 1976; Friend and Threlfall, 1976). There are also instances in which differences in appressorium formation are related to differences in susceptibility between cultivars or between species. Such a relationship has been described for *Rhizoctonia solani*, in which compound appressoria (infection cushions) are formed on susceptible hosts but not on resistant hosts. Although the controlling factors have not been defined, it appears that both the nature of the host surface and host exudates are involved (Dodman, 1970). However, there are many studies in which no such relationship has been found (Matta, 1971), and it is likely that many more will be added with the increased interest in the behavior of pathogens on plants that are normally regarded as nonhosts. Further detailed investigations, such as those carried out by Heath (1977), are needed to determine the importance of appressorium formation and penetration in these types of interactions.

## IV. CONCLUSIONS AND OUTLOOK

It has long been recognized that the activities of pathogens prior to penetration represent an important part of the process of disease development. Although early emphasis was focused on the role of appressoria, there is now an awareness that there are many stages involved in the infection process and also that there is a great diversity of factors that may influence these stages. In recent years many new techniques and procedures have become available and considerable progress has been made in understanding many of the processes involved, especially in the early stages of the interactions between host and pathogen. These investigations have helped to clarify the way in which these processes are controlled, but our understanding of this aspect of pathogenesis is still incomplete. An important aspect of research in the future will be the integration of information on the different stages of penetration and their controlling factors. In this way, some progress can be made in explaining the behavior of pathogens in the field, and opportunities for disease management can be explored.

## References

Blakeman, J. P. (1971). The chemical environment of the leaf surface in relation to growth of pathogenic fungi. In "Ecology of Leaf Surface Microorganisms" (T. F. Preece and C. H. Dickinson, eds.), pp. 255–268. Academic Press, New York.

Borges, O. O., Stanford, E. H., and Webster, R. K. (1976). The host-pathogen interaction of alfalfa and Stemphylium botryosum. Phytopathology 66, 749–753.

Brown, W. (1936). The physiology of host–parasite relations. Bot. Rev. 2, 236–281.

Christou, T., and Snyder, W. C. (1962). Penetration and host–parasite relationships of Fusarium solani f. phaseoli in the bean plant. Phytopathology 52, 219–226.

Clark, C. A., and Lorbeer, J. W. (1976). Comparative histopathology of Botrytis squamosa and B. cinerea on onion leaves. Phytopathology 66, 1279–1289.

Coley-Smith, J. R., and Holt, R. W. (1966). The effect of species of Allium on germination in soil of sclerotia of Sclerotium cepivorum. Ann. Appl. Biol. 52, 273–278.

Cook, R. J., and Papendick, R. I. (1970). Soil water potential as a factor in the ecology of Fusarium roseum f. sp. cerealis 'Culmorum.' Plant Soil 32, 131–145.

Cook, R. J., and Snyder, W. C. (1965). Influence of host exudates on growth and survival of germlings of Fusarium solani f. phaseoli in soil. Phytopathology 55, 1021–1025.

Dickinson, C. H., and Preece, T. F., eds. (1976). "Microbiology of Aerial Plant Surfaces." Academic Press, New York.

Dodman, R. L. (1970). Factors affecting the prepenetration phase of infection by Rhizoctonia solani. In "Root Diseases and Soil-Borne Pathogens" (T. A.Toussoun, R. V. Bega, and P. E. Nelson, eds.), pp. 116–121. Univ. of California Press, Berkeley.

Dodman, R. L., Barker, K. R., and Walker, J. C. (1968). A detailed study of the different modes of penetration by Rhizoctonia solani. Phytopathology 58, 1271–1276.

Emmett, R. W., and Parbery, D. G. (1975). Appressoria. Annu. Rev. Phytopathol. 13, 147–167.

Endo, B. Y. (1975). Pathogenesis of nematode-infected plants. Annu. Rev. Phytopathol. 13, 213–238.

Flentje, N. T. (1957). Studies on Pellicularia filamentosa (Pat.) Rogers. III. Host penetration and resistance, and strain specialization. Trans. Br. Mycol. Soc. 40, 322–336.

Friend, J., and Threlfall, D. R., eds. (1976). "Biochemical Aspects of Plant Parasite Relationships." Academic Press, New York.

Gerlach, W. W. P., Hoitink, H. A. J., and Schmitthener, A. F. (1976). Phytophthora citrophthora on Pieris japonica: Infection, sporulation, and dissemination. Phytopathology 66, 302–308.

Godfrey, B. E. S. (1976). Leachates from aerial parts of plants and their relation to plant surface microbial populations. In "Microbiology of Aerial Plant Surfaces" (C. H. Dickinson and T. F. Preece, eds.), pp. 433–439. Academic Press, New York.

Harvey, I. C. (1977). Studies of the infection of lupin leaves by Pleiochaeta setosa. Can. J. Bot. 55, 1261–1275.

Hayman, D. S. (1970). The influence of cotton seed exudate on seedling infection by Rhizoctonia solani. In "Root Diseases and Soil-Borne Pathogens" (T. A. Toussoun, R. V. Bega, and P. E. Nelson, eds.), pp. 99–102. Univ. of California Press, Berkeley.

152     R. L. DODMAN

Heath, M. C. (1977). A comparative study of non-host interactions with rust fungi. *Physiol. Plant Pathol.* **10**, 73–88.

Hickman, C. J. (1970). Biology of *Phytophthora* zoospores. *Phytopathology* **60**, 1128–1135.

Hickman, C. J., and Ho, H. H. (1966). Behaviour of zoospores in plant-pathogenic Phycomycetes. *Annu. Rev. Phytopathol.* **4**, 195–220.

Ho, H. H., and Hickman, C. J. (1967). Factors governing zoospore responses of *Phytophthora megasperma* var. *sojae* to plant roots. *Can. J. Bot.* **45**, 1983–1994.

Ho, H. H., and Zentmyer, G. A. (1977). Infection of Avacado and other species of *Persea* by *Phytophthora cinnamomi. Phytopathology* **67**, 1085–1089.

Husain, S. S., and McKeen, W. E. (1963). Interactions between strawberry roots and *Rhizoctonia fragariae. Phytopathology* **53**, 541–545.

Ingram, D. S., Sargent, J. A., and Tommerup, I. C. (1976). Structural aspects of infection by biotrophic fungi. In "Biochemical Aspects of Plant Parasite Relationships" (J. Friend and D. R. Threlfall, eds.), pp. 43–78. Academic Press, New York.

Irwin, J. A. G. (1976). Observations on the mode of infection of lucerne roots by *Phytophthora megasperma. Aust. J. Bot.* **24**, 447–451.

Johnson, T. (1934). A tropic response in germ tubes of urediospores of *Puccinia graminis tritici. Phytopathology* **24**, 80–82.

Kamal, M., and Weinhold, A. R. (1967). Virulence of *Rhizoctonia solani* as influenced by age of inoculum in soil. *Can. J. Bot.* **45**, 1761–1765.

Khew, K. L., and Zentmyer, G. A. (1974). Electrotactic response of zoospores of seven species of *Phytophthora. Phytopathology* **64**, 500–507.

Kosuge, T., and Hewitt, W. B. (1964). Exudates of grape berries and their effects on germination of conidia of *Botrytis cinerea. Phytopathology* **54**, 167–172.

Kraft, J. M. (1977). The role of delphinidin and sugars in the resistance of pea seedlings to Fusarium root rot. *Phytopathology* **67**, 1057–1061.

Lewis, B. G., and Day, J. R. (1972). Behaviour of uredospore germ tubes of *Puccinia graminis tritici* in relation to the fine structure of wheat leaf surfaces. *Trans. Br. Mycol. Soc.* **58**, 139–145.

McKeen, W. E. (1974). Mode of penetration of epidermal cell walls of *Vicia faba* by *Botrytis cinerea. Phytopathology* **64**, 461–467.

Malajczuk, N., and McComb, A. J. (1977). Root exudates from *Eucalyptus calophylla* R. Br. and *Eucalyptus marginata* Donn. ex Sm. seedlings and their effect on *Phytophthora cinnamomi* Rands. *Aust. J. Bot.* **25**, 501–514.

Matta, A. (1971). Microbial penetration and immunization of uncongenial host plants. *Annu. Rev. Phytopathol.* **9**, 387–410.

Mayama, S., and Pappelis, A. J. (1977). Application of interference microscopy to the study of fungal penetration of epidermal cells. *Phytopathology* **67**, 1300–1302.

Moller, W. J., and DeVay, J. E. (1968). Insect transmission of *Ceratocystis fimbriata* in deciduous fruit orchards. *Phytopathology* **58**, 1499–1508.

Mount, M. S., and Slesinski, R. S. (1971). Characterization of primary development of powdery mildew. In "Ecology of Leaf Surface Microorganisms" (T. F. Preece and C. H. Dickinson, eds.), pp. 301–322. Academic Press, New York.

Patton, R. F., and Johnson, D. W. (1970). Mode of penetration of needles of eastern white pine by *Cronartium ribicola. Phytopathology* **60**, 977–982.

Pirone, T. P., and Harris, K. F. (1977). Non persistent transmission of plant viruses by aphids. *Annu. Rev. Phytopathol.* **15**, 55–73.

Pope, A. M. S., and Jackson, R. M. (1973). Effects of wheat field soil on inocula of

*Gaeumannomyces graminis* (Sacc.) Arx & Olivier var. *tritici* J. Walker in relation to take-all decline. *Soil Biol. Biochem.* 5, 881–890.

Preece, T. F., and Dickinson, C. H., eds. (1971). "Ecology of Leaf Surface Microorganisms." Academic Press, New York.

Preece, T. F., Barnes, G., and Bayley, J. M. (1967). Junctions between epidermal cells as sites of appressorium formation by plant pathogenic fungi. *Plant Pathol.* 16, 117–118.

Rathaiah, Y. (1976). Infection of sugarbeet by *Cercospora beticola* in relation to stomatal condition. *Phytopathology* 66, 737–740.

Rathaiah, Y. (1977). Stomatal tropism of *Cercospora beticola* in sugarbeet. *Phytopathology* 67, 358–362.

Rishbeth, J. (1951). Observations on the biology of *Fomes annosus* with particular reference to East Anglian pine plantations. II. Spore production, stump infection, and saprophytic activity in stumps. *Ann. Bot. (London)* 15, 1–22.

Robinson, P. W., and Hodges, C. F. (1977). Effect of nitrogen fertilization on free amino acid and soluble sugar content of *Poa pratensis* and on infection and disease severity by *Drechslera sorokiniana*. *Phytopathology* 67, 1239–1244.

Rohde, R. A. (1972). Expression of resistance in plants to nematodes. *Annu. Rev. Phytopathol.* 10, 233–252.

Rovira, A. D. (1965). Plant root exudates and their influence upon soil microorganisms. *In* "Ecology of Soil-Borne Plant Pathogens—Prelude to Biological Control" (K. F. Baker and W. C. Snyder, eds.), pp. 170–186. Univ. of California Press, Berkeley.

Royle, D. J. (1976). Structural features of resistance to plant diseases. *In* "Biochemical Aspects of Plant Parasite Relationships" (J. Friend and D. R. Threlfall, eds.), pp. 161–193. Academic Press, New York.

Royle, D. J., and Hickman, C. J. (1964). Analysis of factors governing in vitro accumulation of zoospores of *Pythium aphanidermatum* on roots. I. Behavior of zoospores. *Can. J. Microbiol.* 10, 151–162.

Royle, D. J., and Thomas, G. G. (1973). Factors affecting zoospore responses towards stomata in hop downy mildew (*Pseudoperonospora humuli*) including some comparisons with grapevine downy mildew (*Plasmopara viticola*). *Physiol. Plant Pathol.* 3, 405–417.

Schroth, M. N., and Snyder, W. C. (1961). Effect of host exudates on chlamydospore germination of the bean root rot fungus, *Fusarium solani* f. *phaseoli*. *Phytopathology* 51, 389–393.

Schroth, M. N., Toussoun, T. A., and Snyder, W. C. (1963). Effect of certain constituents of bean exudate on germination of chlamydospores of *Fusarium solani* f. *phaseoli* in soil. *Phytopathology* 53, 809–812.

Shaykh, M., Soliday, C., and Kolattukudy, P. E. (1977). Proof for the production of cutinase by *Fusarium solani* f. *pisi* during penetration into its host, *Pisum sativum*. *Plant Physiol.* 60, 170–172.

Van den Ende, G., and Linskens, H. F. (1974). Cutinolytic enzymes in relation to pathogenesis. *Annu. Rev. Phytopathol.* 12, 247–258.

Weinhold, A. R., Dodman, R. L., and Bowman, T. (1972). Influence of exogenus nutrition on virulence of *Rhizoctonia solani*. *Phytopathology* 62, 278–281.

Wynn, W. K. (1976). Appressorium formation over stomates by the bean rust fungus: Response to a surface contact stimulus. *Phytopathology* 66, 136–146.

Zentmyer, G. A. (1961). Chemotaxis of zoospores for root exudates. *Science* 133, 1595–1596.

*Chapter 8*

# How the Beachhead Is Widened

RICHARD D. DURBIN

## I. INTRODUCTION

Having entered the host by conquering its peripheral defenses, the pathogen is now ready to broaden its attack. These attack mechanisms constitute the subject of this chapter. Some of what will be said about them is arguable, part will be conjectural, but it is hoped that most will be provocative.

Although pathogens obviously interact with their hosts prior to penetration, it is during and just after this event that the most profound interactions occur. Before penetration the pathogen has existed saprophytically, subsisting on its reserves and/or what it could assimilate from the external environment, where it is often at a competitive disadvantage and nutrients are limiting.

In entering the parasitic phase of its cycle the pathogen commits itself to a radically different environment—that of the host itself. Once inside, the pathogen is confronted with a vast array of novel compounds, many of which require it to synthesize new enzymes before the compounds can be acted upon. In addition, this environment changes as the pathogen advances, for the plant body composed of many diverse cell types that differ in physical and chemical makeup and respond differently to the pathogen.

The strategies, if we may call them that, of the pathogen are clear.

155

PLANT DISEASE, VOL. IV
ISBN–0–12–356404–2

First, it must catabolize these host substances and incorporate them into its own body in sufficient amounts and kinds that it may ultimately reproduce. Second, it must at the same time neutralize one, or usually more, defense mechanisms of the host as rapidly as possible. However, it should be pointed out that metabolically these two facets are not clearly different from each other.

The host–parasite interaction is so complex and dynamic that it is very difficult to separate the role that each plays. Nevertheless, using a variety of operations and experimental approaches—histochemical, physiological, and biochemical—researchers have made advances in many instances. However, in none do we have all the pieces in the puzzle. Nonetheless, patterns are beginning to emerge, and I believe that in the near future they can be woven into a whole.

## II. ATTACK MECHANISMS

### A. General Considerations

Pathogens collectively possess multiple attack mechanisms ranging from those that seek to destroy the host to those that are concerned with initiating and maintaining a "balanced," nondestructive relationship, e.g., haustoria and syncytia. For the most part they can be grouped into one of three general categories: (1) those that actively destroy or modify the structural integrity of the host, (2) those that inhibit selective processes or enzymes of the host, and (3) those that interfere with the regulatory systems of the host. Some are constitutive, whereas others are inducible and may require the prior operation of other attack mechanisms before the inducer becomes available. Interactions between individual mechanisms in different categories also occur, varying from additive cooperation, through synergism, to obligation. Thus, the pathogen is sustained by a series of reactions, some occurring in parallel, others in sequence.

Clearly, attack mechanisms are involved in disease production. However, whether they also play a crucial role in parasitism in unclear. For instance, many toxinless strains of phytopathogenic bacteria appear to be just as effective in causing necrosis and in multiplying in the host as their toxin-producing counterparts (Patil, 1974). We need to make the distinction, at least mentally, between what is crucial for parasitism and what is not.

At first glance the attack mechanisms used by the different taxonomic groups of pathogens might seem to vary enormously. There are some specialized cases, e.g., viruses and nematodes; however, if we consider the overall situation, many similarities appear. One is led to conclude that the different types of diseases, such as soft rots, wilts, or leaf spots, vary not so much in the kinds of attack mechanisms that are operating as in the quantitative balance among them. Still the question remains as to how much the restriction of a pathogen to selective portions of a host depends on deficiencies in attack mechanisms and how much reflects host defense mechanisms.

With the exception of phytohormones, most of our attention to date has been focused on attack mechanisms that destroy or inhibit. I believe more attention should be given to studying how the host's regulatory processes might be affected by pathogens. In principle, these processes might be altered in any of several ways, including changes in pH at the metabolic level (Bateman and Beer, 1965), redox potential, gaseous constituents, water potential, or amounts or properties of proteins, activators, and inhibitors. Merely changing the environment in one or more of these ways could have a very pervasive effect and, by itself, be responsible for major portions of the disease syndrome. Beyond regulation at the metabolic level is the potential for transcriptional and translational regulation. Besides those cases involving *Agrobacterium tumefaciens* and viruses, very little is known about this possibility. Additional examples, if present, would probably be found among the so-called obligate parasites and in situations involving the inability of host defenses to function (as will be discussed in Volume V of the treatise).

At first, disease development in a susceptible plant is relatively simple and involves only a limited area. However, because of the coordinated nature of plant metabolism, the primary effects of the attack mechanisms on the host have a cascading effect. Each alteration by itself soon induces alterations in related host processes until, at an increasingly rapid rate, all are perturbed. Secondary reactions set in, inducible mechanisms are initiated, and an ever-increasing portion of the host becomes affected or invaded. Ultimately, the interaction reaches a crescendo and the host, or a part of it, dies.

The situation becomes even more complex in the case of pathogens that move, grow, or are passively transported from one portion of the host to another. Since the milieu differs from tissue to tissue, different pathogen responses are elicited. Fungi, for example, become compartmentalized. Specialized organs and reproductive or vegetative resting cells form while the periphery of the thallus is advancing vegetatively. All of these situations could lead to the development of a somewhat different complement of attack mechanisms since the requirements of the pathogen would be expected to vary from state to state.

In general, investigators have attempted to study the primary interactions by simplifying the host–parasite system, particularly with regard to the pathogen. Some progress has also been made on the host side but not to the same extent. Technical problems are numerous. Actually it may be very difficult to pinpoint some primary interactions even when one is dealing with a highly simplified system. This arises because we usually assume that the attack mechanism of the pathogen being studied in the simplified system will still reproduce a portion of the disease syndrome. Indeed, this assumption is explicit in the criteria used for proof of complicity of toxins in disease production as discussed by Wheeler (1975). However, it may be that this assumption is not always valid. Consider the case of an attack mechanism that can act only after some other change in the host has occurred. Such a system may have no effect on an unaltered host. The investigator is faced with a dilemma. He needs to simplify the system in order to study it, but if he simplifies it his research may not yield valid results.

An aspect of the attack mechanisms about which we have little information is the existence of feedback control loops in the host. What sorts of chemical and/or physical sginals are being received by the pathogen, how are they recognized, and what are the pathways by which the pathogen's receptors ultimately trigger or modulate pathogen responses?

In addition to these receptors there is the possibility that other bound complexes can function as attack mechanisms. Essentially all the mechanisms we have information about are extracellular. While this phenomenon may be real, it may simply reflect the way in which we have experimentally approached the study of pathogenesis.

Another major, unknown area involves assessing the relative contribution of the parasite and the host to the production of substances important in pathogenesis, e.g., enzymes and small molecules. Commonly, for example, both members alone may be capable of synthesizing such substances. Alternatively, neither member alone may appear to synthesize the substance, although it is formed in the diseased plant. In neither situation can we discern which of the two members is responsible for synthesis. Occasionally, in the former case, it has been possible to make an assessment when the pathway of synthesis or form of the enzyme (isozyme) differs between parasite and host as discussed by Sequeira (1973).

Substances produced by the host that are self-destructive (such as phytotoxins in the restrictive use of the term) or the abnormally high synthesis of phytohormones might be considered secondary attack mechanisms. The methods by which the pathogen evokes their appearance are still obscure but significant in the context of this chapter in that the pathogen can "persuade" the host to do its work for it.

## B. Toxins

Some have suggested that toxins are ubiquitous in plant diseases. While this may be an overstatement, if defined in a broad sense it does appear that they are widespread among the different taxonomic groups of pathogens. However widespread their occurrence, the key question is whether a particular toxin plays a causal role in a particular disease. Unfortunately, convincing evidence for such a role has been provided in only a few cases. Lack of this evidence has required the setting up of operational definitions establishing different degrees of suspected involvement of toxins based on the kind and amount of information available about them (Wheeler, 1975).

This situation has come about for several reasons. At present we are seriously limited in recognizing and initially studying toxins because the bioassay systems used thus far are capable of monitoring only gross, nonspecific effects such as chlorosis, necrosis, growth abnormalities, wilting, gas exchange, flux of metabolites, and loss of structural integrity. Most of the current systems involve a complex, largely unknown series of reactions along the route to the observable reaction of the host. They are subject to large variations and differences in sensitivity.

In addition, several other important impediments to progress in this area exist: (1) The purity and structure of many toxins are uncertain; (2) their primary mechanism of action and metabolism in the host are unknown; (3) they are present in diseased plants only in small amounts; and (4) their exact level is difficult to measure because many of them are unstable and thus difficult to prepare for analysis. Nevertheless, more and more evidence is accumulating on a wide variety of compounds that implicate them in pathogenesis. How many cases will ultimately be shown to be valid cannot be stated, but the trend is clear and strongly suggests that, overall, toxins probably form an important component of the pathogen's armament.

## C. Enzymes

Enzymes play a prominent and relatively well studied role in the mechanism of attack. Of prime importance here are the hydrolases, which catalyze hydrolysis (that is, the direct addition of the elements of water across the bond that is cleaved), and oxidoreductases, which transfer hydrogen atoms or electrons. Other classes of enzymes are also important but have received relatively little study except for several lyases that are important in the degradation of cell walls.

Many of the enzymes are hydrolases that are involved in the catabolism of complex molecules forming the host cell wall. Damage to this structure is particularly devastating because it is of central importance to cell function. These enzymes are produced by most groups of pathogens, possibly because the cell wall constitutes such an important barrier to pathogen movement (see Volume V). Of equal consideration is the requirement of the pathogen for substrate and energy. Thus, from a teleological viewpoint, the evolution of cell-wall-degrading enzymes would appear to have been strongly favored, at least in the less advanced group of pathogens.

Oxidoreductases, among other things, appear to be causally related to the respiratory rise so common in many diseases. What this rise signifies in terms of attack mechanisms is difficult to assess, however. Several possibilities can be envisioned: (1) The host's energy charge is detrimentally altered; (2) host-produced substances are detoxified; or (3) compounds toxic to the host are produced.

## D. Phytohormones

It is generally agreed that phytohormone imbalances occur in numerous diseases. This contention is supported by several lines of evidence. Symptoms such as hypertrophy, abscission, and epinasty can be mimicked by the exogenous addition of appropriate phytohormones. In some cases elevated hormone levels within diseased tissues have been found. Still other studies have shown that pathogens are capable of synthesizing auxins, gibberellins, cytokinins, and ethylene, as well as other substances such as malformin that can affect plant growth and development. Besides synthesizing them, pathogens also have the capability to metabolize phytohormones. This ability has been little studied except with regard to auxins. It is, however, significant because balances between and among the various classes of phytohormes, as well as absolute amounts, determine host responses.

How much of this hormone imbalance is directly due to the pathogen is largely unclear. For example, many pathogens, fungi, bacteria, and nematodes, are capable of synthesizing indole-3-acetic acid (IAA), but to do this most of them require an exogenous source of tryptophan at a concentration much higher than would be expected to occur in the host. Furthermore, many plants and pathogens can metabolize IAA by one or more common pathways. Similar problems exist with interpreting the role of the other phytohormones.

Regardless of the uncertainty as to the source of synthesis or degradation, phytohormone imbalances appear to constitute an important attack mechanism in which the pathogen either produces or catabolizes the hormone directly or induces the host to do so. Phytohormone imbalances provide metabolic sinks, alter senescence, and, by a variety of means, divert host metabolism away from the production of potential defensive chemicals.

## E. Mechanical Force

Thatcher (1939, 1942, 1943) some years ago showed that the water potential of parasitic fungi is lower than that of their host cells. He developed the hypothesis that this differential allows the pathogen to assimilate water from the host. Furthermore, Thatcher, and later others, showed that there is an apparent correlation between susceptibility and increased permeability (Wheeler, 1975). Whatever the attack mechanism(s) responsible for this effect, it seems probable that leakiness of host membranes would favor the pathogen in terms of both providing a readily accessible supply of water and nutrients and upsetting the host's metabolic environment.

Osmotic differences between pathogen and host, if confined to small areas, also can be used to generate rather substantial mechanical forces. Undoubtedly, such forces could act synergistically with enzymes that are weakening the cell's architecture. The movement of certain fungi through the plant would seem to offer examples of this mechanism. Whether mechanical force is an important method making possible the intra- and intercellular movement of fungi is still unresolved, however (van den Ende and Linskens, 1974).

In contrast, the use of mechanical forces is well developed in nematodes. Here the evolution of a robust buccal stylet concomitant with plant parasitism is quite evident. The mechanical thrusting of the stylet punctures cells and allows the nematode to feed on the contents. Stylet movements facilitate internal movement of migratory endoparasites— *Pratylenchus*, *Radopholus*, and *Ditylenchus*—by forcing cells apart or puncturing them. Coupled with these stylet actions are a series of other neuromuscular events that are responsible for the body movements, enabling the nematode to move through the tissues.

## III. CONCLUDING REMARKS

Looking at pathogenesis from the pathogen's viewpoint, several important generalities can be made: (1) The pathogen employs a series of independent and coordinated attack mechanisms that have quite different effects on the host. (2) Some are constitutive whereas others require induction. (3) The basic types of mechanisms intrinsic to the different groups of pathogens are more similar than different. (4) The interaction between a parasite and host is in a dynamic state of disequilibrium. (5) Changes within the diseased plant are usually quantitative in nature and can be rapid.

In this brief survey I have tried to provide a common, conceptual background against which the more specific chapters to follow can be placed. It should be clear that much remains to be discovered about how pathogens induce disease. An understanding of this process would clearly be of great importance not only to basic biology, but also for practical considerations. Once the "stratagems" of the pathogen are known, preventative measures can be more rationally formulated.

### References

Bateman, D. F., and Beer, S. V. (1965). Simultaneous production and synergistic action of oxalic acid and polygalacturonase during pathogenesis by *Sclerotium rolfsii*. *Phytopathology* **55**, 204–211.

Patil, S. S. (1974). Toxins produced by phytopathogenic bacteria. *Annu. Rev. Phytopathol.* **12**, 259–279.

Sequeira, L. (1973). Hormone metabolism in diseased plants. *Annu. Rev. Plant Physiol.* **24**, 353–380.

Thatcher, F. S. (1939). Osmotic and permeability relations in the nutrition of fungus parasites. *Am. J. Bot.* **26**, 449–458.

Thatcher, F. S. (1942). Further studies of osmotic and permeability relations in parasitism. *Can. J. Res., Sect. C* **20**, 283–311.

Thatcher, F. S. (1943). Cellular changes in relation to rust resistance. *Can. J. Res., Sect. C* **21**, 151–172.

van den Ende, G., and Linskens, H. F. (1974). Cutinolytic enzymes in relation to pathogenesis. *Annu. Rev. Phytopathol.* **12**, 247–258.

Wheeler, H. (1975). "Plant Pathogenesis." Springer-Verlag, Berlin and New York.

*Chapter 9*

# How Fungi Induce Disease

PAUL H. WILLIAMS

## I. INTRODUCTION

Nature has evolved a magnificent array of relationships among organisms. Individuals of different species that share a "common life" for at least a portion of their existence are known as symbionts (Cooke, 1977). Within the Plant Kingdom, the range of symbioses extends from the obligately mutualistic relationships of the lichens to the facultatively antagonistic relations between pathogens and their plant hosts. Among the most intriguing and important studies in biology are those that deal with the complexities of symbiotic relationships. Particularly in the cases of antagonistic symbioses, in which disease is the outcome of the relationship, an understanding of the complexities of the relations will most likely provide insights into methods of therapy and disease management.

163

## II. CONCEPTUALIZATION OF FUNGAL PATHOGENESIS

Disease in plants is the injurious alteration of one or more ordered processes of energy utilization caused by a continued irritation by a primary causal factor or factors (see Volume III, Chapter 3). Although there are many causal factors, both animate and inanimate, that are capable of inducing disease in plants this chapter will be concerned with how fungi induce disease. Bateman (see Volume III, Chapter 3) has suggested that the relationships between symbionts, and in particular those involving pathogens and their suscepts, are sufficiently complex that none of the existing theories of parsitism are adequate to provide a sound conceptual basis for meaningful experimentation. Thus, he has proposed a *multicomponent hypothesis,* which relies on a complete understanding of the environments produced by and surrounding the symbionts in order to determine whether a relationship is mutualistic, commensalistic, neutral, or antagonistic. In the case of disease in plants, an unfavorable environment created by the pathogen confronts a favorable environment in the host, whereas, in resistance, an unfavorable environment produced by the host confronts the pathogen. Bateman has purposely defined the basis on which symbioses occur as broadly as possible, providing room for wide-ranging theoretical and experimental exploration. It is within the context of the multicomponent hypothesis, therefore, that I would like to consider fungal pathogenesis.

The development of fungal diseases in plants is extremely complex and in many ways resembles the intricacies of a Shakespearean tragedy or a Thomas Hardy novel. In the play or novel, the protagonists are placed on stage or in a particular setting in some relationship with one another, and the story is set in motion with a communication of some sort. Whether the communication is verbal, an act, or a deed is immaterial, for once a signal has been received, the drama is under way, and time will lead to a changing and increasingly complex environment as the "plot thickens." Inexorably, the interplay of events and interactions among the protagonists and among others in the story leads to the creation of an antagonistic environment and ultimately to the downfall of at least one protagonist. By skillful use of the pen, our great authors and playwrights have been able to depict the minds and temperaments of the characters in their novels and plays and in doing so have been able to reveal, with a high degree of precision, how and why they interact as they do. The reader or observer, too, is able to perceive the total drama as it unfolds and, in its resolution, to develop insights that give him a fuller understanding of human nature.

So, too, should we be able to perceive the biological dramas that embrace symbioses in which plants are confronted by parasitic and pathogenic fungi that are capable of inducing disease. In these biological dramas, the protagonists, the host and pathogen, are found in a specific spatial relationship with one another such that communications between them leads to the creation of an environment that sooner or later is antagonistic to the host plant and results in disease. The problem of understanding disease in plants, therefore, becomes one of how best to perceive the multitude of interactions that inevitably occur as symbiosis unfolds into disease. Unlike plays or novels, which are written in a language and enacted in a medium that is immediately perceptible to the audience or reader, the dramas of symbioses involve a spectrum of interorganismal communications that we are only beginning to learn how to decipher. The challenge to biologists interested in pathogenesis therefore lies in developing methodologies that will permit an accurate perception of the events occurring between the host and pathogen. Essentially, understanding pathogenesis is understanding interorganismal communications or, as Bateman has proposed, understanding the various components of the interacting environments of the symbionts.

As a prelude to examining particular forms of communication between pathogens and their suscepts, it would be useful to review the general relationships that exist between these symbionts by considering what constitutes the *favorable environment* of the host and the *unfavorable environment* of the pathogen. As chemical heterotrophs, fungi display a diversity of mechanisms for the acquisition of nutrients from plant cells and tissues. Within most of the major groups of fungi can be found both the necrotrophs, whose primary sources of nutrients are moribund cells, and biotrophs, which require metabolically active host cells from which to derive their sustenance. In addition to the necrotrophs and biotrophs, there are a number of other fungal pathogens, often termed hemibiotrophs, which for a portion of their lives are biotrophic and for another portion may be either necrotrophic or even saprotrophic. Superimposed on these various trophic or nutritional relationships may also be the degree of ecological interdependence of the symbionts. Many fungal pathogens have an obligate dependence on their host throughout their life cycle, whereas others are facultatively symbiotic and capable of existing as saprotrophs for extended periods (Cooke, 1977).

Each relationship, regardless of the form of nutritional dependence, also involves rather specialized spatial configurations between the host and the pathogen. In addition to supplying the necessary nutrients for pathogen growth, the host also provides various chemical and physical components of the environment essential to the creation of the particular niche required by the parasite.

## A. Favorable Host Environment

When viewed across the various taxa, the cells and tissues of higher plants present a relatively homogeneous, albeit, highly diverse, source of nutrients. This source of substance and potential energy was, of course, created by the plant for its own purposes of growth and development, but, alternatively, this resource could be used to satisfy the heterotrophic requirements of pathogenic fungi. Plant cells are rich in readily available carbon, much of which is sequestered as structural components in the cell walls or stored as energy reserves in carbohydrates, proteins, and lipids. The cytoplasm of plant cells abound in low molecular weight, metabolic intermediates that may provide ready sources of nutrients to both necrotrophs and biotrophs. The obligate biotrophs may even call directly upon the metabolic machinery of the host cells themselves to provide functions that have been lost in the process of evolutionary adaptation. All of these nutritive constituents comprise part of what may be considered to be the favorable environment of the host. Although we know, in general terms, what host components are utilized by pathogens, we are seriously deficient in our understanding of precisely how and why one pathogen must first kill its host before extracting nutrients, while another must maintain its host in a highly viable and stimulated state.

In passing, it is interesting to speculate that if higher plants were capable of providing only a "favorable environment" to fungal pathogens, the disease condition would prevail, and the life of green plants on earth would rapidly terminate. This issue is raised only to point out that in the vast majority of cases in which pathogens find themselves in proximity to the potential nutrient sources of higher plants, it is an "unfavorable host environment" that confronts the pathogen. It is resistance to pathogens that is the commonly observed phenomenon in the Plant Kingdom (see Volume 5).

## B. Unfavorable Pathogen Environment

Despite the fact that virtually all plant cells possess the nutrients essential for fungal growth, only a relatively few fungi have evolved the diverse arsenal of weapons necessary for the acquisition of these nutrients. The diversity of weapons in the pathogen reflects, to some degree, the diverse locations and forms of the essential nutrients within the host and, to a greater degree, the genetically determined capabilities that pathogens have evolved for obtaining these nutrients in ways that are energetically conservative. Some biotrophs, for example, may be able to utilize only small molecular components directly from the "soluble pool" of the host's cytoplasm, whereas various necrotrophs may also be able to call upon the energy-requiring steps of polymer degradation to gain access to additional host carbon reserves.

## C. Interorganismal Communications

In this chapter, pathogenesis is likened to a drama in which the major protagonists, the host and suscept, each consisting of genetically predetermined environments, the former predominantly favorable, the latter predominantly unfavorable, find themselves in such proximity that they are able to communicate with each other. Initially, a signal is sent by one of the symbionts and is received by the other. The signal may be amplified by the recipient and is followed by a response that may constitute a signal or reply to the other partner. With the response and second signal of the recipient, a full cycle of interorganismal communication has occurred. The cycle is repeated again and again, and, when viewed over time, this exchange of information becomes a progression that leads from initial decisions about compatibility to disease in the host and to growth and reproduction of the pathogen. The progression of disease in plants becomes increasingly complex when one considers that superimposed on the exchanges of information that constitute pathogenesis are alterations in the environment of the host and parasite that represent the ontogenetic changes associated with growth and maturation.

It is extremely important for the biologist interested in pathogenesis to be able to dissect out in the most minute detail *all* aspects of communication and response between host and pathogen. In order to do this, the complete time course of disease progression must be taken into account, from the initiation of the first signals to the full expression of disease and its accompanying propagative cycle of the pathogens. The precision with which our observations are made will, to a large measure, contribute to the validity of our insights. Time sequences that now may be perceived in hours or minutes will ultimately be observed in seconds and milliseconds. Perceptions of spatial relationships between hosts and parasites have already been vastly improved through the use of electron microscopy. Yet still greater physical resolution is needed if we are to understand the chemical relationships of interfacial zones. Likewise, what today is largely imperfect physiological or biochemical understanding of enzyme, toxicant, or regulatory molecule action will ultimately be reduced to a biophysical understanding. Before that time is reached, however, the imperfect outlines of the events of pathogenesis as we now understand them must be continually refined using the most rigorous approaches and effective methodologies available.

When exploring the sequence of signals and responses in any particular fungi–host relationship, it will be useful to categorize those components of the host that are currently thought to be involved in pathogenesis as well as those mechanisms of the pathogen that constitute the fungus arsenal. It is important to realize that components which today we may assign a particular role or function may with improved experimental resolution turn out to have a very different function in pathogenesis.


## III. RECEPTIVE AND RESPONSIVE HOST COMPONENTS

Rather than present an exhaustive treatment of those host components that are likely to be involved in pathogenesis, this section will provide a conceptual rationale for viewing some of them as they may be involved in the chemical and physical dialogue with the pathogen. For a more comprehensive review of what is known about host components and various constituents used by fungi in bringing about disease in the host, the reader is referred to various chapters in Heitefuss and Williams (1976).


### A. Physical Barriers

Physical constituents have long been considered to be important in providing barriers to the ingress and establishment of fungal pathogens. Various highly polymerized nonpolar substances on foliar surfaces, such as waxes, cutins, and suberins, have been considered as relatively inert to fungal penetration. In Chapter 7 of this volume, Dodman has dealt thoroughly with current thinking on the role that cuticular components play in regulating fungal ingress. In some cases, the physical conformation of these hydrophobic surfaces is of great importance in stimulating the production and guiding the hyphae in the process of penetration. The waxy cuticular surface, indeed, may be viewed as the stage on which the earliest exchanges of information in pathogenesis occur. Evidence is accumulating that these surfaces may interact chemically with host exudates and surface constituents, providing initial sources of exogenous carbon to the pathogen prior to penetration. Although the ester and ether linkages of waxes are generally difficult for many fungal pathogens to attack, etching of the cuticle by hyphae of leaf surface pathogens, such as powdery mildews (Staub *et al.*, 1974) and *Venturia inaequalis*, has been reported (Bracker and Littlefield, 1973). This suggests that even the "inert" surface barriers may be actively involved in ingress. Lignin is an important compound conferring physical strength and chemical protection to cellulose microfibrils in cell walls, yet we know that various fungi possess the oxidative enzymes necessary for breaking the complex linkages between the phenolic subunits of lignin. The question that must always be raised with regard to the nature of host barriers to ingress is, Are they functioning as a chemical and physical deterrent to penetration of walls and tissues, or are they constitutive, degradable components capable of contributing to the energy requirements of the pathogen or possibly even of responding actively to a communication from the pathogen? The answer to this question lies very much with our understanding of the nature of the fungal–host communication.

## B. Chemical Constituents

In addition to chemicals that may function as inert constitutive components, plant cells contain a wide array of substances that normally are more readily assimilated and that can serve as a major source of carbon, nitrogen, and nutrients for pathogenic fungi. Such compounds as cellulose, pectins, and proteins are vital components of the host cell structure, whereas starches, lipids, and other proteins constitute major reserves of stored energy for the plant. Along with these degradable polymeric components of the host cell are numerous soluble compounds (sugars, amino acids, peptides, organic acids, etc.), many of which may serve as readily available sources of energy for the pathogen.

Whether the host–pathogen relationship is an obligately parasitic one in which much of the vital environment is within the living host cell itself, or whether it is a facultatively perthophytic relationship in which the pathogen is growing in the chemical milieu of dead tissue, many of the molecules produced by the host will act as signals to which the fungus may respond. Many low molecular weight constituents may play an important regulatory role in synthesis of fungal organelles (Maxwell et al., 1975) or regulate synthesis of hydrolytic enzymes via catabolite repression (Bateman and Basham, 1976).

An important contribution of low molecular weight host constituents to the environments of both host and pathogen is in providing organic buffering, in which hormone regulation, toxin action, and polymer degradation all occur.

## C. Metabolic Components

With obligately parasitic relationships, the anabolic and catabolic pathways of the host cell constitute that part of the favorable environment that is highly responsive to and, indeed, intimately tuned to communications from the pathogen. In order to function as a congenial environment for intracellular and haustorial pathogens, the host must provide a stable interface in a suitably buffered osmoticum to permit appropriate regulation of its metabolism by the parasite. Similarly, from the perspective of the pathogen, its own physiology must be delicately adjusted to achieve a nutritionally and metabolically congenial relationship with the host cells. One of the most intriguing mysteries of obligate parasitism is the nature of the information exchanges, which foster increasing demands on the host's anabolic resources and at the same time extract increasing quantities of essential nutrients from the host cell (Daly, 1976a).

Virtually all of the host's organelle systems may be affected by the parasite. Activity of mitochondria and chloroplasts may be enhanced, biosynthetic pathways for protein, starch, and soluble carbohydrate may be activated or repressed, and membranes of the host-parasite interface may be stabilized or disrupted. Although with the obligate biotrophs, the environment of the pathogen may seem highly favorable toward the host, this period is relatively transitory, and, as increasing demands are made upon the host, the environment quickly changes to an antagonistic one. Storage polymers and host reserves become depleted, the metabolic capabilities of the host cell are unable to sustain the ordered discharge of energy, and tissues senesce prematurely (see Volume III, Chapter 18). Frequently, tissues in proximity to and cells colonized by obligate biotrophs are sustained in a viable condition at the expense of surrounding tissues, giving rise to the metabolic sink and "green island" effects on leaves (see this volume, Chapter 8).

### D. Genetic Components

An integral part of the alteration in overall cell metabolism by invading parasites is the regulation of the transcriptional and translational components (see Volume III, Chapter 17). Numerous biotrophic fungi are able to stimulate RNA synthesis within infected cells (Heitefuss and Wolf, 1976). Enhanced RNA synthesis is often accompanied by the rejuvenation of metabolic pathways, leading to cell growth, dedifferentiation, and replication. Although evidence is lacking, rejuvenation is likely to involve hormonal regulation by compounds produced both within the parasitized cells and by the pathogen itself (Pegg, 1976).

The environment of rejuvenated cells appears to represent a particularly favorable one for many biotrophs. Such cells may be metabolically more "leaky" or physiologically more "pliable" than differentiated cells and, thus, be capable of responding to the demands of the parasite without undergoing lethal shock.

The important concept when one is considering the host as a favorable environment for the pathogen is that virtually every component of the host, be it an apparently inert polymer, a single molecule, or a highly regulated and vital metabolic pathway, must be regarded as potentially reactive and, in many cases, receptive to signals from the pathogen. Thus, virtually all host components are brought into the drama of pathogenesis, and the degree to which they are engaged in an active role remains to be determined.

## IV. THE FUNGAL ARSENAL

Pathogenic fungi possess an intriguing array of mechanisms that permit them to acquire nutrients from the host environment. The fungal arsenal, in essence, constitutes the signals to which the host will respond. Some of the fungal components assist the pathogen with physical aspects of ingress, others act primarily in the dissolution of host components prior to assimilation, whereas others perform a dual role of ingress and assimilation. Toxins act selectively or nonselectively, mainly to destroy or depress vital functions within the cell, whereas growth regulatory molecules may stimulate particular components of the host cell's metabolism. In virtually every disease, numerous fungal mechanisms of attack are brought into play, fulfilling particular functions necessary to elicit the full concert of events in pathogenesis.

### A. Structural Weapons

Dodman (this volume, Chapter 7) has dealt thoroughly with those components of the fungus arsenal that are concerned primarily with overcoming the physical and chemical barriers to ingress of the host. Specialized penetration structures, such as infection cushions, appressoria, and the adhesorium and Stachel of the Plasmodiophorales, all represent evolutionary variations on the theme of concentrating either physical or chemical energy over a very small area so that ingress may be achieved efficiently. Whether the location and establishment of the penetration apparatus are strictly physical processes determined randomly, or whether they represent the product of a number of communicative interchanges between the host and pathogen is not fully understood for any host–pathogen relationship. Careful microscopic investigations of host penetration by a number of fungal pathogens have suggested that host wall penetration involves chemical dissolution as well as some physical distortion of the complex wall components (Aist, 1976). Even in the case of penetration by *Plasmodiophora brassicae*

Wor., where the Stachel is virtually expelled from the adhesorium and propelled through the root hair cell wall, Stahmann (Williams *et al.*, 1973) has suggested that the actual penetration of the cell wall could involve the action of enzymes on the surface of the Stachel acting on the microfibrils and matrix of the cell wall. As with so many other events in pathogenesis, this apparently physical penetration of the host cell wall, which occurs in less than 1 second, may be found to involve chemical processes as well.

An aspect of the physical presence of fungi within the host that may contribute significantly to pathogenesis is the actual mass of the hyphae or thallus. The rapid invasion and replacement of vegetative, floral, and reproductive parts by numerous hyphae and spores of the smut fungi indicate the effectiveness of the fungal mass in pathogenesis. With a number of rust fungi and *Albugo* sp., the growth of mycelium and production of sporophores ruptures the host epidermis and contributes to excessive water losses from the leaves. Massive hyphal growth and yeastlike stages of various vascular pathogens may contribute to reduced water movement in the host.

## B. Enzymes

Among the most important group of constituents functioning in pathogenesis are the fungal enzymes involved in the digestion of polymeric host components. Various hydrolytic enzymes are capable of cleaving pectin, cellulose, lipid, and protein components into their respective monomeric units (see Volume III, Chapter 13). Transeliminases produced by some fungi degrade pectic substances via lytic action, whereas oxidative enzymes have been shown to be active in lignin degradation (Kirk and Chang, 1975).

Among the wall-degrading enzymes, the pectic enzymes are the best understood in pathogenesis. Pectinases are almost universally produced by pathogenic necrotrophic fungi. They play a primary role in tissue maceration of soft-rot diseases of fruits and vegetables. The pectic enzymes function first by splitting the $\alpha$1,4-glycosidic bonds in the galacturonide fraction of cell walls by either endwise or random cleavages. In this way, they render the uronide polymers susceptible to other wall-degrading enzymes. They also function to disrupt the permeability of cell membranes, causing leakage and rapid death of cells. Whether the altered pH or ionic properties of the maceration fluid or molecular components cleaved during maceration are responsible for cell death is not known.

Very little is understood regarding the enzymatic dissolution of waxes, cutins, or proteinaceous components of the host. Cytological observations suggest that enzymatic degradation of these components is an important way of breaching host barriers (Aist, 1976).

Change in cell membrane permeability is another almost universally observed physiological symptom of disease (see Volume III, Chapter 15). Although toxins, along with proteases and phospholipases, have been implicated in altered membrane function, very little systematic evidence has been brought to bear on the role of these enzymes in fungal pathogenesis. Phospholipase activity is quite common *in vitro* among fungal pathogens, and macerated tissues have been found to contain higher phospholipase activities; however, evidence on the role of these enzymes in pathogenesis is largely circumstantial.

The production and deployment of enzymes as a group of attacking mechanisms by the fungi is one of the most important aspects in the study of pathogenesis. Necrotrophs frequently secrete sufficient quantities of extracellular enzymes that tissues are macerated well in advance of the invading hyphal front. Cells are killed prior to penetration. In some diseases, the production of pectinases and cellulases may be regulated by the availability of small molecular weight sugars via catabolite repression, whereas, in others, high concentrations of hydrolytic enzymes may be found in diseased tissues.

In contrast to those necrotrophs that rely on massive production of extracellular enzymes are the haustorial and intracellular biotrophs. Although these parasites apparently require the activity of various enzymes capable of degrading cell walls, they must deploy them in a precisely timed and highly localized way during the penetration of living cells. Such enzymes may be under very rigid and localized forms of regulation or perhaps bound to the surface of hyphae at particular points where their activity is needed. As with other components of the fungus arsenal, there has been relatively little critical experimentation *in vivo* on the role of fungal enzymes during the actual progression of disease.

## C. Toxins

Toxins are a diverse group of fungal metabolites that have debilitating effects on one or more vital functions of host cells. Although toxins are usually considered to function apart from enzymes and growth regulators, the opposite is more likely true. Toxins may very well operate in concert with lytic enzymes or certain growth hormones during pathogenesis.

Some toxins show a high degree of host genotype selectivity—a few even exhibit specificity of action at the single-gene level, whereas others act widely over a range of plant species on most cells and tissues. Although more than 110 different compounds have been identified as toxins produced from over 100 species of fungal pathogens, the mode of action of only a very few is understood (Rudolph, 1976; Scheffer, 1976). Some toxins alter plasma membrane function, resulting in increased loss of electrolyte, while others affect chloroplast function or mitochondrail activity. Some toxins mimic the action of various hormonal growth regulators but usually cause lethal side reactions. Most toxins are of relatively low molecular weight and are released extracellularly by fungal pathogens. They often function in pathogenesis in the debilitation of cells and tissues in advance of the invading pathogen. Toxin-induced loss of solutes from host cells may provide the first readily available sources of carbon to the invading pathogen. Toxins are commonly produced by leaf-spotting fungi and may be an important part of the arsenal of many necrotrophs.

It is no surprise that there have been few reports of toxins produced by biotrophs (Rudolph, 1976); almost by definition it seems unlikely that toxins could play a significant role, except perhaps in the terminal phases, of an obligately parasitic relationship.

## D. Growth Regulators

A wide range of pathogenic fungi have been shown to be capable of producing, *in vitro*, various hormonal growth regulators of higher plants. In a number of instances, higher than normal concentrations of these growth hormones have been associated with fungal infections (Pegg, 1976; Dekhuijzen, 1976). For these reasons, growth regulators have been considered to be among the arsenal of compounds used by fungi in pathogenesis. Although most of the research on the production of auxins, gibberellins, cytokinins, and ethylene has been with necrotrophic fungi, it is interesting that the symptoms of apparent imbalances and excesses of these hormones are observed most commonly in disease of biotrophs. As discussed in (Volume III, Chapters 1 and 8) a great deal of research has been done on the physiology of various plant growth regulators, but their role in both healthy and diseased plants remains unclear. This is due in part to the widespread assumption by both plant physiologists and plant pathologists that plant growth is regulated by the concentration of single growth regulatory substances rather than by the dynamic changing ratios among two or more growth hormones. In many cases, the presence of abnormally high concentrations of growth hormones in diseased tissues associated with abnormal growth has been the strongest evidence of a role for these compounds in pathogenesis.

When a clearer understanding of the function of plant hormones is obtained and when suitable methods have been developed for examining their changing ratios and activity *in situ*, a more critical evaluation of their roles in pathogenesis will be possible. Until then, one can only speculate that these compounds probably constitute powerful regulatory signals that condition or maintain parasitized and diseased cells in an active physiological condition, often permitting massive colonization of cells and tissues by various biotrophs. In the case of the fungi, it is unclear whether the parasites themselves are producing the growth regulators, or whether the fungi are stimulating the host cells to produce excesses of the compounds. It is entirely probable that both the pathogen and the host are producing these regulators.

### E. Metabolic Equilibria

An important physiological aspect of disease, which is presumed to be regulated by imbalances in the ratios of growth hormones, is the metabolic sink created in the infection court. Tissues parasitized by biotrophs frequently draw upon surrounding noninfected tissues, redirecting the flow of nutrients toward the infection and creating a debilitating drain on other parts of the plant. Shrunken grain, premature ripening, and senescence are frequent symptoms brought on by the misdirected flow of assimilates to infection sites. Although the ability of an infected region of tissue to redirect nutrient flow from either contiguous or distant tissues may be mediated through gradients of hormones, within individually parasitized cells, the metabolic capabilities of the parasite may dictate the unidirectional flow of nutrients toward the parasite. Smith *et al.* (1969) pointed out that the conversion of glucose and other sugars obtained from the host to sugar alcohols by various biotrophs shifts the equilibrium of nutrient flow across haustorial boundaries in favor of the parasite. Such directional flow of nutrients is only a part of the very complex controls of host metabolism that biotrophic pathogens exert in host tissue (Kosuge and Gilchrist, 1976). Soon after invasion of cells by obligate biotrophs, host anabolism is stimulated and storage polysaccharides accumulate. After a period of time, often coinciding with the onset of sporulation by the pathogen, the excess reserves may become depleted (Long *et al.*, 1975). Although the activities of some of the enzymes involved in starch synthesis and degradation can be examined through classical procedures of extraction and assay, the precise role and regulation of such enzymes in pathogenesis will require new experimental approaches (Daly, 1976b).

## F. Extracellular Genetic Determinants

The genome of the pathogen itself is the primary determinant of all the factors we have discussed in this chapter as components in the offensive arsenal of fungal pathogens. In recent years, we have come to realize that, at least in the case of bacteria, various extrachromosomal genetic determinants, such as viruses, plasmids, and episomes, can confer pathogenic capabilities on the organism. Recent evidence suggests that the presence of specific genetic information on the plasmid of *Agrobacterium tumefaciens* confers tumor-inducing capabilities on the organism, and that the "tumorogenic code" is likely to be incorporated into the genome of the host plant from the pathogen. At this time, it is a matter for speculation as to whether pathogenic fungi possess extranuclear genetic factors such as a virus that could act as a determinant of pathogenesis. The work of Lindberg (1971) on cytoplasmically transmissible factors that condition virulence or certain isolates of *Helminthosporium oryzae* and the extranuclear transmissible hypovirulence factors in *Endothia parasitica* (Grente and Sauret, 1969) points to the direct involvement of extranuclear components of the fungal genome in pathogenesis.

## G. pH and the Ionic Milieu

Within the arena in which pathogenesis takes place, be it the expanding zone of macerating tissue in rotten fruit or damping-off seedlings, the slowly expanding margin of a necrotic leaf spot, or the massively hypertrophic tissues of a maize smut gall, an important aspect of the environment in which the pathogen and the host are communicating is the ionic medium in which virtually all of the molecular events of pathogenesis take place. The pH of the aqueous medium and its ionic characteristics have an important part in determining the effectiveness of enzyme, toxin, and growth regulator activity. In the case of obligately biotrophic or intracellular parasites, these same components of the inorganic chemical environment are important in metabolic regulation. Little is known concerning the physical and chemical properties of the inorganic milieu of the infection court. Undoubtedly, the degree of buffering by both inorganic and organic constituents varies with the progression of disease. What effect the various inorganic solutes have on the reactive components of the fungus arsenal will prove to be one of the most complex aspects to interpret in pathogenesis.

## V. PATHOGENESIS

Disease induced by the fungal pathogens of higher plants represents the end result of a range of processes that are virtually as diverse as the combination of symbionts involved. Although, for convenience, pathologists have classified diseases into various kinds, depending largely on gross symptoms, each host–suspect combination is likely to differ from every other one in the progression of communications exchanged between the symbionts. In a chapter such as this, to return to the traditional classifications of types of disease and to discuss them in terms of the potential mechanisms would be to return to traditional thought and would most likely lead to the inadequacies of our conventional explanations and understanding of disease causation.

Our knowledge of how fungi induce disease in plants is at best only superficial. This is because of both the complexities and variations in the process of pathogenesis and our limited ability to perceive what is being communicated between host and pathogen. What is needed are painstaking, thorough, and deliberate experimental approaches to the processes by which fungi and hosts interact. Our present methods, although useful in providing us with structural and chemical glimpses of pathogenic components and host responses, are too often viewed out of context of the pathological environment. Every researcher of pathogenesis must be prepared to be an experimental innovator, using new methodologies that he has created or adopted from the best methodologies of allied fields.

Of the full spectrum of events in any given disease, certainly the dissection over time of the initial events, representing the first communications between host and pathogen, stand the best chance of providing new insight into disease processes. In the external environment prior to penetration, events can be followed visually and probably represent environments that are less complex than those that follow as ingress and disease proceed. The light and electron microscopes will continue to be important tools in providing new perspectives on the physical progress of disease. Enzymology and physiology will continue to provide information on the capabilities of the pathogen and responses of the host, but these are the conventional studies of the experimentalist today. Progress during the next few years is most likely to come from studies that couple the electron microscope to techniques in enzymology and physiology at the cellular level. With these combined approaches, we will view more precisely what is happening and where it is happening in disease. To answer how events in pathogenesis are progressing will require new methods and insights into the living cell and higher plant.

## References

Aist, J. R. (1976). Cytology of penetration and infection-fungi. In "Physiological Plant Pathology" (R. Heitefuss and P. H. Williams, eds.), pp. 197–221. Springer-Verlag, Berlin and New York.

Bateman, D. F. (1978). The dynamic nature of disease. In "Plant Disease: An Advanced Treatise" (J. G. Horsfall and E. B. Cowling, eds.), Vol. III. Academic Press, New York.

Bateman, D. F., and Basham, H. G. (1976). Degradation of plant cell walls and membranes by microbial enzymes. In "Physiological Plant Pathology" (R. Heitefuss and P. H. Williams, eds.), pp. 316–355. Springer-Verlag, Berlin and New York.

Bracker, C. E., and Littlefield, L. J. (1973). Structural concepts of host-pathogen interfaces. In "Fungal Pathogenicity and the Plant's Response" (R. J. W. Byrde and C. V. Cutting, eds.), pp. 159–317. Academic Press, New York.

Cooke, R. (1977). "The Biology of Symbiotic Fungi." Wiley, New York.

Daly, J. M. (1976a). Some aspects of host-pathogen interactions. In "Physiological Plant Pathology" (R. Heitefuss and P. H. Williams, eds.), pp. 27–50. Springer-Verlag, Berlin and New York.

Daly, J. M. (1976b). The carbon balance of diseased plants: Changes in respiration, photosynthesis, and translocation. In "Physiological Plant Pathology" (R. Heitefuss and P. H. Williams, eds.), pp. 450–479. Springer-Verlag, Berlin and New York.

Dekhuijzen, H. M. (1976). Endogenous cytokinins in healthy and diseased plants. In "Physiological Plant Pathology" (R. Heitefuss and P. H. Williams, eds.), pp. 526–559. Springer-Verlag, Berlin and New York.

Grente, J., and Sauret, S. (1969). L'hypovirulence exclusive, phénomène original en pathologie végétal. C. R. Acad. Sci., Ser. D 268, 2347–2350.

Heitefuss, R., and Williams, P. H., eds. (1976). "Physiological Plant Pathology." Springer-Verlag, Berlin and New York.

Heitefuss, R., and Wolf, G. (1976). Nucleic acids in host–parasite interactions. In "Physiological Plant Pathology" (R. Heitefuss and P. H. Williams, eds.), pp. 480–508. Springer-Verlag, Berlin and New York.

Kirk, T. K., and Chang, H.-M. (1975). Decomposition of lignin by white-rot fungi. II. Characterization of heavily degraded lignins from decayed spruce. Holzforschung 29, 56–64.

Kosuge, T., and Gilchrist, D. C. (1976). Metabolic regulation in host–parasite interactions. In "Physiological Plant Pathology" (R. Heitefuss and P. H. Williams, eds.), pp. 679–702. Springer-Verlag, Berlin and New York.

Lindberg, G. D. (1971). Disease-induced toxin production in Helminthosporium oryzae. Phytopathology 61, 420–424.

Long, D. E., Fung, A. K., McGee, E. E. M., Cooke, R. C., and Lewis, D. H. (1975). The activity of invertase and its relevance to the accumulation of storage polysaccharides in leaves infected by biotrophic fungi. New Phytol. 74, 173–182.

Maxwell, D. P., Maxwell, M. D., Hänssler, G., Armentrout, V. N., Murray, G. M., and Hoch, H. C. (1975). Microbodies and glyoxylate-cycle enzyme activities in filamentous fungi. Planta 124, 109–123.

Pegg, G. F. (1976). Endogenous auxins in healthy and diseased plants. *In* "Physiological Plant Pathology" (R. Heitefuss and P. H. Williams, eds.), pp. 560–581. Springer-Verlag, Berlin and New York.

Rudolph, K. (1976). Non-specific toxins. *In* "Physiological Plant Pathology" (R. Heitefuss and P. H. Williams, eds.), pp. 270–315. Springer-Verlag, Berlin and New York.

Scheffer, R. P. (1976). Host-specific toxins in relation to pathogenesis and disease resistance. *In* "Physiological Plant Pathology" (R. Heitefuss and P. H. Williams, eds.), pp. 247–269. Springer-Verlag, Berlin and New York.

Smith, D. C., Muscatine, L., and Lewis, D. (1969). Carbohydrate movement from autotrophs to heterotrophs in parasitic and mutualistic symbiosis. *Biol. Rev. Cambridge Philos. Soc.* **44**, 17–90.

Staub, T., Dahmen, H., and Schwinn, F. J. (1974). Light- and scanning-electron microscopy of cucumber and barley powdery mildew on host and nonhost plants. *Phytopathology* **64**, 364–372.

Williams, P. H., Aist, J. R., and Bhattacharya, P. K. (1973). Host–parasite relations in cabbage clubroot. *In* "Fungal Pathogenicity and the Plant's Response" (R. J. W. Byrde and C. V. Cutting, eds.), pp. 141–158. Academic Press, New York.

# Chapter 10

# How Bacteria Induce Disease

ARTHUR KELMAN

## I. INTRODUCTION

The bacteria that initiate disease in higher plants comprise only a tiny fraction of the bacteria to which these plants may be exposed in their normal growth. In contrast to the many bacteria that are pathogenic to man and other animals, the phytopathogenic forms represent very few genera, indeed. In man, approximately 60 species in over 40 genera cause disease; in contrast, fewer than five species of bacteria affect each major species of basic food crops.

Certain groups of green plants (mosses, ferns, and gymnosperms) have few, if any, bacterial diseases of consequence. No serious bacterial disease affects the conifers or hardwood species that are of commercial importance in the forests of the world. The monocots are also relatively free of severe attack by bacterial pathogens. Some bacterial diseases of marine algae have been observed; in general, the bacteria involved are inadequately characterized, and Koch's postulates have not been applied (Andrews, 1976). The small number of bacterial diseases re-

PLANT DISEASE, VOL. IV

ported on plants other than angiosperms may merely reflect lack of research and/or commercial impact.

Unlike many fungi, bacteria lack mechanisms for forcing their way physically through protective barriers, such as cuticle, epidermis, or bark; however, so many natural openings exist that morphological factors can have small importance in resistance to bacterial invasion. Since most saprophytic bacteria can invade and degrade dead plant tissue, phytopathogenicity is not dependent on unique nutritional capabilities. Leaves of living plants are covered by teeming populations of epiphytic bacteria that normally are not phytopathogenic. Below ground, large bacterial populations also compete with other soil microflora and fauna for available nutrients that leak from root surfaces. Thus, plants have evolved in a virtual microbial jungle surrounded by free-living bacteria with the potential to embrace parasitism. It is remarkable, therefore, that so very few phytopathogenic bacterial species have evolved.

Plants appear to be more susceptible to fungi than to bacteria; the number of different fungal species that attack a given crop plant successfully is often 10 times higher than that for bacteria on the same plant species. It is clear that mechanisms for defense in plants against bacteria may differ significantly from those for resistance to fungi. Many antifungal compounds are formed in plants during infections by bacteria as well as fungi (Kuć, 1976). Usually, these compounds have little or no inhibitory effect on bacterial pathogens. One can conclude that the bacterial species that cause disease in plants, few as they are, must share certain distinctive characteristics that are not possessed by fungal pathogens or nonpathogenic bacteria. The physiological functions that set the phytopathogens apart from closely related bacteria have been identified in relatively few instances, however.

Bacteria and their close relatives produce diseases that encompass the complete spectrum of symptoms that can be induced by other biotic agents of disease, including fungi, viruses, mycoplasma-like organisms, and nematodes. These symptoms may result from biochemical events in which the metabolites of the host may play a greater role than those of the pathogen. Also, similar symptoms may result from a series of complex reactions started by very different agents. The primary processes in disease induction are constantly changing during the initial period of growth of the bacterial cells in the host tissue. Conversely, this changing environment undoubtedly influences the metabolism of the invading bacterium. Investigators who attempt to focus on this problem find themselves confronted, therefore, with a constantly moving target.

The objective of this chapter is to consider the nature of interactions between bacteria and plants and the chemical weapons of phytopathogenic bacteria involved in pathogenesis. The literature on bacterial pathogenesis has been surveyed in previous reviews and texts (Starr and Chatterjee, 1972; Goodman *et al.*, 1967, 1976; Heitefuss and Williams, 1976; Kelman, 1970; Wood, 1967). Primary emphasis in this discussion is placed on research completed in the past decade. Aspects of pathways of infection and initial development of bacteria in host plants are covered in this volume, Chapters 7 and 8.

## II. INITIAL INTERACTIONS BETWEEN BACTERIA AND PLANTS

Unlike other pathogens, invading bacteria rarely make direct contact with the living protoplast of host plants. Instead, they must affect the living protoplast through the formidable barrier of the plant cell wall. Thus, bacteria must survive initially on the compounds from the intercellular spaces or, in the case of the few vascular parasites, the xylem elements.

In each of the diseases, at least one essential physiological function of the plant is affected adversely. However, a broad range of effects may be induced at the cellular level. The basic changes in host cell metabolism that occur during the early stages of pathogenesis are similar for most diseases caused by bacteria. These include an altered or increased respiration, the accumulation of phenolic substances, an increased permeability of cell membranes (or at least a significant increase in conductivity) and peroxidase activity.

Bacteria that are not known as pathogens can be isolated from plants that are ostensibly healthy (Hayward, 1974). Usually they are the saprophytic bacteria that are present commonly in the phylloplane and rhizoplane or rhizosphere of the specific host plant. Little is known as to how these bacteria maintain themselves inside plants and the effects that they have on host plants. There are few other examples of the type of symbiotic relationship that exists between species of *Rhizobium* and legumes. In contrast to the small number of different bacteria involved in plant symbioses, many species of fungi in numerous genera form compatible mycorrhizal relationships with a variety of plants.

### A. Hypersensitive Reaction

The hypersensitive reaction in plants to bacteria was first reported by Klement (Klement and Goodman, 1967). He observed that the introduction of large numbers of certain incompatible plant pathogenic bacteria into leaves of tobacco plants resulted in the death of cells in the infiltrated area within 12 hours. Although only a few bacterial plant pathogens (certain soft rotting bacteria, for example) lack the capability of eliciting a hypersensitive reaction, saprophytic bacteria generally cannot induce this response in plants.

In man, the main symptoms of damage from bacterial disease often do not result directly from toxic products of the pathogen. Instead, the body tissues may overreact to the invading organism, and thus the damage may be regarded as self-inflicted. In plants, the hypersensitive reaction may be a similar type of self-damaging response. Thus far, very little has been learned about the specific biochemical reactions of the host that are involved in this injurious process.

## B. Recognition Phenomenon

A major advance in our knowledge of disease induction by bacteria in recent years has been the demonstration of mechanisms by which plants recognize certain incompatible bacteria (Sequeira, 1978). Much of this research received its stimulus from studies of the effects of plant lectins on animal cells. Lectins are glycoproteins that can bind to specific sugars on cell membranes and also cause animal and bacterial cells to agglutinate (Sequeira and Graham, 1977) (see Volume V, Chapter 10).

Most bacteria adhere tightly to many types of surfaces (Costerton *et al.*, 1978). This adhesion is attributed to the formation of interwoven fibers of branching sugar molecules that extend from the cell surface to form a "glycocalyx." The ability of this glycocalyx to adhere to surfaces of the host is a key factor in the initiation of a wide range of diseases of man and animals, ranging from dental caries to pneumonia. The glycocalyx has not been seen in bacterial cells grown in culture. Polysaccharide strands of the glycocalyx can bind to a similar glycocalyx on animal cell surfaces, or binding may be mediated by lectins that act as bridging connections. The specificity of certain bacteria for certain tissues in the body may well be attributable to specificity of these interactions. Unfortunately, no detailed studies of glycocalyx formation by phytopathogenic bacteria have been reported.

The characteristics of the binding of animal pathogens to host cells may have relevance to recent research on *Rhizobium* and *Agrobacterium*. Initial investigations with the *Rhizobium*–plant system (Bohlool and Schmidt, 1974) indicated a possible role for lectins in the specificity of *Rhizobium* species for particular legume hosts. Subsequent studies on the *Rhizobium*–clover interaction indicate that a unique surface antigen on the surface of the *Rhizobium trifolii* cell as well as the clover root hair may be cross-bridged by a specific lectin, trifoliin, which is present on clover roots (Dazzo, 1979). The cross-reactive surface antigen is absent on noninfective *R. trifolii* mutants or other *Rhizobium* species that do not infect clover. Virulent cells of the crown gall bacterium, *Agrobacterium tumefaciens*, also bind to cell walls and show certain similarities to *Rhizobium*–plant interactions (Lippincott and Lippincott, 1976). The specific binding site in the case of *A. tumefaciens* has not been determined, although pectic substances rather than lectins may be involved.

The formation of legume root nodules or induction of plant tumors obviously involves processes very different from those that characterize the majority of bacterial diseases of plants. It is perhaps not surprising that attachment of cells indicates an incompatible rather than compatible reaction for those diseases resulting in necrosis of tissues. In the case of the bacterial wilt pathogen, *Pseudomonas solanacearum*, the interaction between host cell walls and incompatible avirulent bacteria involves recognition, attachment, and envelopment (Sequeira, 1978). Interactions occur between specific constituents of the bacterial cell wall and binding sites on the plant cell wall surface. The envelopment that follows attachment results from movement of amorphous material outward through the host cell wall; this material finally completely surrounds the incompatible cells (Sequeira *et al.*, 1977). The major difference between virulent and avirulent cells of the tobacco strain of *P. solanacearum* is the presence or absence of an extracellular polysaccharide. This extracellular polysaccharide apparently interferes with binding of the bacterium to specific receptor sites (Sequeira and Graham, 1977).

Attachment of incompatible, avirulent, and/or saprophytic bacteria to cell walls and their subsequent envelopment may be a common phenomenon in higher plants (Goodman *et al.*, 1976). This can occur not only in dicots, but also in monocots. In maize leaves, attachment and envelopment of cells of *Erwinia carotovora* (a strain nonpathogenic on maize) is similar to that observed for *P. solanacearum;* cells of a pathogenic strain of *E. chrysanthemi* are not bound (J. I. Victoria, G. Gaard, G. A. de Zoeten, and A. Kelman, unpublished observations). Singh and Schroth (1977) reported that a saprophytic bacterium *(Pseudomonas putida)* introduced into bean leaves became attached and enveloped but that compatible (*P. phaseolicola*) and incompatible (*P. tomato*) bacteria did not. However, Roebuck *et al.* (1978) showed that the incompatible race of *P. phaseolicola* does become attached in the resistant variety in a manner similiar to that described by Sequeira and co-workers for avirulent or incompatible virulent strains of *P. solanacearum.*

## III. OPPORTUNISTIC BACTERIA AND SYNERGISTIC RELATIONSHIPS

Research on disease induction has been concentrated on the six genera in which the major bacterial pathogens are represented. However, there is a growing awareness that normally saprophytic bacteria may form synergistic partnerships with plant pathogens—bacteria, fungi, or nematodes (see this volume, Chapter 4 ).

If a species of *Corynebacterium* is present on carnation plants inoculated with *Pseudomonas carynophylli,* the rate of wilting and severity of basal rot is significantly increased (Brathwaite and Dickey, 1970). Similarly, *Achromobacter* sp. in the inoculum increases the number of lesions on beans that result after inoculations with *Pseudomonas phaseolicola.* Following inoculation of primary bean leaves with *Achromobacter,* the bacterium becomes systemic and can be isolated from other symptomless plant parts (Maino *et al.,* 1974).

In the case of clostridial infections of potato tubers, surface-water films that limit availability of oxygen appear to be a key factor in enhancing development of these anaerobes (Lund, 1972; Lund and Kelman, 1977). Under conditions of oxygen stress, the clostridia are as aggressive in causing decay of potato tissue as the typical soft-rot *Erwinia* sp. Clostridia, in this circumstance, fit the behavioral criteria that are applied to define "opportunistic" bacteria in man; that is, they are organisms that cause disease in compromised hosts whose defense mechanisms have been weakened by one or more stress factors.

In early studies on decay of hardwood in living trees, the presence of bacteria in isolations often was overlooked or even ignored. Recent studies indicate that a wide variety of bacteria, including anaerobes, accompany decay fungi in their invasion of trees. In several instances, species of *Clostridium* have been associated with injurious effects that can cause significant economic loss in certain hardwoods (Shigo, 1972; Shigo *et al.,* 1971).

### Relation of Epiphytic Bacteria to Frost Injury

The recent intriguing discovery that epiphytic bacteria are involved in frost injury to plants (Lindow *et al.,* 1978) warrants discussion under the topic of "opportunistic" bacteria. Frost injury, in this instance, can increase the prospect of successful infection by strains of *Pseudomonas syringae*. This is based on the discovery that cells of certain strains of *P. syringae* demonstrate ice nucleation activity. With the exception of certain strains of *Erwinia herbicola*, numerous other bacteria that have been tested do not demonstrate this unusual trait. Under growth room and field conditions, leaves of plants lacking ice nuclei-active bacteria do not freeze at $-2°$ to $-5°C$; in contrast, if ice nucleation-active strains of *P. syringae* and *E. herbicola* are applied to leaves and plants are cooled to $-5°C$, severe frost damage is induced. Over ten species of so-called "tender" plant species, including maize, have been tested with similar results. Leaves of most plants representing a wide range of species collected from different geographic areas and at different seasons in the United States were found to harbor substantial numbers of ice nucleation-active bacteria. It appears that there has been a real, but unrecognized, basis for the constant association of *P. syringae* infections and frost damage to plants. The potential that the destructive impact of frost injury can be moderated on a practical basis by control of certain epiphytic bacteria is an attractive new concept.

## IV. MECHANISMS OF DISEASE INDUCTION

### A. Tissue Degradation

Unlike fungi, which can penetrate protective barriers such as cuticle or cork cells by mechanical means, bacteria are wholly dependent on macerating enzymes that degrade the formidable barrier presented by a complex plant cell wall. Plant pathogenic bacteria that move through parenchymatous tissue in an intercellular manner presumably accomplish this mainly by hydrolysis of the pectic compounds of the middle lamella. An exception is *E. amylovora*, the fire blight pathogen, which is mainly an invader of parenchymatous tissue; none of the enzymes that are typically associated with degradation of major cell wall constituents are produced *in vivo* or *in vitro* by this pathogen (Seemüller and Beer, 1976). Exactly how this bacterium is able to spread through tissue in the absence of such enzymes has not been explained.

Relatively few phytopathogenic bacteria other than certain xanthomonads can utilize starch. Even the soft-rotting *Erwinia* sp. with their notorious capacity to decay fleshy vegetables such as potato tubers leave the starch grains untouched in well-rotted tissue. Similarly, few phytopathogenic bacteria can degrade cellulose at a rate comparable to that of many cellulolytic fungi. In instances in which cellulose is degraded, it does not follow necessarily that the products of degradation can be utilized. Thus, although cellobiose is the product of the degradation of carboxymethylcellulose by *Pseudomonas solanacearum*, this bacterium cannot use cellobiose as a sole carbon source in a defined medium (Kelman and Cowling, 1965).

Although bacterial degradation of all cell wall components, including lignin, has been reported to occur in wood following extensive exposure to high moisture, conclusive evidence for the lignolytic capabilities of the specific bacteria involved has not been presented (Liese, 1970). Furthermore, no bacterial plant pathogens are known that degrade lignin, the second most abundant of the plant materials on earth (Kirk and Connors, 1977).

Maceration of parenchymatous plant tissue is one of the few cases in which the role of a single enzyme has been delineated in bacterial pathogenesis (Bateman and Basham, 1976). A purified polygalacturonate lyase (transeliminase) of *Erwinia chrysanthemi* has been shown to macerate potato tissue. Other enzymes attacking cell wall constituents, such as cellulose, protein, xylan or mannan-based polymers, may enhance the maceration process, however.

One of the puzzling manifestations of bacterial soft rot has been the rapidity of the death of infected cells associated with infection (Wood, 1976). If it was assumed that pectic substrates in the middle lamella were the primary sites of the tissue-macerating factor, polygalacturonate lyase, it was difficult to explain how such a specific enzyme could injure or kill the entire cell. Treatment of tissues with purified enzymes with high specificity for a single component in the cell wall killed cells in treated tissues. In tissues treated with purified polygalacturonate lyase, the cells showed an increase in electrolyte loss and loss in ability to retain neutral red. When tissue treated with pectin lyase was plasmolyzed, cell injury was minimized; in contrast, cells in nonplasmolyzed turgid tissues were killed rapidly.

The following explanation for cell death has been supported by several investigators. In a turgid cell in polygalacturonate lyase-treated tissue, turgor pressure forces the plasmalemma tightly against the cell wall. Rifts and tears in the cell membrane cause loss of integrity, and this is followed by death of the cell. This explanation of cell death is persuasive; however, direct evidence of effects on the plasmalemma is lacking.

## B. Alteration in Permeability

The specific factors involved in permeability changes in bacterial diseases are not understood at present. Increased membrane permeability is characteristic of bacterial infections, in general, and the hypersensitive reaction, in particular. The symptom of water soaking can be produced by treating leaves with specific products of bacterial growth, as was shown for the causal agent of bacterial spot of cucumber, *Pseudomonas lachrymans* (Keen *et al.*, 1969). Initially, the causative compound was considered to be a proteolytic enzyme. Subsequently, it was shown that a bacterial lipomucopolysaccharide was involved. The specific nature and mechanism of action of this compound have not been determined, however.

The increased membrane permeability associated with invasion by soft-rot bacteria indicates that proteolytic enzymes and phospholipases might be involved in the death of cells in tissue macerated by soft-rot bacteria. Unfortunately, the critical experiments needed to demonstrate the involvement of such enzymes are also lacking.

## C. Obstruction of Water Movement

The major advances in our understanding of the mechanisms by which vascular parasites cause wilting have come from research on the wilts that are caused by species of fungi in the genera *Fusarium* and *Verticillium* (Dimond, 1970). Attempts have been made to use knowledge of fungal wilts to explain bacterial wilts because both are characterized by a remarkable restriction to xylem elements in their initial growth and development (Nelson and Dickey, 1970). The mechanisms by which bacteria induce wilting differ significantly from those that are typical for fungal wilt pathogens, however. Initially, it was concluded that mechanical plugging in bacterial wilts resulted merely from the physical occlusion by masses of bacterial cells. Subsequent studies postulated the presence of a systemic toxin that could disrupt the osmotic controls of the cells in the leaf with subsequent increase in water loss by transpiration. This toxin concept of wilting is no longer tenable (Talboys, 1978). Current evidence supports the view that wilting results mainly from a decrease in the rate of water flow in the xylem tissue of the stem. This cessation of water flow results from direct and indirect effects of products of the pathogen in combination with substances formed by the host in response to invasion by the pathogen.

High molecular weight polysaccharides induce wilting readily when introduced into the xylem of plants, and wilt-inducing polysaccharides are produced by many bacteria that are not involved in wilt induction or disease development in host plants. It is not clear how relevant it is to demonstrate wilt induction with high molecular weight compounds produced by organisms that cause foliage blights, leaf spots, cankers, or rots. In the case of certain foliage-invading bacteria, movement of the pathogen is clearly in the vascular bundles of the veins (for example, *Xanthomonas campestris*, causal agent of black rot of cabbage). In these instances, extracellular polysaccharides can interfere significantly with water transport (Sutton and Williams, 1970).

In the bacterial wilt caused by *P. solanacearum*, wilting has been attributed to the combined effect of different factors disrupting the movement of water in the stems (Kelman, 1970). These factors include (1) formation of a highly viscous extracellular polysaccharide in the vessels by the virulent strain of the pathogen, (2) degradation of cell wall components by pectic enzymes and to a lesser degree cellulases formed by the pathogen, and (3) stimulation of tylose formation by the increase of indoleacetic acid (formed by pathogen and host) and collapse of vessels as a result of stimulation of cell division in tissue surrounding the vascular elements. Of these factors, the highly viscous extracellular polysaccharide is considered to play a primary role in decreasing the rate of water flow in xylem vessels (Wallis and Truter, 1978).

A small amount of high molecular weight polypeptide produced by *Corynebacterium insidiosum*, causal agent of bacterial wilt of alfalfa, can elicit typical wilting symptoms when the compound is introduced into cuttings (Strobel, 1977). The glycopeptide isolated from cultures of *C. insidiosum* could be obtained also from diseased alfalfa plants in amounts sufficient to cause wilting in test plants. Membrane damage to xylem parenchyma was considered to be the major factor involved. These conclusions were questioned by Van Alfen and Turner (1975), who found that the primary effect of the toxin is to decrease stem conductance in alfalfa. However, the rapid decrease in transpiration and stomatal conductance that was correlated with the wilting of alfalfa cuttings was not considered to result from membrane damage. It was postulated that the glycopeptide produces these symptoms by reducing or blocking water movement through pit membranes.

Studies of the effect on tomato stem tissue of a glycopeptide formed by *C. michiganense* also led to the conclusion that wilt induction in the host was attributable to the toxic effect of this compound on cell membranes of xylem parenchyma cells (Strobel, 1977). This concept was based in part on evidence that the amount of glycopeptide absorbed by cuttings was too small to plug vessels and on cytological evidence for membrane damage. In a histopathological study of tomato stems infected with this same pathogen, Wallis (1977) concluded that the bacterial canker organism depends for its pathogenicity largely on possession of a very active enzyme system. No vessel-plugging materials or structures were present in quantities sufficient to produce wilting symptoms. Most of the components of the primary cell wall apparently were degraded completely. In the advanced stage, complete vascular tissue breakdown had occurred.

It has been established that movement of water from vessel to vessel occurs through pairs of bordered pits in their adjacent walls. The pit membranes are very permeable to water, and large areas of overlapping of cell walls are covered with these bordered pits. In view of the importance of these pits in water flow, it is reasonable to postulate that deposition of bacterial polysaccharides or glycopeptides on pit membranes would be a major factor in restriction of water movement. This could well occur prior to the adverse effect of hydrolytic enzymes on cell walls of the vascular system; small amounts of these compounds are sufficient to impede water flow significantly. The extracellular polysaccharide of *P. solanacearum* may also be involved in wilt induction in this manner. Beckman and co-workers (Vander Molen *et al.*, 1977) have presented clear evidence that vascular gelation is a general response in plants to vascular infections by fungal wilt pathogens. The relevance of these observations to bacterial diseases has never been explored fully, however.

Symptoms produced by xylem-limited rickettsia-like bacteria differ markedly from the symptoms induced by the other wilt pathogens (Hopkins, 1977). The rickettsias are rarely found outside the xylem tracheids or vessels, and lysogenous cavities are rarely, if ever, formed. Apparently, rickettsias may have to be introduced directly into xylem vessels by the leafhopper vector for infection to occur. Thus, spread in the host may be limited to upward and downward movement with very little lateral movement from vessel to vessel. Unlike other bacterial wilt pathogens, the ricksettsia-like bacteria seem to lack the enzymes necessary to dissolve the cell wall components that could enable them to migrate between xylem vessels. Hence, they are relatively restricted to the xylem elements. The localization of infection by vascular occlusion may be more important in these types of diseases than in other bacterial wilt diseases.

## D. Disruption of Metabolism by Toxins

Most symptoms of disease reflect such a complex and dynamic interplay of metabolites of the host plant and bacterial cells that it is extremely difficult to determine how a bacterial toxin elicits a specific symptom. Initially, toxins produced by plant pathogens were considered to be compounds of low molecular weight which injured plant cells at relatively low concentrations. This concept has been modified to include high molecular weight compounds such as glycopeptides (Patil, 1974; Strobel, 1977). The role of such compounds in wilt induction has been discussed in the previous section.

In the wildfire disease of tobacco (caused by *Pseudomonas tabaci*), a necrotic leaf spot surrounded by a yellow halo is the characteristic symptom. A toxin formed in culture by *P. tabaci* can induce chlorosis in tobacco and in many other plant species, including plants not susceptible to the pathogen. Since methionine sulfoximine produced chlorotic halos when injected into tobacco leaves and inhibited growth of the alga *Chlorella* in a manner very similar to that of the wildfire toxin, it was postulated by Braun and co-workers (see Patil, 1974) that the toxin was an antimetabolite of methionine. This research focused attention on the prospects of explaining disease phenomena on the basis of specific toxic metabolites of plant pathogens.

Subsequent research by Sinden and Durbin (see Strobel, 1977) indicated that both wildfire toxin and methionine sulfoximine can act by inhibiting glutamine synthetase, which catalyzes formation of glutamine from glutamic acid and ammonia. Methionine neither inhibits glutamine synthetase nor protects the enzyme against methionine sulfoximine inhibition, however. Concentrations of wildfire toxin that inhibited glutamine synthetase *in vitro* also caused chlorosis of tobacco leaves. Treated leaves were characterized by an increase in ammonia levels to approximately seven times that in untreated leaves. Injection of high levels of glutamine into the leaves prevented both chlorosis and an increase in levels of ammonia. Thus, inhibition of glutamine synthetase was a reasonable explanation for the effect of the toxin on leaf tissue. Furthermore, the chlorosis was more likely to be caused by toxic levels of ammonia or other intermediates of nitrate metabolism than by an induced deficiency of glutamine.

A reexamination of the structure of the wildfire toxin revealed that two dipeptides are involved, each containing a tabtoxinine-$\beta$-lactam residue linked to the amino group of either threonine (tabtoxin) or serine (2-serine tabtoxin) (Stewart, 1971). A study of a *Pseudomonas* sp., which induces a chlorotic leaf spot of timothy (*Phleum pratense*), revealed that this bacterium produces a toxin with the two identical analogs that are in the toxin of *P. tabaci* (Taylor and Durbin, 1973). Recently it has been found that free tabtoxinine-$\beta$-lactam is produced in culture. This compound can induce chlorosis and may be the active agent in the plant.

When highly purified tabtoxins were tested, glutamine synthetase was not inhibited; under the same conditions, methionine sulfoximine was inhibitory, however. On this basis, Durbin (personal communication) has concluded that the tabtoxins per se do not inhibit glutamine synthetase in pathogenesis; thus, the exact mechanism of toxin action once again is open to conjecture.

Marked shifts in levels of amino acids and amides can occur in leaf tissue infected by certain of the chlorosis-inducing bacteria. In bean leaves infected with P. phaseolicola, the level of ornithine, in particular, increased markedly; the level of ornithine was increased 20-fold in chlorotic tissue over that of healthy tissue. These changes were attributed by Patil and Walker (see Patil, 1974) to the effect of a toxin since no bacteria were found in the chlorotic area.

Subsequently, the toxin of P. phaseolicola, phaseotoxin (Patil, 1974), was shown to be a potent inhibitor of ornithine carbamoyltransferase (OCT). Chlorosis induced by phaseotoxin could be reversed by addition of citrulline or arginine. Phaseotoxin was separated by Patil et al. (1976) into four components (A, B, C, and D); the A fraction was identified as N-phosphoglutamic acid. The other components may be closely related N-phosphoamino acids.

Mitchell and colleagues have completed a detailed characterization of the toxin of P. phaseolicola, with certain results at variance with those reported by Patil and co-workers. Mitchell (1976) assigned the trivial name phaseolotoxin to the major component, ( N -phosphosulfamyl)ornithylalanylhomoarginine, to distinguish it from phaseotoxin described by Patil. In addition to the major component, an analog, 2-serine phaseolotoxin, has also been identified.

Among other unresolved questions, it was not known whether the chlorosis-inducing factor was the same compound that was introduced by the pathogen or whether it was a degradation product. Mitchell and Bieleski (1977) have now demonstrated that the purified phaseolotoxin can elicit the characteristic halo symptom of bean halo blight when as little as 0.1–1 nmole of toxin per gram fresh weight of leaves is applied. At higher concentrations (10–100 nmoles per gram fresh weight), the systemic symptoms may also appear after 1 or 2 days.

When analysis was made of bean plants infected with P. phaseolicola, only small amounts of phaseolotoxin were found. However, the levels of phosphosulfamylornithine present were high enough to induce the observed chlorosis and increase in ornithine. Thus, the toxic agent functional in eliciting the primary symptoms of chlorosis in the bean halo blight disease is phosphosulfamylornithine. Studies using radioactive sulfur ($^{35}$S) gave further evidence that movement of the toxin occurred in the phloem. The toxin may be involved in the induction of the chlorotic phase of the disease as follows: the bacteria release phaseolotoxin into the intercellular spaces; peptidases in the cell wall or plasmalemma are released from cells damaged by the infection and convert phaseolotoxin to phosphosulfamylornithine. This primary toxin diffuses away from the site of infection and into living cells, resulting in accumulation of ornithine and degradation of the chloroplasts. This explains the appearance of the characteristic chlorotic halo that normally occurs around the point of infection. If growth of the bacterium in the intercellular spaces is rapid enough, some of this phosphosulfamylornithine is transported via the phloem to other areas in the leaf, resulting in accumulation of ornithine, chlorosis, and stunting. Apparently, both compounds can be transported in the phloem.

This research on phaseolotoxin is of interest because it provides evidence that a toxin produced in culture may not be the toxin that is functional in the diseased plant even though it may, when applied to the plant, elicit specific symptoms. In these studies, phosphosulfamylornithine, which is the major product detected in the tissue, has the same toxicity as phaseolotoxin during bioassay. Phaseolotoxin contains three amino acids plus phosphate and sulfate. It is distinct from phaseotoxin even though it is produced by the same organism and has many of the same characteristics. Phaseolotoxin is produced by isolates of *P. phaseolicola* from various geographic sources. Phaseolotoxin and 2-serine phaseolotoxin accounted for all of the chlorosis-inducing activity in the culture filtrates of the different isolates tested (Mitchell, 1978b). Phaseolotoxin causes a marked accumulation of ornithine in bean leaves, and it is a potent inhibitor of ornithine transcarbamylase. The effective part of phaseolotoxin is the *N*-substituted ornithine group. The substituent group in phaseolotoxin, sulfamyl phosphate, can be regarded as a simple analog of carbamyl phosphate in which the carbonyl group has been replaced by a sulfonyl group.

A second component associated with toxicity of filtrates of *P. phaseolicola* is a serine analog of phaseolotoxin. As was noted previously, tabtoxin, which is the toxin of *Pseudomonas tabaci* and other closely related *Pseudomonas* species, is also characterized by the presence of a serine analog. Mitchell and Parsons (1977) have proposed the name 2-serine phaseolotoxin for the minor toxin of *P. phaseolicola*.

The toxin of *P. glycinea*, contrary to other data, is basically indistinguishable from coronatine, which is the chlorosis-inducing toxin obtained from *P. coronafaciens* var. *atropurpurea*, the causal agent of the chocolate spot disease of Italian ryegrass (Mitchell, 1978a). Thus, the toxin of *P. glycinea* differs chemically from *P. phaseolicola* toxin, and different biological effects may be involved for these two toxins.

The hypersensitive reaction elicited by incompatible isolates of *P. phaseolicola* in leaves of halo blight-resistant bean seedlings was found to accompanied by accumulation of antibacterial phytoalexins, phaseollin, phaseollinisoflavan, coumestrol, and kievitone (Gnanamanickam and Patil, 1977). If resistant bean plants were treated with phaseotoxin prior to inoculation with the incompatible strain, the hypersensitive reaction was suppressed and typical symptoms were produced. A significant reduction in accumulation of phytoalexins was also noted, and toxin treatment of leaf tissue enhanced bacterial multiplication.

In a series of definitive papers, Owens and colleagues (see Patil, 1974) described their studies on the unusual example of a normally well-behaved symbiont, *Rhizobium japonicum,* that can produce a potent phytotoxin. This investigation resulted from an attempt to determine the cause of chlorosis in soybean plants that were ostensibly pathogen-free. Nodules of chlorotic plants contained strains of *R. japonicum* that produced a potent toxin; this toxin was characterized and named rhizobitoxine. Since *Salmonella typhimurium* was found to be sensitive to low concentrations of the toxin, this bacterium served as a microbial guinea pig in pinpointing the mode of action of the toxin. The conversion of cystathionine to homocysteine was blocked by the toxin as a result of inhibition of $\beta$-cystathionase. How this inhibitor affects host tissue to cause chlorosis is still not clearly defined, however.

None of the toxins discussed thus far demonstrates host specificity. Thus, considerable interest was aroused when Goodman and co-workers (1974) described amylovorin, a toxin produced by virulent strains of the fire blight bacterium, *Erwinia amylovora,* following growth on slices of pear and apple. Since rosaceous hosts susceptible to fire blight were shown to be more sensitive to amylovorin than nonrosaceous species, and resistant apple varieties exposed to the toxin were observed to wilt less rapidly than susceptible ones, amylovorin was reported to be the first bacterial toxin clearly demonstrating host specificity. Recently, exceptions to this postulated host specificity have been noted; for example, certain species of *Spirea,* which are not susceptible to fire blight, are as sensitive to amylovorin as a susceptible host (Sjulin and Beer, 1976). Originally, this toxin was found only in host tissue. Subsequently, high yields of toxin were obtained from apple cell suspension cultures inoculated with *E. amylovora* (Hsu and Goodman, 1978) and directly from a buffered glucose liquid medium in the absence of any host tissue (Beer *et al.,* 1977).

Pathogenesis by *E. amylovora* has been attributed wholly or in part to the action of amylovorin (Goodman, 1976). Plasmolysis of xylem parenchyma cells and related cytoplasmic damage observed in infected tissue were reproduced by treating apple stems with relatively low concentrations of the toxin. As noted in the section on wilt induction, many high molecular weight polysaccharides similar to amylovorin can cause wilting when introduced into stem tissue.

Although a number of important questions remain unresolved with respect to the mode of action and host specificity of amylovorin, these recent studies have contributed to a new understanding of pathogenesis in the fire blight disease.

Virulent isolates of *Pseudomonas syringae* produce a potent toxin that is injurious to plant tissue and a number of microorganisms. Loss of pathogenicity has been correlated with loss in ability to produce the toxin (Gonzalez and Vidaver, 1977). In the purified state, the toxin, syringomycin, produces lesions in host tissues similar to those caused by the pathogen itself. Syringomycin at concentrations as low as 2 $\mu$g/ml can cause water soaking, chlorosis, and necrosis when applied to leaves of maize and cowpea (Gross and DeVay, 1977). One of the interesting new developments in studies with this toxin is the evidence that the presence of a plasmid is correlated with virulence and production of syringomycin (Gonzalez and Vidaver, 1977). A strain of *P. syringae* that lacked the plasmid induced a hypersensitive reaction rather than a typical leaf spot in maize plants. Genetic studies on this relationship between syringomycin production, virulence, and presence of a plasmid should be very productive.

## E. Hypertrophic Growth

Hypertrophic growth is characteristic of many bacterial diseases. Reviews of the early and recent literature on abnormal growth have stressed the difficulties of determining cause and effect relationships (Daly and Knoche, 1976; Dekhuijzen and Pegg, 1976; Sequeira, 1973). Bacteria are similar to other plant pathogens with respect to their ability to enhance the formation of auxins, gibberellins, cytokinins, ethylene, and growth inhibitors in diseased tissue. Plant pathogenic bacteria can produce these same compounds (with the possible exception of the gibberellins) in host tissues. Although growth regulators produced by plant pathogens are considered to be intimately involved in the disease process, the number of investigators studying these interactions using bacterial diseases as the subjects for their studies has declined markedly in the past decade. In part, this reflects the awareness of the difficulty and complexity of the problem and also a shifting of emphasis on research to new areas.

Since so many interacting, constantly changing systems are involved, the possibility of errors in analysis and interpretation in this research area is intimidating. Probably no meaningful new research will be completed in studies on hormone imbalance unless a coordinated approach is used in which the major growth-regulating compounds are followed in a single disease. A number of bacterial diseases would provide excellent model systems for such studies, but few research groups have the resources necessary for the definitive study that is needed.

In contrast to most other areas in phytobacteriolology, the research effort on the crown gall disease has expanded at a logarithmic rate during the past decade. No other bacterial disease is being studied as intensively as is crown gall at present. Recent, very complete reviews make it unnecessary to reassess the voluminous literature on this unusual disease of plants (Braun, 1978; Merlo, 1978; Lippincott and Lippincott, 1976; Kado, 1976). A major advance in our understanding of the molecular mechanism of plant tumorigenesis came with the finding that a specific large plasmid was present in all virulent strains of *Agrobacterium tumefaciens* (Van Larebeke *et al.*, 1975, cited in Merlo, 1978). Once this was confirmed, researchers quickly demonstrated that the plasmid was essential for virulence. When this plasmid was tranferred to plasmid-less virulent strains of *Agrobacterium*, these strains became virulent. In fact, when the tumor-inducing plasmid was transferred to the closely related bacterium, *Rhizobium trifolii*, cells of this recipient strain became tumorigenic although they retained the capacity to fix nitrogen (Hooykas *et al.*, 1977).

The evidence is now very persuasive that a long-standing presumption is, in fact, correct and that genetic information from the bacterium is passed directly into host cells during the induction process. A contiguous segment of DNA that represents 5–10% of the tumor-inducing plasmid is transferred to the plant cell (Chilton *et al.*, 1977, cited in Merlo, 1978). Furthermore, crown gall cells synthesize RNA transcript that specifically hybridizes to the transferred segment of the tumor-inducing plasmid (Drummond *et al.*, 1977, cited in Merlo, 1978). These data indicate that the final result of infection is the expression in the crown gall cell of this tiny bit of foreign DNA introduced from the tumor-inducing plasmid of *A. tumefaciens*.

It now appears that the mechanism of induction of tumors in plants resembles that which characterizes some virally induced animal tumors; i.e., a eukaryotic cell is transformed by adding new genetic information, which is maintained in subsequent cell divisions. Certain genes in the *A. tumefaciens* cell that are not transcribed in the bacterium are transmitted to the transformed plant cell. One of these genes controls the synthesis of octopine or nopaline (unusual amino acids not formed in normal tissue). The bacterium does not synthesize these amino acids; rather it utilizes them as a source of carbon and/or nitrogen for its growth (Kerr *et al.*, 1977, cited in Merlo, 1978). The hypothesis has been proposed that the transferred region of the tumor-inducing plasmid may not be expressed in the bacterium because it is of plant origin, but once transferred back to the plant, it can be expressed. Although the goal of crown gall research is closer than ever to resolution now, the basic question as to exactly how autonomous growth is triggered is still unresolved.

## V. AREAS FOR FUTURE RESEARCH

Most investigators on bacterial plant pathogens have dealt with a relatively small number of well-recognized species. There is a need to expand current research boundaries to encompass diseases reflecting synergistic relationships among two or more organisms and borderline diseases caused by opportunistic pathogens, including anaerobes such as species of clostridia.

Significant progress has been made in advancing our understanding of bacterial pathogenesis in the six decades since Erwin F. Smith (1920) summarized his assessment of the mechanisms by which bacteria cause plant disease. A large number of different compounds produced by phytopathogenic bacteria have been characterized chemically and shown to elicit profound and complex physiological effects in host plants. These compounds range from simple ones, such as indoleacetic acid to relatively complex high molecular weight materials, such as glycopeptides and extracellular polysaccharides. Correlations between virulence and formation of certain toxins, extracellular polysaccharides, and hydrolytic enzymes have been demonstrated in only a few cases. Although intensive efforts have been made to analyze the specific manner in which some of these compounds are involved in pathogenesis, conclusive evidence as to the precise mode of action of specific compounds in living tissue is still lacking for most bacterial diseases. Further, none of the extracellular products produced by bacteria has been shown to be a primary determinant of the specificity of a pathogen–plant interaction. It is attractive to postulate that chemical interactions between antigens common to certain plants and their bacterial pathogens determine whether the disease process will proceed. The role of lectins and related compounds in the recognition of incompatible bacteria is receiving considerable attention as a relatively new area of research for phytobacteriologists. Intensive analysis of the initial events in the association of bacteria with plant cells must be forthcoming in order for the subsequent steps in pathogenesis to be adequately understood.

A massive reservoir of genetic knowledge and related biochemical data has been amassed by bacterial geneticists and molecular biologists. The tremendous potential for precise genetic analysis that has been recently demonstrated in recombinant DNA research also remains to be exploited with most bacterial plant pathogens. With this information, phytopathogenic bacteria may now offer more possibilities for productive research on host–parasite relationships and pathogenesis than any other type of plant pathogen.

## References

Andrews, J. H. (1976). The pathology of marine algae. *Biol. Rev. Cambridge Philos. Soc.* 51, 211–253.

Bateman, D. F., and Basham, H. G. (1976). Degradation of plant cell walls and membranes by microbial enzymes. In "Physiological Plant Pathology" (R. Heitefuss and P. H. Williams, eds.), pp. 316–355. Springer-Verlag, Berlin and New York.

Beer, S. V., Baker, C. J., Woods, A. C., and Sjulin, T. M. (1977). Production of amylovorin *in vitro* and partial characterization. *Proc. Am. Phytopathol. Soc.* 4, 182–183.

Bohlool, B. B., and Schmidt, E. L. (1974). Lectins: A possible basis for specificity in the Rhizobium-legume root nodule symbiosis. *Science* 185, 269–271.

Brathwaite, C. W. D., and Dickey, R. S. (1970). Synergism between *Pseudomonas caryophylli* and a species of *Corynebacterium*. *Phytopathology* 60, 1046–1051.

Braun, A. C. (1978). Plant tumors. *Biochim. Biophys. Acta* 516, 167–191.

Costerton, J. W., Geesey, G. G., and Cheng, K. J. (1978). How bacteria stick. *Sci. Am.* 238, 86–95.

Daly, J. M., and Knoche, H. W. (1976). Hormonal involvement in metabolism of host–parasite interactions. In "Biochemical Aspects of Plant Parasite Relationships" (J. Friend and D. R. Trelfall, eds.), pp. 117–133. Academic Press, New York.

Dazzo, F. B. (1979). Adsorption of micro-organisms to roots and other plant surfaces. In "Adsorption of Micro-Organisms to Surfaces" (G. Britton and K. C. Marshall, eds.). Wiley, New York (in press).

Dekhuijzen, H. M., and Pegg, G. F. (1976). Endogenous cytokinins in healthy and diseased plants. In "Physiological Plant Pathology" (R. Heitefuss and P. H. Williams, eds.), pp. 526–559. Springer-Verlag, Berlin and New York.

Dimond, A. E. (1970). Biophysics and biochemistry of the vascular wilt syndrome. *Annu. Rev. Phytopathol.* 8, 301–322.

Gnanamanickam, S. S., and Patil, S. S. (1977). Phaseolotoxin suppresses bacterially induced hypersensitive reaction and phytoalexin synthesis in bean cultivators. *Physiol. Plant Pathol.* 10, 169–180.

Gonzalez, C. F., and Vidaver, A. K. (1977). Syringomycin and holcus spot of maize: Plasmid associated properties. *Proc. Am. Phytopathol. Soc.* 4, 107.

Goodman, R. N. (1976). Physiological and cytological aspects of the bacterial infection process. In "Physiological Plant Pathology" (R. Heitefuss and P. H. Williams, eds.), pp. 172–196. Springer-Verlag, Berlin and New York.

Goodman, R. N., Kiraly, Z., and Zaitlin, M. (1967). "The Biochemistry and Physiology of Infectious Plant Disease." Van Nostrand, Princeton, New Jersey.

Goodman, R. N., Huang, J. S., and Huang, P. Y. (1974). Host-specific phytotoxic polysaccharide from apple tissue infected by *Erwinia amylovora*. *Science* **183**, 1081–1082.

Goodman, R. N., Huang, P. Y., and White, J. A. (1976). Ultrastructure evidence for immobilization of an incompatible bacterium, *Pseudomonas pisi*, in tobacco leaf tissue. *Phytopathology* **66**, 754–757.

Gross, D. C., and DeVay, J. E. (1977). Role of syringomycin in holcus spot of maize and systemic necrosis of cowpea caused by *Pseudomonas syringae*. *Physiol. Plant Pathol.* **11**, 1–11.

Hayward, A. C. (1974). Latent infections by bacteria. *Annu. Rev. Phytopathol.* **12**, 87–97.

Heitefuss, R., and Williams, P. H., eds. (1976). "Physiological Plant Pathology." Springer-Verlag, Berlin and New York.

Hooykas, P. J. J., Klapwijk, P. M., Nuti, M. P., Schilperoort, R. A., and Rörsch, A. (1977). Transfer of the *Agrobacterium tumefaciens* Ti plasmid to avirulent agrobacteria and to *Rhizobium ex planta*. *J. Gen. Microbiol.* **98**, 477–484.

Hopkins, D. L. (1977). Diseases caused by leafhopper-borne rickettsia-like bacteria. *Annu. Rev. Phytopathol.* **15**, 277–294.

Hsu, S. T., and Goodman, R. N. (1978). Production of a host-specific, wilt-inducing toxin in apple cell suspension cultures inoculated with *Erwinia amylovora*. *Phytopathology* **68**, 351–354.

Kado, C. I. (1976). The tumor-inducing substance of *Agrobacterium tumefaciens*. *Annu. Rev. Phytopathol.* **14**, 265–308.

Keen, N. T., Williams, P. H., and Upper, C. D. (1969). A re-evaluation of the protease from *Pseudomonas lachrymans*: Isolation of a fraction producing non-catalytic solubilization of proteins in trichloroacetic acid. *Phytopathology* **59**, 703–704.

Kelman, A. (1970). How bacteria damage crops. *In* "How Crops Grow. A Century Later" (P. R. Day, ed.), *Conn. Agric. Exp. Stn., New Haven, Bull.* No 665, pp. 128–154.

Kelman, A., and Cowling, E. B. (1965). Cellulase of *Pseudomonas solanacearum* in relation to pathogenesis. *Phytopathology* **55**, 148–155.

Kirk, T. K., and Connors, W. J. (1977). Advances in understanding the microbiological degradation of lignin. *In* "Recent Advances in Phytochemistry. The Structure, Biosynthesis, and Degradation of Wood" (F. A. Loewus and V. C. Runeckles, eds.), Vol. 2, pp. 369–394. Plenum, New York.

Klement, Z., and Goodman, R. N. (1967). The hypersensitive reaction to infection by bacterial plant pathogens. *Annu. Rev. Phytopathol.* **5**, 17–44.

Kuć, J. (1976). Phytoalexins. *In* "Physiological Plant Pathology" (R. Heitefuss and P. H. Williams, eds.), pp. 632–652. Springer-Verlag, Berlin and New York.

Liese, W. (1970). Ultrastructural aspects of woody tissue disintegration. *Annu. Rev. Phytopathol.* **8**, 231–258.

Lindow, S. E., Arny, D. C., Upper, C. D., and Barchet, W. R. (1978). The role of bacterial ice nuclei in frost injury to sensitive plants. *In* "Plant Cold Hardiness and Freezing Stress" (P. H. Li and A. Sakai, eds.), pp. 249–263. Academic Press, New York.

Lippincott, J. A., and Lippincott, B. B. (1976). Morphogenic determinants as exemplified by the crown-gall disease. *In* "Physiological Plant Pathology" (R. Heitefuss and P. H. Williams, eds.), pp. 356–388. Springer-Verlag, Berlin and New York.

Lund, B. M. (1972). Isolation of pectolytic clostridia from potatoes. *J. Appl. Bacteriol.* **35**, 609–614.

Lund, B. M., and Kelman, A. (1977). Determination of the potential for development of bacterial soft rot of potatoes. *Am. Potato J.* **54**, 211–225.

Maino, A. L., Schroth, M. N., and Vitanza, V. B. (1974). Synergy between *Achromobacter* sp. and *Pseudomonas phaseolicola* resulting in increased disease. *Phytopathology* **64**, 277–283.

Merlo, D. J. (1978). Crown gall—a unique disease. *In* "Plant Disease: An Advanced Treatise" (J. G. Horsfall and E. B. Cowling, eds.), Vol. III, pp. 201–213. Academic Press, New York.

Mitchell, R. E. (1976). Isolation and structure of a chlorosis-inducing toxin of *Pseudomonas phaseolicola. Phytochemistry* **15**, 1941–1947.

Mitchell, R. E. (1978a). *Pseudomonas glycinea* toxin: Isolation and characterization as coronatine. *Phytopathol. News* **12**, 201. (Abstr.)

Mitchell, R. E. (1978b). Halo blight of beans: Toxin production by several *Pseudomonas phaseolicola* isolates. *Physiol. Plant Pathol.* **13**, 37–50.

Mitchell, R. E., and Bieleski, R. L. (1977). Involvement of phaseolotoxin in halo blight of beans: Transport and conversion to functional toxin. *Plant Physiol.* **60**, 723–729.

Mitchell, R. E., and Parsons, E. A. (1977). A naturally occurring analogue of phaseolotoxin (bean halo blight toxin). *Phytochemistry* **16**, 280–281.

Nelson, P. E., and Dickey, R. S. (1970). Histopathology of plants infected with vascular bacterial pathogens. *Annu. Rev. Phytopathol.* **8**, 259–280.

Patil, S. S. (1974). Bacterial phytotoxins. *Annu. Rev. Phytopathol.* **12**, 259–279.

Patil, S. S., Youngblood, P., Christiansen, P., and Moore, R. E. (1976). Phaseotoxin A: An antimetabolite from *Pseudomonas phaseolicola. Biochem. Biophys. Res. Commun.* **69**, 1019–1027.

Roebuck, P., Sexton, R., and Mansfield, J. W. (1978). Ultrastructural observations on the development of the hypersensitive reaction in leaves of *Phaseolus vulgaris* cv. Red Mexican inoculated with *Pseudomonas phaseolicola* (race 1). *Physiol. Plant Pathol.* **12**, 151–157.

Seemüller, E. A., and Beer, S. V. (1976). Absence of cell wall polysaccharide degradation by *Erwinia amylovora. Phytopathology* **66**, 433–436.

Sequeira, L. (1973). Hormone metabolism in diseased plants. *Annu. Rev. Plant Physiol.* **24**, 353–414.

Sequeira, L. (1978). Lectins and their role in host-pathogen specificity. *Annu. Rev. Phytopathol.* **16**, 453–481.

Sequeira, L., and Graham, T. L. (1977). Agglutination of avirulent strains of *Pseudomonas solanacearum* by potato lectin. *Physiol. Plant Pathol.* **11**, 43–54.

Sequeira, L., Gaard, G., and de Zoeten, G. A. (1977). Interaction of bacteria and host cell walls: Its relation to mechanisms of induced resistance. *Physiol. Plant Pathol.* **10**, 43–50.

Shigo, A. L. (1972). Successions of micro-organisms and patterns of discoloration and decay after wounding in red oak and white oak. *Phytopathology* **62**, 256–259.

Shigo, A. L., Stankewich, J., and Consenza, B. J. (1971). *Clostridium* sp. associated with discolored tissues in living oaks. *Phytopathology* **61**, 122–123.

Singh, V. O., and Schroth, M. N. (1977). Bacteria–plant cell surface interactions: Active immobilization of saprophytic bacteria in plant leaves. *Science* **197**, 759–761.

Sjulin, T. M., and Beer, S. V. (1976). Evidence that amylovorin and *Erwinia amylovora* infection induce shoot wilt by different mechanisms. *Proc. Am. Phytopathol. Soc.* **3**, 260.

Smith, E. F. (1920). "Bacterial Diseases of Plants." Saunders, Philadelphia, Pennsylvania.

Starr, M. P., and Chatterjee, A. K. (1972). The genus *Erwinia:* Enterobacteria pathogenic to plants and animals. *Annu. Rev. Microbiol.* **26**, 389–426.

Stewart, J. M. (1971). Isolation and proof of structure of wildfire toxin. *Nature (London)* **229**, 174–178.

Strobel, G. A. (1977). Bacterial phytotoxins. *Annu. Rev. Microbiol.* **31**, 205–224.

Sutton, J. C., and Williams, P. H. (1970). Relation of xylem plugging to black rot lesion development in cabbage. *Can. J. Bot.* **48**, 391–401.

Talboys, P. W. (1978). Dysfunction of the water system. In "Plant Disease: An Advanced Treatise" (J. G. Horsfall and E. B. Cowling, eds.), Vol. III, pp. 141–162. Academic Press, New York.

Taylor, P. A., and Durbin, R. D. (1973). The production and properties of chlorosis-inducing toxins from a pseudomonad attacking timothy. *Physiol. Plant Pathol.* **3**, 9–17.

Van Alfen, N. K., and Turner, N. C. (1975). Changes in alfalfa stem conductance induced by the *Corynebacterium insidiosum* toxin. *Plant Physiol.* **55**, 559–561.

Vander Molen, G. E., Beckman, C. H., and Rodehorst, E. (1977). Vascular gelation: A general response phenomenon following infection. *Physiol. Plant Pathol.* **11**, 95–100.

Wallis, F. M. (1977). Ultrastructural histopathology of tomato plants infected with *Corynebacterium michiganense*. *Physiol. Plant Pathol.* **11**, 333–342.

Wallis, F. M., and Truter, S. J. (1978). Histopathology of tomato plants infected with *Pseudomonas solanacearum* with emphasis on ultrastructure. *Physiol. Plant Pathol.* **13**, 307–317.

Wood, R. K. S. (1967). "Physiological Plant Pathology." Blackwell, Oxford.

Wood, R. K. S. (1976). Killing of protoplasts. In "Biochemical Aspects of Plant Parasite Relationships" (J. Friend and D. R. Threlfall, eds.), pp. 105–116. Academic Press, New York.

*Chapter 11*

# How Mycoplasmas and Rickettsias Induce Plant Disease

KARL MARAMOROSCH

## I. INTRODUCTION

Understanding this chapter requires a definition of the terms "mycoplasma" and "rickettsia," as well as of the abbreviations MLO and RLO. Since 1967 a large group of plant diseases, earlier believed to be caused by viruses, have become known as plant mycoplasma diseases. Doi *et al.* (1967) discovered mycoplasma-like bodies in the phloem of diseased plants. Ishiie *et al.* (1967) described the temporary remission of the diseases in plants that were treated with tetracycline antibiotics. Nasu *et al.* (1967) found mycoplasma-like bodies not only in diseased plants but also in leafhopper vectors. The causative agents of witches'-broom and yellows-type diseases were therefore recognized as nonviral. As work progressed in various laboratories around the world, it became obvious that the plant disease agents do not belong to the family Mycoplasmataceae or Acholeplasmataceae of the order Mycoplasmatales (Razin, 1973, 1978). The terms "mycoplasma-like organisms" (MLO) (Maramorosch, 1976) and "rickettsia-like organisms" (RLO) (Maramorosch *et al.*, 1975) therefore came into use. Neither of these terms is satisfactory, and they are reminiscent of the long-abandoned term "pleuropneumonia-like" organisms (PPLO), which are now properly classified as mycoplasmas (Maramorosch, 1974).

PLANT DISEASE, VOL. IV

Mycoplasmas are wall-less prokaryotes that differ from bacteria in having an exterior membrane instead of a cell wall. The mycoplasmas are the smallest prokaryotes. They contain RNA ribosomes and a double-stranded DNA genome. In 1967 the mycoplasmas were placed in the newly created class Mollicutes (Tully and Razin, 1977). The trivial name mycoplasmas covers all species of the Mollicutes, including the plant pathogenic spiroplasmas (Saglio et al., 1973). The RLOs are very different from mycoplasmas because they are walled microorganisms, Two RLO groups have been recognized: one confined to the phloem, the other to the xylem, of diseased plants. The members of the latter group are now considered as bacteria (Auger et al., 1974). The agents of ratoon-stunting disease of sugarcane (Plavsic-Banjac and Maramorosch, 1972) and of Pierce's disease of grapes belong to this group. The phloem-restricted RLOs probably represent small bacteria or, perhaps, chlamydia or bedsonia but not rickettsia, as originally thought. The agent of clover club leaf disease belongs to this group.

Although MLOs and RLOs differ taxonomically, they have certain features in common. Most seem to require insect vectors for their transmission to plants (Maramorosch, 1963, 1969). Vectors act also as alternate hosts and reservoirs of MLOs and RLOs. Therefore, the induction of disease has to be considered not only as the interaction between pathogen and plant, but as a complex interaction between the pathogen, the plant, and the insect vector. The pathogens usually require an intrinsic incubation period in the vectors, starting at the time of oral acquisition from diseased plants and ending at the time the insects are able to transmit the disease agents to susceptible host plants during feeding. The MLOs or RLOs enter the insect's gut, penetrate the peritrophic membrane, are carried by the hemolymph to various sites and organs, reach the salivary gland, and exit through the vehicle of saliva. A protracted incubation period is usually interpreted as indicative of multiplication of the disease agent in an insect vector. Actual multiplication has been demonstrated in several instances. The same species of vector can sometimes act as a transmitter of MLOs and of viruses, and both can multiply in the same vector simultaneously. Before it was recognized that yellows disease agents differ from viruses, it was often assumed that the vector specificity is so high that a disease can be diagnosed by ascertaining the proper vector. Two examples showed that this is not a valid criterion: the transmission by Dalbulus maidis and D. elimatus (corn leafhoppers) of both the rayado fino virus and the corn stunt spiroplasma (Wolanski and Maramorosch, 1979), and the transmission of the oat blue dwarf virus and the aster yellows spiroplasma by Macrosteles fascifrons (six-spotted leafhoppers) (Banttari, 1979).

## II. ETIOLOGY

Plant diseases are malfunctioning processes. During the first decade since the discovery of MLOs and RLOs, the study of plant yellows diseases has concentrated on the characterization of the pathogens and neglected the study of disease processes. A notable exception is the work at the John Innes Institute in Norwich, England, by Daniels and Meddins (1974), where the mechanism of action has been explored. The tendency to equate the pathogen with the disease is often evident in the terminology and the misuse of such terms as "insect transmission of diseases." Diseases are not transmissible, even though we are faced with such well-entrenched terms as the Communicable Disease Center, the name for one of the leading medical research institutions in the world, which is engaged in the study of diseases caused by infectious "communicable" agents.

The causal agents lumped together as MLOs and RLOs are necessary to cause "mycoplasma diseases," but the agents by themselves do not cause these diseases, nor are they necessarily deleterious. The term "thoroughgoing etiology," as used by Bateman (1978), clearly defines the complex nature of diseases. His term applies to MLOs and RLOs as well. For example, the diseases of plants caused by MLOs and RLOs require specific insect vectors for their introduction into plants, but that is only one of the numerous factors needed in addition to the specific agents. Resistance to disease is not well understood, and we do not know why certain species, or even certain individual members of a single plant species, are susceptible or resistant to MLOs and RLOs.

The successful survival of MLOs and RLOs in the ecosystem, as well as their successful transmission to plants, requires not only susceptible plants but also susceptible insect vectors. The so-called efficient vector is, in fact, a highly susceptible arthropod animal, which readily acquires the microbial agents and usually becomes infected itself systemically. In instances in which multiplication of the disease agents in vectors has been studied in more detail, an increase in titer has been ascertained, but the actual sites of proliferation have been studied in less detail than those of certain vector-borne viruses (Maramorosch, 1969).

Not all MLOs have deleterious effects in plants, and only a few cause mild insect diseases. Color changes in plants might sometimes be rather attractive, and the same pertains to certain distortions, as in the case of the ornamental cactus *Opuntia tuna monstrosa*, which is widely grown and propagated but is merely a spiroplasma-infected *Opuntia tuna* (Maramorosch *et al.*, 1972). The spiroplasma of aster yellows induces sterility, stunting, and chlorosis and often kills affected plants.

However, in one instance, diseased and discolored gladiolus flowers, green instead of red, were actually preferred by clients of funeral parlors who felt that the pale color was proper to adorn the coffins! An instance of a "beneficial effect" of the aster yellows spiroplasma was accidently discovered when *D. maidis* leafhoppers, confined to aster yellows-diseased plants, acquired the ability to feed on various species of plants that ordinarily were unsuitable for these insects (Maramorosch, 1958). Apparently, the acquisition of corn stunt spiroplasma became beneficial to corn leafhoppers by changing their feeding or digestive ability (Orenski, 1964).

## III. ECOLOGICAL FACTORS

The influence of environmental factors on disease outbreaks has to be considered not only with respect to plant hosts and the diverse disease agents discussed in this chapter, but also with respect to the insect vectors responsible for the transmission of MLOs and RLOs. Some MLOs are extremely sensitive to elevated, naturally occurring summer temperatures. The MLO of peach yellows is inactivated at 29°C, and the aster yellows spiroplasma at 32°C. The spread of aster yellows in nature is therefore curtailed in many states in North America during the summer months. Peach yellows is geographically limited to areas where summers are comparatively cool. A sharp temperature optimum for *Spiroplasma citri* probably accounts for its erratic spread in certain years. As for the vectors, severe winters can decrease the surviving overwintering stages and thus account for a reduced vector population. In recent years, unusually warm summers in central and northern Europe have brought about an increase in stolbur potato and tomato outbreaks, most likely because the warmer period favors an increase in the vector population. Temperature has a direct effect on the length of the intrinsic incubation of MLOs and RLOs in their vectors. Very long incubation periods at low temperature can result in a decrease in the spread of disease agents. There is inadequate information on the effect of environmental factors on plant susceptibility to infection by MLOs and RLOs.

## IV. INSECT VECTORS

The interactions between disease agents and vectors of MLOs and certain RLOs are reminiscent of interactions between viruses transmitted biologically by vectors (Maramorosch, 1969). Most likely the agents that multiply in their vectors proliferate first in the insect gut and later in other organs, including the salivary glands. Actual multiplication was demonstrated in several instances when the agents were still diagnosed as plant viruses. The RLO of clover club leaf disease was the first microbial agent in this category that was shown to multiply in its leafhopper vector (Black, 1950). Ironically, its microbial nature was recognized three decades later by the original discoverer, and the classic transovarial passage of a "plant virus in an insect vector" became the passage of a penicillin-susceptible microbe! The same irony pertained to the demonstration that the aster yellows "virus" multiplied in leafhopper vectors during ten mechanical serial passages (Maramorosch, 1952). Yet another example of correct observation but incorrect interpretation was the first classic case describing the deleterious effect of a plant "virus" on a leafhopper vector (Jensen, 1959). The Western X disease "virus" has now been recognized as an MLO.

Since insect vectors play an essential role in the transmission of the MLO and RLO disease agents, their susceptibility to infection also has a crucial effect on the way in which plant disease is caused, irrespective of whether the insects are or are not affected by the MLOs and RLOs. A comparatively small number of genera, belonging to a few families, are known to act as vectors of more than 90% of all phloem-restricted agents (Nielson, 1979). Susceptible (active) and nonsusceptible (inactive) races of leafhoppers have been known since the classic studies of Storey (1933). Among the progeny of a single female vector some individuals might be highly susceptible while others are highly refractive. The vector's susceptibility to infection explains, in part, the efficient and nonefficient vectorial capacity.

Plants differ not only in their susceptibility to MLOs and RLOs, but also in their susceptibility to insect vectors (Maramorosch, 1979). In the past, the attention of breeders has been directed primarily toward resistance to the infectious agents. More recently, efforts have been made to incorporate plant resistance to vectors in breeding programs. If vectors can be prevented from feeding on MLO- or RLO-susceptible plants, the plants cannot become infected. In practice, this principle has been used by growers of China aster seed, who routinely cover the plants with cheesecloth or special netting.

We can describe the interactions among plants, insect vectors, MLOs, and RLOs as follows. In nature the disease syndrome requires the interaction of host plants, specific insect vectors, and disease agents. There is no evidence to suggest that the disease agents must undergo different developmental cycles in their plant and insect hosts, as is the case with certain protozoan parasites of animals. For example, the vector can be bypassed by inducing the disease by grafting. In turn, the plant can be bypassed by inoculating the vector mechanically, using extracts from the other vectors. It seems strange that mechanical inoculation of plants has not been achieved, and we can only speculate at the reasons why all attempts to inoculate plants with extracts from either vectors or diseased plants have failed. Perhaps the introduction of the infectious agents into the phloem or the xylem cells is done more precisely and delicately by the proboscis of a vector than by a rigid micropipette or needle. We know that spiroplasmas and MLOs are easily destroyed by changes in pH, osmotic pressure, and oxidation. Cells that are injured by needle puncture might no longer provide the necessary environment for the survival and proliferation of the pathogens.

Vector specificity is usually high, and sometimes related species are unable to act as transmitters of MLOs, even though they can acquire the disease agents by feeding. The phloem-restricted MLOs are usually transmitted by one or a few related leafhopper species. Xylem-confined RLOs are less specific in their insect vector interactions. Recently the artificial production of nonspecific vectors succeeded when cultured *Spiroplasma citri* was injected into *Euscelis plebejus* (Markham and Townsend, 1974) and *Euscelidius variegatus* (Markham and Townsend, 1979). This success is important for vector experimentation and laboratory tests. It raises hopes that eventually mechanical inoculation of plants with cultured spiroplasmas might prove feasible.

MLOs and RLOs can be maintained in insect vectors artificially, in serial passages, without the alternating plant hosts. However, more important for the actual mechanism by which disease is caused, the syndrome can be reproduced in the absence of the disease agent by the use of MLO-produced toxins. At present, the knowledge of the mechanism of action is limited to a single MLO, *S. citri,* and to the pioneering work of Daniels and Meddins (1974).

## V. MODE OF ACTION

How can the virulence of MLOs and specifically the virulence of spiroplasmas be explained? The mechanism by which plant cells control the exogenous cholesterol incorporated into their plasma membranes is of great biological significance, and the events of this process have been unveiled, in part, by studying cholesterol incorporation by mycoplasmas (Razin, 1978). Cholesterol uptake does not depend on metabolic energy, and it is due to a physical adsorption process. The invading spiroplasmas use cholesterol for their own membranes, thus upsetting the balance of the plant's metabolism. The spiroplasmas also interfere with the disposition of lipids. However, these interferences with cellular metabolism are probably a minor part of the disease syndrome. The very severe disturbance in the photosynthetic functions, growth, and hormonal changes that are manifested in the induction of axillary buds, shoot proliferations, witches'-broom production, virescence, big buds and leafy flowers, seed sterility, and breaking of dormancy are indicative of the action of potent antimetabolites or toxins produced by the pathogens.

Daniels and Meddins (1974) reported that one, or perhaps two, toxins are produced by the cultured *S. citri*, the agent of citrus stubborn disease. The toxin, produced *in vitro*, inhibits growth of unicellular algae as well as the growth of clover seedlings and causes cell injury in leaves. Partial reproduction of the yellows syndrome was actually achieved using the toxic substance. The toxin is a low molecular weight compound with hydroxyl groups and one or more nitrogen atoms, perhaps an amino sugar. The chemical identification is expected to be completed soon, and this will help to understand the mode of action. Surprisingly, the same group at the John Innes Institute tried but failed to isolate a toxin from the cultured corn stunt spiroplasma. Perhaps not only dialyzable, but also protein-type, membrane-bound, nondialyzable toxins can be produced by MLOs.

No information is available on receptors for spiroplasmas and MLOs or the chemical nature of binding sites on plant cell membranes. The intimate association between adhering MLOs and RLOs and plant host cells probably provides the proper conditions for toxic metabolites to cause cell damage. It has been proposed that close adherence leads to fusion of the eukaryote cell membrane with that of the host (Gabridge *et al.*, 1977).

## VI. VIRUSES ASSOCIATED WITH MLOs AND RLOs

In 1973 Cole *et al.* reported the first occurrence of viruses in cultured *S. citri*. Two phage-type tailed viruses and a rod-shaped virus have been reported to be associated with the citrus stubborn organism. More recently an indicator strain for one of the viruses has been discovered (Cole, 1977; Maniloff *et al.*, 1977), and this has permitted quantitative studies and definitive infectivity tests. Calavan and Gumpf (1974) described viruses in their *S. citri cultures*. A rod-shaped particle resembling the filamentous SV-C1 virus of Cole *et al.* (1973) was found in the cultured aster yellows spiroplasmas in our laboratory (Kondo *et al.*, 1977; Maramorosch and Kondo, 1978). Cole (personal communication) detected virus particles in the cultured corn stunt spiroplasma.

The first notion that MLOs might be associated with viruses came from electron micrographs of thin sections of infected plants and insect vectors (J. Giannotti, 1970, personal communication; Allen, 1972). In recent years, such associations have been reported in several MLO and RLO diseases, including aster yellows, stolbur, corn stunt, and others. We described polyhedral viruslike particles in a phloem-restricted RLO agent (Maramorosch *et al.*, 1975). We also noted that plants sometimes recovered spontaneously, and in such plants no RLOs could be detected by electron microscopy techniques. We speculated that the RLO virus destroyed the infectious agent and caused a spontaneous cure.

When *Macrosteles fascifrons* leafhoppers were confined to older leaves of systemically infected plants, they were unable to acquire the aster yellows spiroplasma (Maramorosch, 1964). Electron microscopy revealed that the older leaves of systemically infected plants contain deteriorating MLOs, often associated with rod-shaped, viruslike particles (Hirumi and Maramorosch, 1972). Since the spiroplasma culture of aster yellows contains similar particles, and these particles increase in successive subcultures, their role is of particular interest. We have found that some of the subcultures are "lost," without visible contamination, and electron microscopy has revealed that spiroplasma membranes are being disrupted and destroyed in virus-containing cultures. Could spiroplasmas' own viruses provide a spontaneous elimination of the disease agents from living plants? Perhaps future artificial manipulation of the interrelationships between spiroplasmas and their viruses could provide means for biological control. Virus particles have been reported to be associated with corn stunt spiroplasma in the salivary glands of *Dalbulus maidis* (Granados, 1969). These observations might help solve the puzzling mechanism by which spiroplasmas cause plant disease. It is possible that pathogenicity and antigenic variations among spiroplasmas depend on their infection by spiroplasma viruses. If viruses are required for toxin production, the removal of such viruses would provide a possible means of control. The role of viruses of spiroplasmas with respect to virulence and toxin production is conjectural at present, and the speculation might seem farfetched, but the subject deserves special attention. A team of Chinese investigators in Shanghai (Anonymous, 1974) expressed the view that mulberry dwarf disease might be caused by the interaction of MLO and a rod-shaped virus always found in association with MLO.

## VII. UNSOLVED PROBLEMS

Only a few aspects of the new groups of pathogens, MLOs, and RLOs have been studied during the first decade since their recognition, and the unsolved problems will provide fertile ground for plant pathologists and entomologists for many years to come. New and precise identification methods are needed not only for practical purposes but also for the theoretical distinction between these agents. The specificity of insect vectors and the symptomatology in affected plants have been used in the past to identify yellows-type diseases, but mycoplasma taxonomists frown on the use of hosts to determine the species or genus. The identification of recently isolated spiroplasmas seems to require criteria other than those used customarily for members of the Mycoplasmatales (Tully et al., 1973). The plant and vector host range might be among the future criteria. The ELISA test, which proved of great value in the identification of viruses (Clark and Adams, 1977; Voller et al., 1976), has also been of considerable value in distinguishing spiroplasmas (Bar-Joseph, personal communication). The study of relatedness among spiroplasmas might be among the high-priority subjects in the near future. Polyacrylamide gel electrophoresis has already provided a means of distinguishing certain species (Padhi et al., 1977a,b; Padhi and McIntosh, 1976).

The motility of MLOs and RLOs might be closely linked with their pathogenicity, and the study of movement will therefore receive close attention. The motility of spiroplasma filaments can be nontranslational, as seen in viscous media and on contact with solid surfaces (Davis, 1979). Nontranslational rotation, swimming, flexing, and crawling are characteristic of spiroplasmas. In addition, translational locomotion occurs in viscous media and on surface contact (Davis, 1979). Surprisingly, spiroplasmas without spiral forms also occur (Townsend et al., 1977).

The artificial cultivation of MLOs and RLOs is receiving much attention. Different spiroplasmas require different media for isolation and propagation, and we can expect that several additional spiroplasmas will be grown in artificial cell media soon. It is difficult to predict whether, and when, the phloem-restricted RLOs will be grown outside of cells. Perhaps the accurate establishment of growth requirements in vitro and the analysis of phloem and xylem components will provide the clues to culturing MLOs and RLOs. Attempts to use phloem exudate, coconut water, and plant extracts have been made during the past decade, but the results have been disappointing.

The proper study of viruslike particles associated with MLOs and RLOs requires indicator strains. New methods of manipulating the reactions between viruses and MLOs, RLOs, and spiroplasma are needed. We can hope that the cure of plants can be achieved through the destruction of the microbial agents within living plants.

Is there a relation between the shape of mycoplasmas and their mode of infection? We shall restrict our speculation to spiroplasmas grown in cell-free media. Their ability to move might perhaps facilitate their penetration into insect salivary glands and the movement from cell to cell of the phloem through sieve pores. Their restriction to the phloem is probably linked to the presence of plant sterols in the phloem sap, which are necessary for the proliferation of spiroplasma. All of this is, at present, speculation at best, but speculation must precede experimentation.

## VIII. ORIGIN AND EVOLUTION OF MLOs AND RLOs

The MLOs and RLOs are a heterogenous group of several types of microorganisms belonging to diverse classes, Mollicutes, and Schizomycetes. They share a common feature—their dependence on insect vectors for transmission to plants. Long before the microbial nature of these pathogens became known, I postulated that they might have originated in insects, as insect pathogens, and that their evolutionary association with plants is a more recent development (Maramorosch, 1955). The speculation was based on the high degree of vector specificity and narrow insect host range combined with the absence of disease of the arthropod hosts. This is in striking contrast to the severe, often fatal, plant diseases. The theory postulated by Theobald Smith that long association between parasite and host results in an almost symbiotic relationship, while more recent associations are less balanced, influenced this speculation. There are exceptions to the general rule, but most spiroplasmas, MLOs, and RLOs fit the hypothesis.

## IX. CONCLUSIONS AND SUMMARY

The training of plant pathologists is diverse, as pointed out by Bateman (Volume III), and they view their work from the perspective of a mycologist, a bacteriologist, a virologist, or an agronomist–horticulturist, without focusing on their activities from the perspective of pathology as a biological discipline. This is especially true in the new area of plant mycoplasma diseases. Those who entered this study in or after 1967 were previously trained as plant virologists or entomologists but not as bacteriologists, and practically none of them had any idea of the techniques specially developed for mycoplasma research. Consequently, progress was still not achieved in the cultivation of MLOs because the expertise of bacteriologists and mycoplasmatologists was lacking. Also in the area of antibiotic tests, there was a lack of proper background, and some of the earlier reports describing the antibiotic spectrum should never have been accepted for publication. The authors assumed that anything sprayed or poured on the soil around diseased plants actually entered plants; without checking for antibiotic activity in plant tissues they concluded that some antibiotics were, or were not, acting on MLOs. Some authors concluded that plant MLOs differ in their antibiotic sensitivity from animal mycoplasmas because plants did not show signs of recovery when sprayed with the compounds. Furthermore, the unfortunate lack of collaboration between entomologists and plant pathologists, who during and after training at universities seldom have close contact with each other, hampered rapid progress in the study of MLO interactions with insect vectors.

It was fortunate that veterinarians and human pathologists working with mycoplasmas became interested in the plant-infecting MLOs and that their advice was sought and followed. In fact, the discovery of mycoplasmas in plants was due to a fortunate accident that occurred in an electron microscopy laboratory in Tokyo. There, a veterinarian, Dr. Kaoru Koshimizu, happened to see the electron micrographs of Dr. Doi, and the resemblance of the MLOs in plant phloem to mycoplasma in cells of diseased birds. Dr. Koshimizu pointed out this resemblance to plant pathologists (Maramorosch, 1976) and also suggested the use of tetracyclines to cure plants. Although no cure was achieved, temporary remission and disappearance of MLOs from treated plants provided the additional evidence needed to make the story irrefutable. While no acknowledgement of this important, although accidental, collaboration between a veterinarian and a plant pathologist was made in Japan, the fact is of more than historical importance. It points out vividly the value of collaboration between diverse disciplines.

The discovery of spiroplasmas also occurred by accident when Dr. Robert F. Davis, meeting Dr. Shmuel Razin, one of the world's foremost mycoplasma experts, asked how one could study MLOs without having access to an electron microscope. Razin suggested phase-contrast and dark-field microscopy. Shortly afterward, Davis discovered spiral forms in crude extracts from corn stunt-diseased plants. When he examined the cultured MLOs of citrus stubborn disease, he detected the same type of forms and named them spiroplasmas.

Needless to say, numerous pitfalls have been avoided by consulting experts who have worked with mycoplasmas for many years. In our laboratory, the so-called pseudocolonies of MLOs were observed on agar; were it not for Dr. Ruth G. Wittler, who called our attention to the earlier reports of such artifacts, their actual nature would have been misinterpreted. The successful cultivation of the plant spiroplasmas from the cactus monstrosity disease and from lettuce infected with aster yellows was achieved by Dr. Fusao Kondo in our laboratory, thanks to his training in animal mycoplasmatology and veterinary science, not in plant pathology or entomology.

Insect vectors, such as leafhoppers, planthoppers, and psyllids are the main but not the only transmitters of MLOs and RLOs. Grafting, both natural and artificial, also accounts for transmission. Several plant parasitic species of Cuscuta (dodder) and the plant parasitic *Santalum* (sandal) can act as transmitters, and they themselves can contract disease. It is of historical interest that a report in *Phytopathology* (Narasimhan, 1933) describing the sandal spike disease and the observation that not only sandal, but also the tree onto which sandal is grafted contracted the yellows-type syndrome provided the impetus to several plant pathologists to experiment with dodder grafts. Consequently, and independently, C. W. Bennett, Folke Johnson, and Lindsay M. Black (personal communication) used dodder to transmit viruses and MLOs. The technique became a standard tool for experimental transmission whenever the proper vectors were unknown.

Present knowledge of how MLOs and RLOs cause plant diseases is inadequate. The complex nature of interactions among plants, vectors, and disease agents, the factors that influence susceptibility to infection, the production of toxins, the interrelationships with bacteriophage-type and other viruses, and environmental influences all play a role in disease induction. Proper understanding of the basic mechanisms involved will eventually lead to new means of controlling these diseases.

### References

Allen, T. C. (1972). Bacilliform particles within asters infected with a western strain of aster yellows. *Virology* **47**, 491–493.

Anonymous (1974). Studies on the pathogen of mulberry dwarf disease. I. Virus-like particles and mycoplasmalike bodies associated with mulberry yellow dwarf disease. *Sci. Sin.* **17**, 421–427.

Auger, J. G., Shalla, T. A., and Kado, C. I. (1974). Pierce's disease of grapevine: Evidence for a bacterial etiology. *Science* **184**, 1375–1377.

Banttari, E. (1979). Interactions of mycoplasmalike organisms and viruses in dually infected leafhoppers, planthoppers and plants. *In* "Leafhoppers as Vectors of Plant Disease Agents" (K. F. Harris and K. Maramorosch, eds.), pp. 327–348. Academic Press, New York.

Bateman, D. (1978). The dynamic nature of disease. *In* "Plant Disease" (J. G. Horsfall and E. B. Cowling, eds.), Vol. 3, pp. 53–83. Academic Press, New York.

Black, L. M. (1950). A plant virus that multiplies in its insect vector. *Nature* (*London*) 166, 852–853.

Calavan, E. C., and Gumpf, D. J. (1974). Studies on citrus stubborn disease. *Colloq. INSERM* 33, 181–185.

Clark, M. F., and Adams, A. N. (1977). Characteristics of the microplate method of enzyme-linked immunosorbent assay for the detection of plant viruses. *J. Gen. Virol.* 34, 475–483.

Cole, R. M. (1977). Spiroplasmaviruses. *In* "The Atlas of Insect and Plant Viruses" (K. Maramorosch, ed.), pp. 451–457. Academic Press, New York.

Cole, R. M., Tully, J. G., Popkin, T. J., and Bove, J. M. (1973). Morphology, ultrastructure, and bacteriophage infection of the helical mycoplasma-like organism (*Spiroplasma citri* gen. nov., sp. nov.) cultured from "stubborn disease" of citrus. *J. Bacteriol.* 115, 367–386.

Daniels, M. J., and Meddins, B. M. (1974). The pathogenicity of *Spiroplasma citri*. *Colloq. INSERM* 33, 195–200.

Davis, R. E. (1979). Spiroplasmas. Newly recognized arthropodborne pathogens. *In* "Leafhoppers as Vectors of Plant Disease Agents" (K. F. Harris and K. Maramorosch, eds.), pp. 451–488. Academic Press, New York.

Doi, Y., Terenaka, M., Yora, K., and Asuyama, H. (1967). Mycoplasma or PLT group-like microorganisms found in the phloem elements of plants infected with mulberry dwarf, potato witches' broom, aster yellows, or paulownia witches' broom. *Nippon Shokubutsu Byori Gakkaiho* 33, 259–266.

Gabridge, M. G., Stahl, Y. D. B., Polisky, R. B., and Engelhardt, J. A. (1977). Differences in the attachment of *Mycoplasma pneumoniae* cells and membranes to tracheal epithelium. *Infect. Immun.* 16, 766–772.

Granados, R. R. (1969). Electron microscopy of plants and insect vectors infected with corn stunt disease agent. *Contrib. Boyce Thompson Inst.* 24, 173–187.

Hirumi, H., and Maramorosch, K. (1972). Natural degeneration of mycoplasmalike bodies in an aster yellows infected host plant. *Phytopathol. Z.* 75, 9–26.

Ishiie, T., Doi, Y., Yora, K., and Asuyama, H. (1967). Suppressive effects of antibiotics of tetracycline group on symptom development of mulberry dwarf disease. *Nippon Shokubutsu Byori Gakkaiho* 33, 267–275.

Jensen, D. D. (1959). A plant virus lethal to its insect vector. *Virology* 8, 164–175.

Kondo, F., Maramorosch, K., McIntosh, A. H., and Varney, E. H. (1977). Aster yellows spiroplasma: Isolation and cultivation *in vitro*. *Proc. Am. Phytopathol. Soc.* 4, 190–191.

Maniloff, J., Das, J., and Christensen, J. R. (1977). Viruses of mycoplasmas and spiroplasmas. *Adv. Virus Res.* 21, 343–380.

Maramorosch, K. (1952). Direct evidence for the multiplication of aster-yellows virus in its insect vector. *Phytopathology* 42, 59–64.

Maramorosch, K. (1955). Multiplication of plant viruses in insect vectors. *Adv. Virus Res.* 3, 221–249.

Maramorosch, K. (1958). Beneficial effect of virus-diseased plants on non-vector insects. *Tijdschr. Plantenziekten* 64, 383–459.

Maramorosch, K. (1963). Arthropod transmission of plant viruses. *Annu. Rev. Entomol.* 8, 369–414.

Maramorosch, K. (1964). Interrelationships between plant pathogenic viruses and insects. *Ann. N.Y. Acad. Sci.* 118, 363–370.

Maramorosch, K., ed. (1969). "Viruses, Vectors, and Vegetation." Wiley (Interscience), New York.

Maramorosch, K. (1974). Mycoplasma-like bodies in plants. *In* "Bergey's Manual of Determinative Bacteriology", 8th Ed., (R. E. Buchanan and N. E. Gibbons, eds.) The Williams and Wilkins Co., Baltimore, pp. 954–955.

Maramorosch, K. (1976). Plant mycoplasma diseases. *In* "Plant Physiology" (P. H. Williams and R. Heitefuss, eds.), Vol. 4, pp. 150–171. Springer-Verlag, Berlin and New York.

Maramorosch, K. (1979). Insects and plant pathogens. *In* "Breeding Plants for Insect Control" (F. Maxwell and P. Jennings, eds.). Wiley (Interscience), New York (in press).

Maramorosch, K., and Kondo, F. (1978). Aster yellows spiroplasma: Infectivity and association with a rod-shaped virus. *Proc. IOM Conf., 2nd Freiburg 1976.*

Maramorosch, K., Klein, M., and Wolanski, B. S. (1972). Beitrag zur Aetiologie der Hexenbesenkrankheit der Kaktee *Opuntia tuna* (+ *tuna monstrosa*). *Experientia* **28**, 362.

Maramorosch, K., Hirumi, H., Kimura, M., and Bird, J. (1975). Mollicutes and rickettsia-like plant disease agents (Zoophytomicrobes) in insects. *Ann. N.Y. Acad. Sci.* **266**, 276–292.

Markham, P. G., and Townsend, R. (1974). Transmission of *Spiroplasma citri* to plants. *Colloq. INSRM* **33**, 201–206.

Markham, P. G., and Townsend, R. (1979). Experimental vectors of spiroplasmas, *In* "Leafhopper Vectors of Plant Disease Ageints" (K. F. Harris and K. Maramorosch, eds.), pp. 485–514. Academic Press, New York.

Narasimhan, M. J. (1933). Cytological investigations on the spike disease of sandal (Santalum album). *Phytopathology* **23**, 191–202.

Nasu, S., Sugiura, M., Wakimoto, T., and Iida, T. T. (1967). On the etiologic agent of rice yellow dwarf disease. *Nippon Shokubutsu Byori Gakkaiho* **33**, 343–344.

Nielson, M. W. (1979). Taxonomic relationships of leafhopper vectors of plant pathogens. *In* "Leafhopper Vectors of Plant Disease Agents" (K. F. Harris and K. Maramorsch, eds.), pp. 3–28. Academic Press, New York.

Orenski, S. W. (1964). Effects of a plant virus on survival, food acceptability, and digestive enzymes of corn leafhoppers. *Ann. N.Y. Acad. Sci.* **118**, 374–383.

Padhi, S. B., and McIntosh, A. H. (1976). Comparison of two plant spiroplasmas by electrophoresis of their proteins. *In Vitro* **12**, 334–335.

Padhi, S. B., McIntosh, A. H., and Maramorosch. K. (1977a). Characterization and identification of spiroplasmas by polyacrylamide gel electrophoresis. *Phytopathol. Z.* **90**, 268–272.

Padhi, S. B., McIntosh, A. H., and Maramorosch, K. (1977b). Polyacrylamide gel electrophoresis distinguishes among plant pathogenic spiroplasmas. *Proc. Am. Phytopathol. Soc.* **4**, 194.

Plavsic-Banjac, B., and Maramorosch, K. (1972). Electron microscopy of xylem of ratoon-stunted sugarcane. *Phytopathology* **62**, 498–499.

Razin, S. (1973). Physiology of mycoplasmas. *Adv. Microb. Physiol.* **10**, 1–80.

Razin, S. (1978). The mycoplasmas. *Microbiol. Rev.* **42**, 414–470.

Saglio, P., L'Hospital, M., Lafleche, D., Dupont, G. Bove, J. M., Tully, J. G., and Freundt, E. A. (1973). *Spiroplasma citri* gen. and sp. n.: A mycoplasma-like organism associated with "Stubborn" disease of citrus. *Int. J. Syst. Bacteriol.* **23**, 191–204.

Storey, H. H. (1933). Investigations of the mechanism of the transmission of plant viruses by insect vectors. I. *Proc. R. Soc. London, Ser. B* **113**, 463–485.

Townsend, R., Markham, P. G., Plaskitt, K. A., and Daniels, M. J. (1977). Isolation and characterization of a non-helical strain of *Spiroplasma citri*. *J. Gen. Microbiol.* **100**, 15–21.

Tully, J. G., and Razin, S. (1977). The mollicutes. *In* "CRC Handbook of Microbiology" (A. I. Laskin and H. A. Lechevalier, eds.), 2nd Ed., pp. 417–459. CRC Press, Cleveland, Ohio.

Tully, J. G., Whitcomb, R. F., Bove, J. M., and Saglio, P. (1973). Plant mycoplasmas: Serological relation between agents associated with citrus stubborn and corn stunt diseases. *Science* **182**, 827–829.

Voller, A., Bidwell, D. E., and Bartlett, A. (1976). Enzyme immunoassays in diagnostic medicine: Theory and practice. *Bull. W.H.O.* **53**, 55–65.

Wolanski, B. S., and Maramorosch, K. (1979). Rayado fino virus and corn stunt spiroplasma: Phloem restriction and transmission by *Dalbulus elimatus* and *D. maidis*. *Fitopatologia Brasileira* **4**, 47–54.

*Chapter 12*

# How Nematodes Induce Disease

## VICTOR H. DROPKIN

## I. INTRODUCTION

The study of plant parasitic nematodes (phytonematodes) and of diseases associated with them has developed more recently than other branches of phytopathology. Most phytonematodes affect roots or underground stems hidden from view in the soil; thus, we become aware of them only after large populations are present and crop damage is obvious. Nematodes reproduce more slowly than bacteria and fungi; the number of offspring per female ranges from tens to hundreds or in a few cases to a thousand or so. Consequently, nematode-induced diseases tend to remain localized and often escape detection for years.

Since plants and nematodes coexist in an environment that includes other pathogens, and since disease may result from the interactions of more than one kind of pathogen, it is not a simple matter to determine the precise role of nematodes in the causation of plant disease. The object of this chapter is to examine the mechanisms by which nematodes damage plants.

Nematodes are small, elongate, tubular animals that live on the moist

219

PLANT DISEASE, VOL. IV
Copyright © 1979 by Academic Press, Inc.
All rights of reproduction in any form reserved
ISBN–0–12–356404–2

surfaces in the soil, in fresh and marine waters, and in the interior of animals and plants. They feed in various ways. Many soil-inhabiting species ingest whole bacteria, which they macerate in a muscular structure equipped with three sets of comblike teeth. Carnivorous nematodes are equipped with teeth, by which they capture small animals, such as other nematodes and mites. These carnivores feed on the body contents removed by suction from their prey. Phytonematodes have a piercing and sucking tube, the stylet, through which they withdraw the fluid contents of cells by applying suction generated by a muscular pump.

Damage to plants by nematodes cannot be understood solely on the basis of cell response to puncture and withdrawal of contents. A growing plant is a dynamic system through which energy flows in continually shifting patterns as the plant progresses through its life cycle. Since the plant's structure and physiology result from many processes that interact in complex ways, a disturbance in one part of the system can have consequences throughout the integrated whole. For example, *Trichodorus* sp. are associated with a disease called "stubby root." The parasites feed principally at root tips; root growth halts when the nematode injuries are numerous, and side roots emerge behind the injured tips. These side roots in turn are injured. The inadequate root system that develops fails to supply adequate water, growth regulators, reduced nitrogen, and other products to the top. Consequently, the plants grow poorly, flowering is delayed or absent, and yield is reduced. Another example of the effect of nematodes on tissues at a distance from the site of feeding is the influence of infection with *Meloidogyne* sp. on predisposition to fungus infection. The presence of these nematodes in one part of a root may predispose the entire root to attack by fungi to which it is genetically resistant. For further information on this fascinating interaction, see this volume, Chapter 6.

## II. CYTO- AND HISTOPATHOLOGY

When one watches phytonematodes in root-observation chambers, one soon realizes that these small animals seem to know what they are doing. Each species has its own distinctive mode of life. For example, ring nematodes (*Criconemoides* sp. and related genera) insert their long stylets into cortex tissues while the body remains external to the root. Lesion nematodes (*Pratylenchus* sp.) jab and wriggle their way into plants; they penetrate completely and move about within the root cortex. They do not usually invade vascular tissues. They also migrate freely between root and soil. On the other hand, root-knot nematodes (*Meloidogyne* sp.) move directly to the region of differentiating vascular tissues,

## TABLE I
### Parasitic Habits of Phytonematode Genera

*Ectoparasites*—Genera containing species of nematodes that remain outside the plant and penetrate with only a small portion of their bodies
  A. Feeding on surface tissues: *Trichodorus, Paratylenchus, Tylenchorhynchus, Tylenchus, Tetylenchus*
  B. Feeding on deeper tissues: *Belonolaimus, Hoplolaimus, Criconemoides* (and related genera), *Longidorus* (G),[a] *Xiphinema* (G), *Hemicycliophora* (G)
*Endoparasites*—Genera containing species of nematodes that enter tissues completely or with a large portion of their bodies
  C. Migratory in herbaceous plants
     Roots: *Pratylenchus*
     Stems and leaves: *Ditylenchus*
     Buds and leaves: *Aphelenchoides, Anguina* (G)
  D. Migratory in perennial plants: *Rhadinaphelenchus, Bursaphelenchus, Radopholus*
  E. Sessile
     Partly within roots: *Tylenchulus, Rotylenchulus*
     Entirely within roots: *Nacobbus* (G), *Heterodera* (and related genera), *Meloidogyne* (and related genera) (G)

[a] The letter G indicates that galls are often induced.

where they remain in one location. The head moves within a limited area, but the rest of the body is fixed in position. Some nematodes (*Aphelenchoides* sp.) ascend stems, enter stomata, and feed on leaf mesophyll (see Table I).

Each species of nematode has its own characteristic behavior which influences the cytopathology and histopathology of infected tissues. Nematodes have developed a great variety of parasitic associations with host plants; these are summarized in Table I. Genera are listed in accordance with the behavior of most species within the genus. However, the classification cannot be absolute, since the particular parasitic relationship between a nematode and its host depends on both partners. For example, although the genus *Hoplolaimus* is listed under ectoparasites, some species of *Hoplolaimus* penetrate tissues of certain hosts completely. *Aphelenchoides* is listed under endoparasites, but there are species that feed on the outer tissues of developing leaves enclosed within buds.

## A. Surface Feeders

*Trichodorus, Paratylenchus, Tylenchorhynchus, Tylenchus,* and *Tetylenchus* are nematodes that feed on cells of the root epidermis or on root hairs. In general, they do not penetrate the cortex. *Trichodorus* sp.

are the most damaging of the group, both because they severely stunt root growth and because they transmit viruses. Pitcher (1967) found that these nematodes accumulated in large numbers around the growing root tips of *Malus sylvestris* (apple) roots growing against the glass walls of an underground root-observation chamber. The area behind the tips became necrotic, and the roots ceased to grow. Others have observed *Trichodorus* sp. feeding very briefly (5–10 seconds) or more persistently (3–5 minutes) on an individual cell. Mitosis is inhibited in root tip cells so that "stubby root" symptoms develop. An individual of *Paratylenchus* sp., however, has been observed to remain in position for up to 1 week, feeding on one epidermal cell with intermittent periods of stylet thrusts and rests, and with no apparent damage to the host.

## B. Ectoparasitic Feeders on Deeper Tissues

In contrast to surface feeders, these nematodes feed on cells in the outer cortex or more deeply, close to the stele. The stylets and sometimes a small portion of the anterior body penetrate into root tissues. The distinction between partial and complete entry is not always clear, and a single species may behave in both ways. This group includes some genera of destructive nematodes (*Belonolaimus, Hoplolaimus, Rotylenchus, Criconemoides* and other ring nematodes, *Longidorus, Xiphinema,* and *Hemicycliophora*). *Belonolaimus* sp. nematodes kill cells as they feed; narrow-necked lesions develop, and affected tissues rot as bacteria and fungi enter. More restricted lesions are characteristic of infections with *Hoplolaimus* sp. and *Rotylenchus* sp. *Criconemoides* and related genera have very long stylets with which they reach cells deep within the root cortex. These nematodes destroy individual cells by removing their contents.

Three genera of this group (*Hemicycliophora, Longidorus,* and *Xiphinema*) induce galls with enlarged and proliferated cells. We can speculate that nematode secretions stimulate cells to enlarge or divide at a distance from those penetrated. *Hemicycliophora arenaria* also stimulates the proliferation of side roots, a sign of altered growth regulator activity. In addition, cells at the feeding sites of *Hemicycliophora* and *Xiphinema* enlarge, lose their vacuoles, and may become multinucleate. These changes suggest that the cells are adapting to the partial removal of their contents by increasing their metabolism.

## C. Migratory Endoparasites

Some nematodes habitually penetrate plant tissues and move about actively. This group includes *Pratylenchus* sp., *Radopholus* sp., *Aphelenchoides* sp., and *Ditylenchus* sp. Roots of many plants infected with

*Pratylenchus* sp. have elongate, narrow, brown streaks at the surface. These enlarge to form more extensive necrotic areas, which often coalesce into extensive discolored lesions harboring bacteria, fungi, and free-living nematodes. The nematode feeds on a cell, killing it, and moves on to an adjacent cell, usually in a longitudinal direction. Adult females deposit eggs within the cortex, and emerged larvae feed in the same region of the root so that extensive regions of the root may be destroyed. In addition, both young and adult nematodes migrate freely between root and soil. The color of the lesion depends on the species of both *Pratylenchus* and the host.

*Radopholus similis* seems to be much more active than *Pratylenchus*. In banana it makes extensive tunnels through the cortex. The large lesions become dark, and most of the nematodes move into fresh tissues. Another biotype of this species is a pathogen of citrus roots. It invades the stele as well as the cortex. In addition to destroying cells, this parasite stimulates cell enlargement and multiplication of vascular parenchyma tissues. Thus, it is not only a hungry parasite but a host-irritating pathogen.

The genus *Aphelenchoides* includes species of both ecto- and endoparasitic habit. Large populations of these long, thin nematodes severely damage leaves of strawberries, chrysanthemums, begonias, and many other plants. *Aphelenchoides* nematodes infecting chrysanthemums survive the winter in temperate climates within fallen leaves around the base of the host. In the spring they emerge during warm, wet periods and migrate to the growing points of adjacent young host plants. As the plants enlarge, the parasites find their way to new leaves. On the leaf surface, this eelworm wriggles its way through stomata into mesophyll tissues, where it moves about actively in the interveinal areas. Cells are destroyed, and as the population increases, characteristic sectors of brown host tissue develop.

*Ditylenchus dipsaci* severely damages plants. It invades nonvascular tissues of stems, leaves, and cotyledons. Cells in the vicinity of the nematode separate, and large empty spaces develop. Consequently tissues are destroyed. In addition, the entire growth of an infected host is disturbed. Stems have abnormally short internodes and may be galled; leaves and flowers are malformed.

## D. Migratory Endoparasites in Perennial Plants

In the red-ring disease of palms, a palm weevil (*Rhynchophora palmivora*) transmits nematodes that it has acquired from diseased trees. When the insect burrows into the tops of healthy palms, nematodes (*Rhadinaphelenchus cocophilus*) are carried along on the body of the insect. They enter the plant through the insect burrows and invade the stem parenchyma over the length of the trunk and on into the roots. Cells in a ring around the trunk are killed, and the tree eventually dies. In the pine wilt disease in Japan, *Bursaphelenchus lignicolus* populations develop within resin canals of pines, where they damage epithelial cells. The tree dies within weeks of the time the nematodes enter. A wood-boring beetle (*Monochamus alternatus*) is the agent of the dispersal (Mamiya, 1976). In a diseased tree, when the adult stage of the insect emerges from the pupa, nematodes enter its tracheal passages. They are carried along in the beetle, multiply rapidly in the resin canals, and kill the infected tree in a few weeks.

## E. Sessile Parasites

Several genera show striking adaptations to sedentary life within host plants. All induce the formation of specialized cells, which survive repeated removal of portions of their contents. All adult females are sessile and swollen, while the males are elongate and active forms.

*Tylenchulus semipenetrans* penetrates and grows partly into citrus roots, leaving the posterior portion of the adult female external to the root, where it swells into a saclike structure. The nematode feeds on a set of normal-sized cells, called "nurse cells," in the root cortex. These have enlarged nuclei, no central vacuoles, and dense cytoplasm. They remain as individual cells not joined together in a syncytium (Van Gundy and Kirkpatrick, 1964).

*Rotylenchulus reniformis* also penetrates with its anterior portion as far as the endodermis; the posterior remains external to the root and swells into a kidney shape. A few cells immediately adjacent to the nematode's head enlarge and fuse by loss of part of their cell walls. A curved sheet of altered pericycle develops in the outer stele of the host with cells of normal or somewhat enlarged size but with abnormal morphology. The host cytoplasm contains a superabundance of plastids, and the central vacuoles disappear (Jones and Dropkin, 1975; Rebois *et al.*, 1975).

*Nacobbus* sp. invade the stele and stimulate both cell division and the fusion of cell protoplasts after partial dissolution of cell walls. A spindle-shaped mass of cells with partially dissolved walls develops as a syncytium, and the nematode feeds at one end of this group of interconnected cells. In contrast to normal host cells, the syncytium cells are very rich in starch. The feeding site is at the center of a conspicuous gall resulting from the hyperplasia of cortex and vascular tissues (Jones and Payne, 1977).

Infective second-stage larvae of *Heterodera* sp. move into host cortex, where they induce cells to enlarge and the walls between them to partially break down. The resulting syncytium invades the stele by incorporation of cells distally from the nematode. Syncytium walls next to xylem vessels develop fingerlike ingrowths, and the central vacuoles disappear. Nuclei of incorporated cells enlarge but do not divide.

*Meloidogyne* sp. are sessile endoparasites that induce the formation of large multinucleate syncytia ("giant cells"). Differentiating xylem elements around the head of the nematode enlarge, undergo repeated nuclear divisions, lose their central vacuoles, and develop wall ingrowths. The extensive dissolution of cell walls characteristic of *Heterodera*-induced syncytia does not occur in giant cells induced by *Meloidogyne*.

The syncytia that result from infections with *Heterodera* and *Meloidogyne* have characteristics in common with normal "transfer" cells observed at certain locations in healthy plants. The cytoplasm appears dense, the central vacuole is lost, the nucleus is larger than normal, and numerous wall ingrowths are formed, usually adjacent to a vessel. These "transfer cells" develop in situations in which there is very active transport within plants. They occur in places with heavy but temporary demands for nutrients in such regions as leaf traces in stems, or in flower buds. In normal host tissues, these transfer cells are believed to facilitate the movement of solutes, such as amino acids, from a vessel to growing tissues. They are also found in nitrogen-fixing nodules of legume roots and in leaves of most plants, where they may facilitate movement of nitrogenous compounds and photosynthate into vessels. The sessile nematodes convert plant nutrients into their substance together with a large population of eggs and accessory gelatinous proteins. Thus, syncytia may be considered as large, specialized transfer cells that develop in relation to the nematode's utilization of plant resources.

The discussion of cellular pathology thus far emphasizes the role of the pathogen. However, in every host–pathogen association, we observe the dual action of both host and pathogen. An example of the influence of the host on pathology of a nematode infection is found in the association of *Pratylenchus scribneri* with snap or lima beans. Lesion nematodes often induce narrow, elongate, dark lesions distributed generally throughout the roots. These are regions of necrotic cortex tissues. According to histochemical evidence, the brown color results from oxidation of phenols with the formation of a brown pigment. In some hosts, quantities of phenols are low, and the browning reaction is slight. However, the presence of brown pigment in diseased tissue is not a reliable guide to the pathogenicity of the nematode. *Pratylenchus scribneri* is a serious pest of snap bean in California, but it fails to induce elongate brown lesions. In lima beans, however, such lesions are conspicuous, but *P. scribneri* does not have much influence on top growth. In greenhouse experiments, nematode populations increased about sevenfold in 50 days in inoculated snap beans. The weight of infected plants was less than half that of noninoculated plants. The cortex was almost destroyed at 32 days after inoculation, but no conspicuous brown lesions appeared. In lima beans, however, nematode populations barely increased, groups of cells around each invader died and became brown, but few eggs were deposited, and the plant tolerated the local tissue destruction. Large numbers of nematodes are apparently necessary for sufficient root destruction to damage the host plant significantly (Thomason *et al.*, 1976).

## III. CHEMICAL AND PHYSIOLOGICAL PATHOLOGY

Metabolic alterations occur in diseased plants, whether the causal agents are nematodes or other pathogens. Accelerated respiratory enzyme activity (ascorbic acid oxidase, cytochrome *c* oxidase, catalase, and peroxidase) has been observed in rice with the white-tip disease induced by *Aphelenchoides besseyi*. Nonspecific increases in enzyme activity occur in cells on which nematodes feed (Endo and Veech, 1970). Various metabolites in addition to enzymes also change in plants infected with nematodes. Tomato root galls induced by *Meloidogyne* have decreased concentrations of polysaccharides and increased concentrations of nucleic and amino acids. They also synthesize more protein than uninfected tissues (Owens and Specht, 1966). The amino acid composition of infected tissues may depart from the normal condition. Sugar beet roots of susceptible plants with *Heterdera schachtii* have more total free amino acids and in particular increased amounts of aspartic and glutamic acids as well as glutamine in comparison with resistant roots (Doney *et al.*, 1970). Galls induced by *Longidorus* sp. in bur marigold also have more free amino acids, and especially a greater concentration of proline than do healthy roots (Epstein and Cohn, 1971). Perhaps high proline concentrations are a sign of stress rather than a specific indicator of nematode damage.

Nematodes do not invariably induce symptoms of disease in plants. A plant growing under favorable conditions may not show signs of stress from low or moderate populations of nematodes. Indeed, some plants are stimulated to grow more vigorously by small numbers of nematodes probably because these parasites stimulate additional root growth. However, when stress from nematodes is added to that from other agents, such as drought, wet soils, insect damage, or inadequate nutrition, nematode damage is often detected.

## IV. MECHANISMS OF PATHOGENICITY

### A. Interactions with Other Pathogens and Root Symbionts

Many species of phytonematodes predispose plants to infection by fungi. Most of the reports in the literature simply record that plants suffer greater damage from concurrent infections with *Meloidogyne* sp. and fusarium wilt fungi than from either pathogen alone; the increased severity of disease is synergistic; and total damage is more than the sum of the effects of each pathogen.

A few explorations of the mechanisms of interactions have been made. Cotton, tobacco, and tomatoes lose genetic resistance to fusarium wilt fungi in the presence of root knot. When invasion by nematodes precedes invasion by fungi, the synergistic effect is more pronounced than with simultaneous invasion by both pathogens. Hyphae grow vigorously in giant cells and galled tissues. The giant cells are killed by the fungus, which also invades vascular tissues and grows vigorously into the stem. Data from split-root experiments suggest that metabolic disturbances induced by nematodes extend to the entire root system. Sidhu and Webster (1977) showed that amino acid auxotrophs of *Fusarium oxysporum* f. sp. *lycopersici* were nonpathogenic to susceptible tomato cultivars without simultaneous infection by nematodes. In the presence of *M. incognita* nematodes, however, they regained pathogenicity. These authors believe that changes in amino acid content of plants infected with root-knot nematodes are partly responsible for increased growth of the fungal pathogens.

The stress imposed by nematode invasion may be nonspecific in some associations. Any similar stress from whatever source may have the same effect. Perhaps nematode-induced stress reduces a plant's defense reactions to other diseases. Alterations in cell walls may render them more readily penetrable by fungi, or altered conditions in the rhizosphere may shift the populations of fungi surrounding roots in favor of pathogens.

For years it has been known that fluids leak out of roots of growing plants. "Virtually all types of compounds commonly found in plant cells have been detected in root exudates, organic acids, sugars, polysaccharides, amino acids, nucleotides and flavonones, to list a few" (Mitchell, 1976). *Meloidogyne* nematodes cause increased leakage of electrolytes from galls. *Pratylenchus* sp. in one portion of a root predisposes the unin-fected portion to invasion by *Verticillium* fungus. Both observations sug-gest that infection with nematodes alters root exudation from the whole root. This no doubt influences the germination, attraction, and penetra-tion by other pathogens.

One thorough study has been made of the mechanism by which *Meloidogyne incognita* alters the response of tomato to a fungus, *Rhizoc-tonia solani*. The experiments were based on the observation that both pathogens together induce a severe root rot, but each by itself does not (see tabulation below).

| Agent | Root rot |
|---|---|
| *Rhizoctonia* | No |
| *Meloidogyne* | No |
| *Rhizoctonia* plus *Meloidogyne* | Yes |
| *Rhizoctonia* plus leachate from healthy plant | No |
| *Rhizoctonia* plus leachate from nematode-infected plant | Yes |
| *Rhizoctonia* plus *Meloidogyne* plus continual irrigation to remove leachate | No |

As determined by means of radioactive tracers and chemical analysis, carbohydrates accumulate in galled roots, especially during the first two weeks of infection. At this time also, root exudates contain increased amounts of carbohydrates in comparison with those from noninfected roots. In subsequent weeks, exudates from galled roots contain increased quantities of protein and amino acids. These changes in concentrations and composition of organic substances emanating from roots doubtless influence the attraction and growth of the fungus and enhance its ability to penetrate into roots. Further, once inside the galled root, the fungi have access to a rich nutritional base from which to colonize the rest of the plant (Van Gundy *et al.*, 1977). Both fungus and exudates from roots infected with nematodes were necessary for *Rhizoctonia* root rot to occur. When the root leachates from infected roots were removed by trickle irrigation, root rot did not develop.

Microorganisms of the rhizosphere feed on exudates from roots and on tissues sloughed from root surfaces. When tomato plants are infected with *Meloidogyne javanica*, for example, there are changes in the micro-flora around roots. Infective units of a wilt-inducing fungus (*Fusarium*) are increased, but the numbers of an actinomycete are reduced. Addition of a soil amendment (chitin) increases the actinomycetes and protects plants from the combined action of soil microorganisms and root-knot nematodes in root pathology (Bergeson et al., 1970).

Growing roots must contend with a great variety of organisms; when pathogenic nematodes are present, the hazards of infection by other pathogens are increased. Not all species of *Meloidogyne* are equally effective in breaking genetic resistance to fusarium wilt. In tomato, *M. hapla* apparently is less effective than *M. incognita* in this respect. Genetic resistance to wilt breaks down in cotton, tobacco, and peas in the presence of root knot. Resistance also is ineffective in the combination of root knot and *Phytophthora* fungus in the black-shank disease of tobacco. However, genetic resistance to black shank persists in infections with *Pratylenchus* sp. and the fungus. Another effect of root knot is that certain fungi not normally pathogenic may become pathogens when these nematodes are present. Powell lists seven fungi common in soil but not normally pathogenic to tobacco that induce extensive decay of roots bearing *Meloidogyne* nematodes. This nematode and *Fusarium* together also predispose tobacco to infection by the foliar fungus pathogen, *Alternaria alternata* (Powell, 1971).

In addition to *Meloidogyne*, other nematodes also are important in multiple-pathogen interactions with fungi. *Pratylenchus* sp. predisposes roots of peppermint to infection; if the nematode is introduced to one part of a split root, the nematode-free portion is also predisposed to infection by *Verticillium* fungus. Extracts of nematode-infected plants stimulate growth of fungi in culture more than extracts of noninfected plants.

All the examples given thus far indicate that plants infected with nematodes are more readily damaged by concurrent fungus pathogens. However, each pathogen in a host triggers a chain of responses that influence the reaction of the host to other pathogens; for this reason, we might expect that the presence of fungal infection will alter nematode growth and reproduction. Premature destruction of giant cells by fungi leads to smaller populations of *Meloidogyne* sp. in cabbage roots infected with the club root fungus (*Plasmodiophora*). The same is true in chrysanthemums infected with *Pythium*. However, *Belonolaimus* sp. was not affected in these experiments, perhaps because it is an ectoparasite and does not depend on the development of an adaptive cell as a source of food. Populations of *Heterodera* spp., on the other hand, may be suppressed in tomato and beets by concurrent fungal infections. In the case of *Pratylenchus* sp., the populations of nematodes increase in the presence of *Verticillium* sp. and *Aphanomyces* sp. fungi. This is probably related to metabolic changes, since *Pratylenchus* nematodes increase more rapidly in roots of plants whose tops are periodically cut than in undisturbed hosts. Fungus infections may cause nutritional changes in host plants similar to those induced by periodic removal of tops. Additional evidence of metabolic change is that temperature optima for nematode reproduction and for wilt disease development are shifted when both *Verticillium dahliae* and *Pratylenchus minyus* are both present in peppermint.

The great importance to agriculture of nitrogen fixation by bacteria makes the role of nematodes of great interest. Nematodes, legumes, and bacteria are all interconnected in complex ways. In the field, soybeans infected with *Heterodera glycines* usually have few bacterial nodules. However, not all populations of this nematode depress nodulation equally. "Race 1" in the United States reduces nodulation more severely than three other "races" tested. On white clover, neither *H. trifolii* nor *Meloidogyne javanica* has much effect on nodulation. Both endoparasites reproduce well in nodules. However, nodules containing *M. javanica* deteriorate more rapidly than nodules free of nematodes, and thus the concurrent nematode infection reduces the total benefit from the nitrogen-fixing bacteria.

Since roots of legumes are parasitized by many species of nematodes, and since there are differences among host species in interactions among nematodes, host plants, and nitrogen-fixing bacteria, we can expect a rich variety of effects of nematodes on nitrogen fixation. Not only phytoparasitic nematodes but also nematodes that feed on symbiotic bacteria destroy tissues within nodules and depress the fixation of nitrogen.

Another system in which nematodes are important is the mycorrhizal association between certain fungi and plant roots. Most forest trees depend on symbiotic fungi living in close association with juvenile feeder roots. Certain fungi "penetrate juvenile feeder roots intercellularly, partially replace the middle lamellae between cortical cells . . . and eventually produce a continual hyphal network over the exterior of the root" (Ruehle, 1973). Other fungi penetrate intracellularly and form vesicles and complex branching structures within cortex cells. These two kinds of mycorrhizal fungi benefit trees by increasing the supply of essential elements to roots. In effect, they increase the root surface available for uptake of soil nutrients. In the case of pine seedlings, successful growth is impossible in the forest without mycorrhizae. Plants other than trees also have fungal symbionts that are important for their normal development (see Volume III, Chapter 11).

Nematodes appear to have two kinds of interactions with the fungal symbionts of roots. Many species of nematodes feed on fungi. Large

populations of such nematodes (e.g., *Aphelenchus avenae*) are sometimes collected from soils. In laboratory experiments, this nematode can
prevent the formation of mycorrhizal associations. Second, when roots
are subject to attack by phytoparasitic nematodes, they may be damaged sufficiently to block the symbiotic plant–fungal association. This
action has been observed with species of *Pratylenchus, Xiphinema,* and
*Meloidogyne.* Mycorrhizal fungi often protect plants from pathogenic
fungi, but they do not seem to protect against parasitic nematodes.

## B. Induction of Cellular Pathology

### 1. Esophageal Glands

Plant parasitic nematodes have relatively enormous esophageal glands
in comparison with species that feed on bacteria. We assume that secretions from these glands play a role in the induction of certain abnormalities in infections of plants by nematodes. Since cytoplasm must be
ingested through a stylet with an internal diameter of about 0.2–0.3 $\mu$m,
the nematode must have a system to keep its stylet unplugged, and
perhaps also to lower the viscosity of a plant cell's cytoplasm. Phytonematodes maintain a high internal hydrostatic pressure. The pharyngeal
pump (median bulb) has well-developed muscles to dilate the lumen
of the pharynx in a rhythmic fashion during feeding. This presumably
provides enough suction to remove the cell contents. The esophageal
glands of phytonematodes are prominent structures with ducts that open
into the lumen of the pharynx at several positions. The dorsal gland
duct of most species enlarges into an ampulla that empties close to the
base of the stylet. Ducts of the two subventral glands, on the other hand,
empty into the lumen of the pharynx in the median bulb. In species of
the superfamily *Aphelenchoidea,* both the dorsal and subventral gland
ducts open in the median bulb. Is there any significance to these positions? We can expect that the exudate from the dorsal gland is injected
into plant cells, while that from the subventral glands mingles with the
ingested food en route to the intestine (Jones, 1978). Perhaps dorsal
gland products act as a type of "anticoagulant" to keep the lumen of the
stylet open during feeding. If so, how do the nematodes without this
anatomy keep their stylets open? Perhaps gland products emptying into
the midpharynx are passed forward to the stylet while dorsal gland
secretions join food on its way to the intestine.

There is one published example of passive feeding by a nematode
(*Hexatylus*) on fungal cell contents. Doncaster and Seymour (1974)
described the sequence of feeding from films of living specimens. This

species has no muscular pump in its pharynx. Subventral gland secretions flowed from their entry into the midpharyngeal lumen forward to the stylet and into the hyphal cell. Dorsal gland secretions were not seen to enter hyphal cytoplasm, and the authors believe that they are swallowed with ingested food. From the film analysis, Doncaster and Seymour concluded that the internal hydrostatic pressure of this nematode is below that of the hyphal contents. During feeding, hyphal cytoplasm is forced into the nematode by the fungus. Gland contents are forced into hyphal cells by pressure generated by the glands themselves. This does not appear to operate in the case of phytonematodes feeding on cells of higher plants, however.

It would be of great interest to analyze gland contents chemically. Two papers have been published on stylet exudates of *Meloidogyne* sp. Living adult females of this sedentary endoparasite survive removal from roots and incubation in water or salt solution. In a favorable specimen, the stylet continues to thrust in and out of the mouth for long periods—sometimes several hours. Small accumulations of exudate may be observed at the stylet tip. Bird (1968) maintained specimens in a flow chamber and performed histochemical tests on the exudate. He concluded that basic proteins but not nucleic acids were present. Hussey and Sasser (1973) found peroxidase activity in stylet exudates of adult females.

## 2. Cell Wall Alteration

Alterations of cell walls occur in certain infections by phytonematodes. Wall dissolution is associated with some phytonematodes. Riedel and Mai (1971a,b) demonstrated that aqueous extracts of *Ditylenchus dipsaci* contain several pectinases, of which endopolygalacturonase is the active component in maceration of plant tissues. When the scales of onion bulbs were incubated with extracts of nematodes, the walls of epidermal cells lost some of their pectin. Callus tissues infected with nematodes contain pectinases; upon removal of the nematodes, the enzyme activity falls and is restored by replacing the nematodes. From these observations we can conclude that nematodes produce pectinases and that these enzymes probably play a role in maceration of host tissues.

In soybean roots infected by *Heterodera glycines*, cells around the head of the nematode enlarge, and extensive holes appear in the location of pit fields. The syncytium enlarges by continued incorporation of additional cells in an orderly fashion. The dissolution of cell walls appears to be controlled and not random.

Giant cells induced by *Meloidogyne* sp. do not undergo the extensive wall dissolution that is characteristic of infections with *Heterodera* sp. The cells enlarge to 500 to 1000 times the normal volume, nuclei undergo synchronous mitoses and become polyploid, and cytokinesis does not occur. Perhaps the nematode injects a compound similar to caffeine, which is known to inhibit normal cell wall synthesis in plant cells. The enlargement may indicate a reduction in tensile strength of the cell wall together with increased wall synthesis.

### 3. Reduction of Photosynthesis

Nematodes probably have an effect on photosynthetic rates of infected plants. Chlorosis is a symptom of infection of soybeans with *Heterodera glycines*. In Japan the name "yellows disease" is applied to the condition. These nematodes are known to reduce nodulation of soybean roots, and this causes malfunction of leaves. Loveys and Bird (1973) found that seedlings of tomato each inoculated with 30,000 larvae of *M. javanica* had reduced rates of photosynthesis. During the experimental period of 22 days, rates of $CO_2$ exchange increased fivefold in non-inoculated plants but only twice in infected plants. Part of this difference resulted from reduced growth of infected seedlings. In a short-term experiment, reduced photosynthetic rates were observed on the third and fourth days after inoculation with 50,000 larvae per seedling. Reduced photosynthesis was apparent when calculations were made either on the whole plant or on the basis of leaf area, chlorophyll content, or fresh weight. It is difficult to relate these data to field conditions, since seedlings exposed to very high levels of inoculum in laboratory experiments probably suffer much greater root damage than seedlings exposed to less intense attack in the field.

### 4. Growth Regulators

The simplest hypothesis for the induction of host responses by certain nematodes might be that nematodes synthesize plant growth regulators and introduce them into host tissues. Only one comprehensive demonstration of the presence of a chemically defined growth regulator in a plant parasitic nematode has been made (Cutler and Krusberg, 1968). There is no evidence that the compound found (methyl ester of indole-acetic acid in *Ditylenchus dipsaci*) causes the drastic growth abnormalities seen in infected plants. Several investigators have extracted materials from galls and nematodes with growth-promoting effects in standard tests. Without rigorous chemical identification their identity is uncertain. Further, the cource of growth-stimulating compounds may be the plant rather than the nematodes. Tissue damage per se may lead to changes in concentration of growth regulators. *Ditylenchus dipsaci* has enzymes that inactivate auxin. This inactivation may inhibit the development of internodes and lead to the dwarfing of infected plants.

In addition, nematodes destroy plant tissues that are known to be sites for synthesis of growth regulators. *Trichodorus* sp. attack root tips—a source of cytokinins. *Aphelenchoides* sp. in buds may alter quantities of auxin translocated from buds down to the rest of the plant.

## 5. Permeability of Membranes

Nematodes of the genus *Pratylenchus* invade root cortex tissues and destroy cells during feeding. In many hosts, these nematodes induce brown lesions visible at the root surface. However, there may be deleterious effects in addition to lesion formation. The electrical resistance of lightly infected sunflower roots is lower than normal before macroscopic lesions are visible. Perhaps this indicates generalized damage to root cell membranes, resulting in increased quantities of electrolytes in the intercellular spaces. In corn roots, however, *Pratylenchus* nematodes do not induce brown lesions, and no change in the electrical resistance of infected corn roots is detectable (Kaplan *et al.*, 1976b).

## 6. Water Stress

One of the first indications of root malfunction is decreased tolerance to water stress. It is well known that nematode-infected plants wilt earlier than healthy plants in the same field. Measurements of the quantitative changes in water potential of leaves resulting from root infections with nematodes are not available to my knowledge (see Volume III, Chapter 1). Stomata of sunflowers infected with *Pratylenchus* nematodes close before those of noninfected plants as the plant is subjected to increased water stress (Kaplan *et al.*, 1976a). This indicates that movement of water to leaves is impaired. Perhaps some of the changes in metabolites recorded in the phytonematological literature as the specific result of nematode action are simply secondary effects of increased water stress. Even mild water stress may have profound effects on the physiology of plants. As the soil dries out and plants begin to lack water, a series of plant responses is observed. In drying conditions at less than −5 bars water potential, cell growth, wall synthesis, protein synthesis, chlorophyll formation, and nitrate reductase activity are impaired. Water stress from −5 to −10 bars elicits increased abscissic acid accumulation, decreased concentration of cytokinin, closing of stomata, decreased $CO_2$ assimilation, and changes in respiration. At about −10 bars, great increases in proline accumulation and changes in sugar accumulation occur (Hsiao, 1973).

## V. A VISION OF THE FUTURE

Very little attention is currently being given to quantative changes induced by nematodes in the physiological functions of their host plants. The thorough study of such changes should lead to a much deeper understanding of nematode damage to plants. The work of Pate and his associates on the physiology of lupine offers an excellent guide for analyses of physiological changes in nematode-induced diseases of plants. He measured the flow of certain organic solutes through xylem and phloem of *Lupinus albus* in relation to the requirements of the developing fruit. Phloem contents were collected from bleeding tips of cut fruit and xylem sap from upper parts of shoots under mild vacuum. The developing fruit was enclosed for studies of gas exchange and transpiration. Data on transport and accumulation were obtained from pulse-labeling experiments followed by subsequent analysis of xylem and phloem. By combining data on transport with those on accumulation, Pate constructed balance sheets for each week of pod and seed formation and maturation. He suggests that each organ of the plant differs in its patterns of utilization of C, N, and water. He conjectures that the plant has regulatory devices of great complexity and subtlety (Pate, 1976).

One of the regulatory devices seen in plants is the "transfer cell". These cells occur in locations where great quantities of solutes are presumed to move across cell boundaries. Transfer cells are found in healthy plants in regions where there is a heavy, temporary demand for nutrients, such as along leaf traces in stems or in flower buds. The morphology of these specialized cells suggests an active metabolism as well as a great capacity for transport. In comparison with adjacent cells, the cytoplasm has an increased population of organelles, the central vacuole is lost, and the nucleus is enlarged. In addition, numerous wall ingrowths are formed, usually adjacent to a vessel. These are fingerlike projections of the cell wall directed inward. They are covered with plasmalemma. Thus, the cell surface available for movement of solutes is increased. Transfer cells are believed to facilitate the movement of solutes such as amino acids *from* a vessel to growing tissues. They are also found in nitrogen-fixing nodules of legumes where they probably aid the movement of nitrogenous compounds *into* vessels (Pate, 1976). In leaves, transfer cells are presumed to facilitate movement of photosynthates into phloem.

A sessile phytonematode creates a center of demand within a plant for nutrients to support its growth. Not surprisingly, transfer cells develop in association with some kinds of such nematodes. Both *Meloid-ogyne* sp. and *Heterodera* sp. induce the development in roots of special-ized cells with the characteristic appearance of transfer cells. Nematodes of certain other genera (*Nacobbus, Rotylenchulus, Tylenchulus,* and *Xiphinema*) also induce the development of specialized cells, but these do not have the wall ingrowths characteristic of transfer cells.

The entire pattern of flow into and out of tomato roots is altered in plants with root-knot nematodes. Xylem composition and root exudates change during nematode infections, but few data have been recorded. Wang and Bergeson (1974) analyzed xylem sap and root exudates of tomato in relation to infection with *Meloidogyne incognita* (root-knot nematode). The nematodes had little influence on the proportions of 18 amino acids in xylem sap, but the total quantity of these compounds was reduced in sap of infected plants. The quantity of xylem sap was also reduced in plants with galled roots. In contrast with the reduction in amounts of amino acids, the concentration and total quantity of car-bohydrates were greater in xylem sap of infected plants, while the sugar composition was comparable in both infected and noninfected plants. Exudates from roots of infected plants contained fewer amino and or-ganic acids but more carbohydrates than exudates from noninfected plants. Total carbohydrate content of daily exudate from infected roots was more than twice as much as that from noninfected roots. Moreover, a water-soluble polysaccharide consisting mostly of glucose made up more than half the carbohydrate content of the exudate from galled roots. This represented an 18-fold increase over the quantity of polysaccharide released from noninfected roots.

Gommers and Dropkin (1977) found four times more glucose in transfer cells induced by *Meloidogyne incognita* in soybean and *Im-patiens balsamina* than in root tip cells of uninoculated plants. The same increase in glucose concentration occurred in infections of soybeans with the cyst nematode, *Heterodera glycines*. These profound alterations in the physiology of root galls and cyst nematode feeding sites must be connected with changes in the movement of solutes between stems, leaves, and roots. There is little information on such alterations of host physiology in infections with other nematodes.

The data on altered physiology of plants infected with nematodes are too fragmentary to permit any general statements about the influence of these organisms on root physiology and on the physiology of the whole plant. When a thorough study is made (in the style of Pate), we shall have some understanding of the mechanisms by which nematodes influence crop yields.

## VI. SUMMARY

Plant parasitic nematodes comprise over 2000 species. Among these, there are different modes of behavior, and different structural and physiological adaptations to parasitic life. The pathological effects of nematodes within plants are varied, as indicated in this chapter. The entire plant is an integrated unit—alterations of one part lead to other alterations, which eventually affect the whole. The challenge for future research is to discover how nematodes, fungi, and plants respond to each other, what molecular messages pass back and forth between host and parasites, what general changes in host and pathogen physiology take place, and how we can manage crops to minimize the damage that nematodes induce.

### References

Bergeson, G. B., Van Gundy, S. D., and Thomason, I. J. (1970). Effect of *Meloidogyne javanica* on rhizosphere microflora and *Fusarium* wilt of tomato. *Phytopathology* **60**, 1245–1249.

Bird, A. F. (1968). Changes associated with parasitism in nematodes. IV. Cytochemical studies on the ampulla of the dorsal esophageal gland of *Meloidogyne javanica* and on exudations from the buccal stylet. *J. Parasitol.* **54**, 879–890.

Cutler, H. G., and Krusberg, L. R. (1968). Plant growth regulators in *Ditylenchus dipsaci*, *Ditylenchus triformis* and host tissues. *Plant Cell Physiol.* **9**, 479–497.

Doncaster, C. C., and Seymour, M. K. (1974). Passive ingestion in a plant nematode, *Hexatylus viviparus* (*Neotylenchidae: Tylenchida*). *Nematologica* **20**, 297–307.

Doney, D. L., Fife, J. M., and Whitney, E. D. (1970). The effect of the sugarbeet nematode *Heterodera schachtii* on the free amino acids in resistant and susceptible *Beta* species. *Phytopathology* **60**, 1727–1729.

Endo, B. Y., and Veech, J. A. (1970). Morphology and histochemistry of soybean roots infected with *Heterodera glycines*. *Phytopathology* **60**, 1493–1498.

Epstein, E., and Cohn, E. (1971). Biochemical changes in terminal root galls caused by an ectoparasitic nematode, *Longidorus africanus*: Amino acids. *J. Nematol.* **3**, 334–340.

Gommers, F. J., and Dropkin, V. H. (1977). Quantitative histochemistry of nematode-induced transfer cells. *Phytopathology* **67**, 869–873.

Hsiao, T. C. (1973). Plant responses to water stress. *Annu. Rev. Plant Physiol.* **24**, 519–570.

Hussey, R. S., and Sasser, J. N. (1973). Peroxidase from *Meloidogyne incognita*. *Physiol. Plant Pathol.* **3**, 223–229.

Jones, M. G. K., and Dropkin, V. H. (1975). Cellular alterations induced in soybean roots by three endoparasitic nematodes. *Physiol. Plant Pathol.* **5**, 119–124.

Jones, M. G. K., and Payne, H. L. (1977). The structure of syncytia induced by the phytoparasitic nematode *Nacobbus aberrans* in tomato roots, and the possible role of plasmodesmata in their nutrition. *J. Cell Sci.* **23**, 299–313.

Jones, R. K. (1978). The feeding behavior of *Helicotylenchus* spp. on wheat roots. *Nematologica* **24**, 88–94.

238     VICTOR H. DROPKIN

Kaplan, D. T., Rohde, R. A., and Tattar, T. W. (1976a). Leaf diffusive resistance of sunflowers infected by *Pratylenchus penetrans*. *Phytopathology* 66, 519–570.

Kaplan, D. T., Tattar, T. A., and Rohde, R. A. (1976b). Reduction of electrical resistance in sunflower roots infected with lesion nematodes. *Phytopathology* 66, 1262–1264.

Loveys, B. R., and Bird, A. F. (1973). The influence of nematodes on photosynthesis in tomato plants. *Physiol. Plant Pathol.* 3, 525–529.

Mamiya, Y. (1976). Pine wilting disease caused by the pine wood nematode, *Bursaphelenchus lignicolus*, in Japan. *Jap. Agr. Res. Quart.* 10, 206–211.

Mitchell, J. E. (1976). The effect of roots on the activity of soil-borne plant pathogens. *In* "Physiological Plant Pathology" R. Heitefuss and P. Williams, eds.), Vol. 4, pp. 104–128. Springer-Verlag, Berlin and New York.

Owens, R. G., and Specht, H. N. (1966). Biochemical alterations induced in host tissues by root-knot nematodes. *Contrib. Boyce Thompson Inst.* 23, 181–198.

Pate, J. S. (1976). Nutrient mobilization and cycling: Case studies for carbon and nitrogen in organs of a legume. *In* "Transport and Transfer Processes in Plants" (I. F. Wardlaw and J. B. Passioura, eds.), pp. 447–462. Academic Press, New York.

Pitcher, R. S. (1967). The host–parasite relations and ecology of *Trichodorus viruliferous* on apple roots, as observed from an underground laboratory. *Nematologica* 13, 547–557.

Powell, N. T. (1971). Interactions between nematodes and fungi in disease complexes. *Annu. Rev. Phytopathol.* 9, 253–274.

Rebois, R. V., Madden, P. A., and Eldridge, B. J. (1975). Some ultrastructural changes induced in resistant and susceptible soybean roots following infection by *Rotylenchulus reniformis*. *J. Nematol.* 7, 122–139.

Reidel, R. M., and Mai, W. F. (1971a). Pectinases in aqueous extracts of *Ditylenchus dipsaci*. *J. Nematol.* 3, 28–38.

Reidel, R. M., and Mai, W. F. (1971b). A comparison of pectinases from *Ditylenchus dipsaci* and *Allium cepa* callus tissue. *J. Nematol.* 3, 174–178.

Ruehle, J. L. (1973). Nematodes and forest trees—Types of damage to tree roots. *Annu. Rev. Phytopathol.* 11, 99–118.

study the predisposition phenomena in the root-knot wilt fungus disease com-
Sidhu, G. S., and Webster, J. M. (1977). The use of amino acid fungal auxotrophs to plex of tomato. *Physiol. Plant Pathol.* 11, 117–127.

Thomason, I. J., Rich, J. R., and O'Melia, F. C. (1976). Pathology and histopathology of *Pratylenchus scribneri* infecting snap bean and lima bean. *J. Nematol.* 8, 347–352.

Van Gundy, S. D., and Kirkpatrick, J. D. (1964). Nature of resistance in certain citrus rootsticks to citrus nematode. *Phytopathology* 54, 419–427.

Van Gundy, S. D., Kirkpatrick, J. D., and Golden, J. (1977). The nature and role of metabolic leakage from root-knot nematode galls and infection by *Rhizoctonia solani*. *J. Nematol.* 9, 113–121.

Wang, E. L. H., and Bergeson, G. B. (1974). Biochemical changes in root exudate and xylem sap of tomato plants infected with *Meloidogyne incognita*. *J. Nematol.* 6, 194–202.

Chapter 13

# How Insects Induce Disease

DALE M. NORRIS

## I. INTRODUCTION

The objective of this chapter is to consider insects as inducers, not just of injury, but of disease in plants. The task of writing about this subject was accepted because the author, who is both an entomologist and a plant pathologist, believes that it addresses the most common major "void" in contemporary considerations of plant disease situations. The lingering tendency of entomologists and phytopathologists to underestimate the roles of insects in the induction of plant disease is especially disconcerting in view of Leach's (1940) and Carter's (1973) now historical interdisciplinary analyses of the topic.

By addressing this subject, one automatically stimulates debate on the semanitics of "disease" versus "injury." However, our mission cannot afford to be bogged down in such discourse. If a reader is concerned about these semantics we urge him to trace the evolution of terminology that has accompanied the consideration of nematodes as pathogens instead of as pests and to make the obvious analogies between these animals and the insects of present concern.

Because I am identified more commonly as an entomologist than as a plant pathologist, many entomologists may conclude that I am personally forfeiting the Insecta that utilize plant hosts to "those plant

PLANT DISEASE, VOL. IV

pathologists." However, our real concerns are simply better biological understanding and less "pigeonholing" of facts and judgments in those long-overused human artifacts called disciplines of science (e.g., entomology and plant pathology). It is time to acknowledge positively and internationally that many thousands of insect species are inducers of plant disease.

This chapter focuses on two realms of insect induction of disease: (1) insect regurgitation, excretion, secretion, or injection of nonliving materials onto, or into, plant tissues, resulting in the development of symptoms, and (2) insect symbioses with microbes, resulting in plant disease. Neither of these subjects has been ignored entirely in pertinent scientific investigations, but the magnitude of their involvement in inducing plant disease has not been adequately recognized or studied. This chapter seeks to stimulate research in these fascinating areas of plant disease.

## II. INSECT–DERIVED PHYTOALLACTINS

Insect secretions, excretions, or regurgitations that cause changes in plants have generally been considered as *phytotoxins* which cause *phytotoxemias* (Carter, 1973). However, some insect-produced compounds at given concentrations cause hyper- or hypoplasia or hyper- or hypotrophy and do not kill cells. The latter chemicals thus function by regulating cell division, growth, or differentation and have properties of *phytohormones* or *growth regulators*.

The full range of insect-derived compounds that alter plants thus encompasses more than necrosis-producing phytotoxins. The term *phytoallactin* is proposed here to include all insect-produced chemicals that cause changes in plants. *Allact* is the Greek root meaning *change*. Phytoallactin thus is defined as a plant-altering chemical. Phytotoxins and phytohormones are types of phytoallactins. This terminology is proposed to convey more thoroughly the spectrum of actions that insect-derived molecules have in plants. There are obvious situations in which a given molecule at one concentration is a phytotoxin and at a different level is a phytohormone. Such situations further emphasize the usefulness of the new, more broadly defined term "phytoallactin."

### A. Phytotoxins

DeBary in 1886 presented the concept that a pathogen may produce a toxin that causes disease symptoms in a plant. Although debates over the semantics and specific roles of phytotoxins continue, the major importance of such toxic chemicals in causing plant disease is no longer contested (Strobel, 1974; Heitefuss and Williams, 1976). In this discussion, the toxin-producing agent (i.e., pathogen) is an insect or mite. The majority of species responsible for phytotoxemias belong to the orders Hemiptera and Homoptera. These are the so-called sucking insects. Some Acarina (mites) also are sufficiently important *toxicogenic* arthropods to warrant inclusion here even though they are not insects. Correlaries will be drawn between the toxicogenic actions of Homoptera, Hemiptera, and Acarina when appropriate.

All Homoptera, but only some Hemiptera, are phytophagous. The latter insects are usually referred to as "plant bugs." The feeding processes of these two groups of insects are important in influencing the type of phytotoxemia produced. Most Homoptera and the Pentatomorpha (pentatomid plant bugs) secrete viscous material just before and during feeding. This secretion gels to form a sheath around the mouthparts regardless of their route in the plant tissue. The sheath is mainly protein and becomes stabilized through keratinization. It contains about 10% phospholipid, a small amount of carbohydrate, and perhaps some tannins (Miles, 1968, 1972). Sheath-forming insects leave the sheath trace in the pierced tissues, but gross lesions directly attributable to the sheath are not usually formed. However, besides the solidifying sheath material, Homoptera and Pentatomorpha also inject a more watery saliva, which contains a variety of materials and which can spread significantly from the sheathed area. Such watery saliva is secreted from and sucked, at least partially, back into the food canal. It is quite clear that the contents of this saliva contribute to the phytotoxemias induced by the feeding of some Homoptera and Pentatomorpha.

Mirid bugs and other Cimicomorpha do not form sheaths, and the mirids (capsids) are especially notorious for causing severe phytotoxemia. The feeding by these types of bugs causes a breakdown and partial homogenization of a large pocket of cells. The contents of these cells are flushed out of ruptured tissues by excesses of watery saliva.

The contents of watery saliva vary among insects, and the components that cause phytotoxemia also probably differ. The insect-derived causes of phytotoxemias remain ill defined, but three major theories concerning such toxicogenesis exist.

The amino acid theory purports that amino acids in the watery saliva elicit the phytotoxemia (Kloft, 1960; Anders, 1961). Known effects of free amino acids in watery saliva when injected artificially into plants include streaming of cytoplasm; increased cell permeability, respiration rate, and transpiration; and decreased photosynthesis. These effects may account for some toxic reactions in the plant. Lysine, histidine, trypto-

phan, glutamic acid, and valine are amino acids that have been especially considered as phytotoxins. However, the reported effects of such
acids have been duplicated with injections of potassium phosphate
solution (Miles, 1968). Thus, the amino acid theory remains highly
questionable. Perhaps the strongest support for this theory came from
Schäller's (1968) demonstration that the greater the overall concentration of amino acids in the saliva of *Viteus vtifolii*, the grape phylloxera,
the greater the ability of the insect to induce phytotoxemia. The concentration of indoleacetic acid (IAA) in the saliva also was positively
correlated with the insect's ability to cause phytotoxemia. Schäller
subsequently concluded that IAA must interact with free amino acid
to produce phytotoxemia. Such possible interactions will be discussed
later in this section in conjunction with the polyphenol oxidase (PPO)
theory.

The second theory concerning causation of phytotoxemias involves
the pectinase enzymes [e.g., salivary pectin polygalacturonase (Strong,
1970)]. Such enzymes have been found in the salivary glands of several
species of mirids, and it is proposed that these enzymes significantly
contribute to the development of disease symptoms through the destruction of meristematic tissues. Pectinases dissolve the middle lamella
of cells, and this allows the more extensive spread of salivary secretions
into surrounding plant cells during the lacerate and flush feeding by
mirids. The theorized net result is a lesion that is much larger than
would result from the feeding of a nonmirid sap-sucking insect of comparable size.

The third theory is that salivary phenols (e.g., IAA) or PPO directly
interfere with basic subcellular processes in plant tissues. The disease
may result from either an excess or a shortage (i.e., an imbalance) of
such chemicals. Tissue damage from insect feeding may initiate a series
of wound-response reactions in the plant tissues, and in some circumstances (e.g., a hypersensitive reaction) phenols in the damaged and
surrounding cells are oxidized to highly toxic quinones. These quinones
may directly kill plant cells or they may transform tryptophan, by PPO-
catalyzed oxidative deamination, to toxicogenic amounts of the phytohormone IAA. Thus, the possible combined roles of free amino acids
[e.g., dihydroxyphenylalanine (dopa) and tryptophan], and phenols
(e.g., IAA) in the causation of phytotoxemias seem to deserve much
more study. The probably unique PPO system(s) in the salivary glands
of each species of insect (Miles, 1968) would seem to allow for significant species-specific phytotoxemias. The effects of an insect's watery
saliva on pierced and surrounding plant cells probably are greatly influenced by whether the insect's or plant's PPO system dominates the

disease scene. If the plant's PPO system prevails, the highly toxic quinones are likely to be produced. If the insect's PPO system dominates, it may prevent or reduce oxidation of the accumulated plant phenolics to toxic quinones or may rapidly detoxify existing quinones. Either of the latter actions may reduce hypersensitivity reactions in the plant cells, and in some cases cell death may not occur.

As our knowledge of insect-induced phytotoxemias increases in the future, chemicals that fit in each of the above-theorized categories of phytotoxins probably will be found in the watery saliva of several, if not most, insect species that cause phytotoxemias. Thus, significant common chemical principles of insect-induced phytotoxemias are likely to emerge.

Representative examples of insect-induced phytotoxemias are given in Tables I and II. The cited toxemias have been divided into *local-lesion* versus *systemic* groupings. However, it is important that these rather arbitrary categories do not significantly inhibit our considerations of the true spectrum of such toxemias. Some species of insects even induce a combination of local lesions and various systemic symptoms. An example is a mirid, *Adelphocorus lineolatus*, which causes a range of toxemias in alfalfa. Symptoms initially are localized around the feeding puncture but later spread to other parts of individual flowers. The final range of toxemias caused by *A. lineolatus* includes bud blast, cell disintegration in the ovary and ovules, flower fall, and pod necrosis.

## B. Phytohormones

The functional components of plants usually develop in an extraordinarily coherent and flexible, but definitely limited, pattern. In the normal organism, morphogenetic laws are strictly obeyed, and the processes of metabolism, growth, cellular differentiation, and organogenesis are precisely regulated. Processes start and stop in harmony and finally yield a plant of certain proportions, which remain more or less constant throughout many generations. However, the harmony of form or the limit on size may be abruptly disrupted. A prime cause of such abnormalities is an alteration of the amount of and/or balance among various growth-regulating hormones in the affected plant. These abnormalities are designated in this discussion as *growth and differentiation disorders.*

Some insect-induced malformations in plants involve relatively harmonious (i.e., proportional) changes, but the growth responses are either increased or decreased. Examples of such disorders are cercopid-induced reductions of the length and diameter of canes and increases in the development of adventitious buds of sugarcane; spittle bug-induced

**TABLE I**

Representative Insect-Induced Local-Lesion Phytotoxemias

| Inducing insects | Toxemia | References |
| --- | --- | --- |
| Cicadellidae (leafhoppers) | | |
| *Empoasca bifurcata, E.* *erigeron, E. filamenta,* *E. abrupta, E. maligna* | Leaf spotting and stippling | Smith and Poos (1931) |
| *Eutettix strobi* (nymphs) | Crimson spots on *Chenopodium album* | Fenton (1925), Carpenter (1928) |
| *Graminella nigrifrons* | Chlorotic spots on rice plants | Deemsteegt *et al.* (1969) |
| *Erythroneura comes,* *Zyginia rhamni,* *Arboridia dalmatia* | Depigmentation of vines | Vindane (1967) |
| Aphididae (aphids) | Yellow and red | Wadley (1929) |
| *Toxoptera graminum* | spots of grasses | |
| *Myzus ornatus,* *M. persicae* | Chlorotic spots on several plants | Jensen (1954) |
| Coccidae (scales) | | |
| *Pseudococcus adonidum* | Circular chlorotic spots on pineapple plants | Carter (1942) |
| *Dysmicoccus brevipes* | Irregular chlorotic spots on pineapple plants | Carter (1933, 1939) |
| *D. neobrevipes* | Green spotting of pineapple plants | Beardsley (1959) |
| *Melanaspis bromeliae* | Circular green spotting of pineapple plants | Carter (1949) |
| *Lecanium coryli* (adult females) | Dieback of ash | Komarek (1946) |
| Pentatomidae (stinkbugs) *Euschistus variolarius* | White cottony pockets on d'Anjou pears | Wilks (1964) |
| Acarina (mites) *Vasates fockeui* | Chlorotic fleck of Myrobalan plum | Gilmer and McEwen (1958) |
| *V. cornutus* | Yellow spot of peach | Wilson and Cochran (1952) |
| *Aceria ficus* | Leaf distortion, russetting, and chlorosis | Flock and Wallace (1955) |
| *Brevipalpus phoenicis* | Diffuse chlorotic spotting of orange trees | Knorr and Denmark (1970) |

## TABLE II
### Representative Insect-Induced Systemic Phytotoxemias

| Inducing insects | Toxemia | References |
|---|---|---|
| Miridae (capsids) | | |
| *Helopeltis bergrothi* | Gnarled stem canker | Leach and Smee (1933) |
| *H. schoutedeni* | Systemic necrosis | Schmitz (1958) |
| *Sahlbergella singularis* | Blast, dieback, and staghead of cacao | Goodchild (1952) |
| *Distantiella theobroma* | Blast, dieback, and staghead of cacao | Gibbs and Leston (1970) |
| *Lygus hesperus, L. elis* | Pitting of lima beans, shedding of blossoms and pods | Baker *et al.* (1946) |
| *Adelphocorus lineolatus* | Bud blast, flower fall, and pod injury | Hughes (1943) |
| Coreidae (coreids) | Flower and nut | Way (1953) |
| *Pseudotheraptus wayi* | destruction on coconuts | |
| *Amblypelta cocophaga,* | Pod distortion and | Brown (1958) |
| *A. theobromae* | dieback of cacao | |
| Aphididae (aphids) | Chlorotic spotting of | Stace-Smith (1954) |
| *Amphorophora rubitoxica* | black raspberry | |
| Coccidae (scales) | Resin excretion | Carle *et al.* (1970) |
| *Matsucoccus feytaudi* | and tree death | |

rosetting and shortening of internodes on alfalfa and clover; leafhopper (i.e., *Empoasca*) stunting, rosetting, and proliferation of dwarfed shoots on hosts; and mirid-caused growth of multiple stems (i.e., witches'-broom effect) instead of single stems in many young trees.

The specific insect-induced causes of relatively harmonious malformations in plants are still ill defined, but they apparently involve imbalances in auxins in meristematic tissues. These imbalances may result directly from (1) withdrawal of such chemicals from tissues along with the ingested liquefied contents of the disrupted cells, (2) injection of auxins into meristematic cells during feeding or oviposition, or (3) injection of auxin inhibitors during feeding or oviposition. Such auxin imbalances also may result indirectly from altered levels of (1) amino acids, (2) oxidative enzymes, (3) sugars, (4) protein–carbohydrate ratio, (5) phenols, (6) auxin precursors, or (7) combinations of the previous factors.

At the opposite end of the spectrum of insect-induced growth and differentiation disorders in plants are the amorphous changes in growth pattern. These can be usefully divided into self-limiting overgrowths (e.g., galls) and non-self-limiting overgrowths (e.g., tumors) which involve hyperplasia or hypertrophy or both.

It is highly interesting and biologically significant that each species of gall-forming insect produces a unique self-limiting overgrowth (i.e., zoocecidium). The initial gall development involves a general stimulation of the growth of meristematic plant cells. Cell nuclei enlarge, and the nucleic acid content of such cells increases. Chloroplasts degenerate, and starch is converted to sugar. Cell permeability is increased, and cells enlarge. Young parenchymatous cells may become meristematic, and the rate of divisions among such cells increases. New vascular tissues appear in the gall, and these join the plant's vascular system. The polarity of the tissues of the gall is related to the insect, and not to the rest of the plant. However exotic the final structure of the gall may be, it is composed of cells that are modifications of normal plant cells. The variously differentiated cells of the gall simply have a unique relationship with each other and with the cells in the rest of the plant.

The lingering central question regarding insect-induced galls in plants (i.e., *cecidogenesis*) is, Does the gall-forming insect produce its specific chemical organizers of the plant tissues, or does it duplicate nonspecific hormonal (i.e., growth regulator) controls of the plant but with a polarity and timing controlled by the insect? Current knowledge does not provide the answer to this intriguing question, but existing evidence suggests that most insects probably utilize both specific and general chemicals in their overall induction of a unique gall. It is known that some cecidogenic (gall-forming) insects contain the general natural phytohormone IAA in pertinent tissues (e.g., salivary glands of coccids and aphids and ovipositional glands of tenthredinid Hymenoptera). Actual biosynthesis of IAA from tryptophan also has been demonstrated in the salivary glands of cecidogenic insects. Thus, current information strongly suggests that IAA is a principal chemical inducer of cecidogenesis. Specific mixtures of amino acids, peptides, and proteins also occur in the pertinent tissues of gall-forming insects; species-characteristic enzymes or mixtures also are involved.

The utilization of these various chemicals in cecidogenesis also seems unique in time and space for each insect species. Thus, an insect probably applies quite precise amounts of each compound to finite tissues or individual cells according to an inherently programmed timetable. The more elaborate the gall is in terms of distinctly differentiated cells and tissues, the more complex the inherently programmed gall-inducing behavior of the insect is likely to be.

Efforts to induce cecidogenesis under experimental conditions without the insect have yielded erratic results. Amorphous plant cell calluses are easily produced by means of such experimental methods, but "successful" imitations of cecidogenesis have been obtained only when accurate amounts of growth regulator (e.g., IAA) have been applied to the "correct" cells at the "proper" times. More spectacular results have been achieved with synthetic auxins. This is true perhaps because such growth regulators are not as subject to inhibition by natural inhibitors in plant tissues as is the naturally occurring IAA.

The major past problems in research on cecidogenesis strongly suggest that some future investigations should emphasize the use of cecidogenic insects on plant cells in tissue culture conditions in the laboratory. The general field of *in vitro* cell, tissue, and whole-plant culture has advanced sufficiently to allow such methodologies to be effectively adapted to the chemical analysis of specific cecidogenesis. Knowledge of insect nutrition and continuous aseptic culture in laboratory conditions also is sufficiently adequate for such research to be undertaken. If the culturing of cecidogenic insects on properly grown meristematic cells in defined environments under aseptic conditions is combined with qualitative and quantitative analyses of the associated chemical constituents over time and in space, definitive data on the molecular bases of such gall development should be obtained.

Non-self-limiting overgrowths (tumors) involve new growth, which is composed of persistently altered, more or less randomly proliferating cells that reproduce true to type. There is no adequate control mechanism against this growth in the host. Affected cells thus acquire, as a result of their transformation, a capacity for autonomous growth. Acquisition of the capacity for autonomous growth theoretically requires something newly activated and distinctive that urges such cells to continue the abnormal and essentially unregulated proliferation. The tumors involve cells whose energies are directed largely toward synthesis of substances required specifically for cell growth and division. These molecules include nucleic acids, histones, and other substances involved in mitosis. Some basic questions to be answered regarding nonlimiting tumor development thus are the following: (1) what entities initiate the altered cell conditions, and (2) what biochemical mechanism(s) in such plant cells are responsible for the permanent switch in biosynthetic metabolism?

Regarding insect-induced tumors in plants, physical irritations (e.g., feeding punctures or ovipositional wounds) or chemical compounds (e.g., certain electron donors) produced by the insect may, jointly or singularly, induce tumors. These *incitants* also may be termed *proximate causes;* they render the plant cell neoplastic. Incitants condition the cell and induce tumorogenesis, but they frequently play no significant role in the continued abnormal and autonomous proliferation of the tumorous cell. Causes of continued abnormal and autonomous proliferation of the tumorous cell correspondingly may be termed *continuing causes.*

The abnormal plant growth induced by the balsam woolly aphid (*Adelges picae*) is an example of an insect-induced tumor. This tumorous growth has been partially duplicated by applying IAA in lanolin to scarified bark surface. Thus, the roles of the aphid in this tumorogenesis seem largely proximate in nature and include cell conditioning and induction. The continuing causes perhaps are hormonal imbalances induced in the involved cells by the feeding activities of the aphids. The permanent biochemical bases for the continuation of such a tumorogenesis might evolve from the induced imbalance in phytohormones in such plant cells resulting in the production of a unique progeny cell genome which lacks the molecular constraints required for regulated cell growth and divisions. A specific biochemical mechanism might be (1) an irreversibly altered histone–nucleic acid relationship that results in the portion of the genome involved with cell division and growth being selectively and repeatedly replicated to yield new tumorous cells; (2) omission of an enzyme that prevents continuous replication of the portion of the genome involved in cell division; or (3) addition of an enzyme that promotes such continuous cell divisions.

## III. INSECTS AS COMPONENTS OF PHYTOPATHOGENIC SYMBIOTIC COMPLEXES

Senescence and death of plants are sequential processes usually elicited by a complex of causal factors. Each factor thus inflicts some damage or incites a certain disease, and the host then may become more susceptible to other parasitic agents. The attack sequence of the agents usually is quite precise, and the success of each parasitic species depends greatly on the conditioning of the host by preceding species.

The discipline of plant pathology thus far has largely emphasized single species as causal agents and has made great use of the pure (i.e., single-species) culture technique. However, the frontiers of plant pathology are no longer amenable to analysis through the sole use of the pure culture technique. Leading experimentation now must consider the effects of complexes of agents on plants (see also this volume, Chapter 6).

Insects that attack plants commonly bear several microbial symbiotes, which also inflict some disease in the plant. In numerous situations the insect cannot survive on the host plant without the symbiotic microbes being present; thus, the symbiotic complex including the insect is the truer inducing unit. This situation strongly suggests that the effects of insects and microbes in plants are difficult to clearly or accurately separate. Few entomologists or plant pathologists have yet adequately recognized this prevalent situation.

## A. Ectosymbiotes of Phytopathogenic Insects

Microbial symbiotes of insects can be classified as either *ectosymbiotes* or *endosymbiotes*. Ectosymbiotic microbes are borne by insects in ectodermal structures. Specially evolved structures are termed *mycangia, mycetangia, esophageal bulbs*, etc. These structures may occur in various developmental forms of the insect, but they are most common in larvae or nymphs, and adults. Mycangia in the immature insect seem especially important in enabling the individual to spread ectosymbiotes along the walls of its tunnels in the plant tissue. This ability to spread ectosymbiotes makes such immature insects very active participants in the extension of pathogenesis in plants.

Mycangia in adults may also enable them to inoculate a complex of phytopathogenic microbes throughout feeding scars, tunnels, brood galleries, etc., in plant tissues. However, these structures in adults are especially important for the perpetuation of a phytopathogenic complex of microbes from plant to plant. During the movement of the insect from a diseased to a healthy plant, the symbiotic phytopathogenic microbes are protected and variously nourished in these specialized ectodermal organs. In some instances the quantity of microbes in a mycangium increases more than 100-fold (e.g., symbiotes in the mycangium of the ambrosia beetle, *Trypodendron retusum*) during the time when the adult insect is not on, or in, a plant (e.g., tree). This marked multiplication of the phytopathogenic microbes occurs at the nutritional expense of the insect, and the required nutrients are secreted into the organ from specialized epithelial cells that line the hemocoel side of the involved insect cuticle. It seems that the evolution of these ectodermal "culture chambers" for the insect's phytopathogenic symbiotes has enabled it to maintain a dependable supply of required microbial inoculum for use in pathogenesis in plants without allowing the microbes to invade its body as typical parasites.

Current knowledge of the interdependencies between phytopathogenic insects and their phytopathogenic ectosymbiotic fungi, bacteria, protozoa, and nematodes clearly demonstrates that many such symbioses are incredibly intimate behaviorally, anatomically, nutritionally, and biochemically. Thus, the historical attitude that such interrelationships are "strictly mechanical," "crude," or "biologically simple" now clearly qualifies as faulty dogma. In several significant aspects the complexities of molecular interdependencies that exist in these ectosymbioses exceed those between phytopathogenic insects and their ectosymbiotic viruses. In the latter situations, the causal entities apparently are only borne on the external surface of the insect's mouthparts.

Knowledge of the molecular interdependencies between the phytopathogenic ambrosia beetle, *Xyleborus ferrugineus,* and its many microbial ectosymbiotes (Norris, 1975) aptly reveals the extremely intimate and obligatory specific relations that exist between such insects and their complexes of ectosymbiotes. This beetle has coevolved with ectosymbiotic phytopathogenic fungi and bacteria such that its (1) nutrition, (2) growth, (3) metamorphosis, (4) reproduction, and (5) life span are regulated by symbiotic-derived chemicals. The symbiote-free insect simply is not an organism in the sense of being a reproducing entity. The concept of such insects being basically an embodiment (e.g., host or home) for a multispecies pathogenic complex thus seemingly deserves our most critical consideration and experimental evaluation.

Many ectosymbiotic microbes in these multispecies pathogenic complexes apparently are not as dependent on the insect as it is on them. However, in most investigated cases, the microbes appear to have relinquished some otherwise valued microbial traits while evolving as symbiotes of insects. Many, if not most, filamentous fungal ectosymbiotes reproduce only asexually in the host insect. The elaborate behaviors and mechanisms that the insects have evolved to perpetuate the asexual forms of such fungi perhaps have reduced the benefits from sexual forms to a trivial status. Ectosymbiotic fungi and bacteria also have evolved to grow on the nutrients available in the host insect's mycangia as well as on the nutrients available in the insect's tunnels in plant tissues. Such fungi frequently are extremely pleomorphic, and their appearances in mycangia may have few, if any, major characteristics common to their forms in the insect's tunnels in plants. Their growth characteristics on standard laboratory culture media also may not resemble any of the above-discussed forms. Thus, association with the host insect markedly alters the appearance of such ectosymbiotes.

Ectosymbiotic bacteria seem to show much less change in growth and form when associated with the host insect than do the fungi. However, their growth rates may be critically regulated by insect excretions or secretions.

Ectosymbiotic nematodes frequently complete sexual reproduction in plant tissue while somewhat removed from the influences of their insect hosts. Protozoan ectosymbiotes are still poorly understood, but existing knowledge indicates that they have distinctive forms as "free livers" in plant tissues compared to when they are intimately associated with their insect hosts.

The ectosymbiotic forms of phytopathogenic viruses or mycoplasms transmitted by insect hosts have been studied extensively, and numerous texts (e.g., Maramorosch, 1977) on these agents are available. Viruses and mycoplasms in plant cells do have appearances that distinguish them from those seen on insect mouthparts.

The extent to which the phytopathogenic ectosymbiotes of an insect engage in pathogenesis in plant tissues while significantly removed from the insect's feeding, tunneling, or ovipositional sites apparently varies greatly among the involved microbes. In some cases (e.g., *Fusarium* sp. ectosymbiote of *Xyleborus* beetles), the fungus may spread systemically in the plant and generate pathology far from the limits of the insect's activities. However, ectosymbiotic yeasts commonly are restricted in their phytopathogenesis to plant cells in the immediate area of insect attack on the plant. *Pseudomonas* sp. of bacterial ectosymbiotes of several phytopathogenic fruit flies spread extensively beyond the mines of the fly larvae and may rot whole plant organs (e.g., apples, olives, or cherries). Viruses and mycoplasms also may move far from the pathology immediately associated with the insect host and vector.

Nematodes that cause red-ring disease of coconut and oil palms are prime examples of such ectosymbiotes of phytopathogenic insects (i.e., the palm weevil) that cause plant-wide systemic pathogenesis. Other nematode ectosymbiotes (e.g., *Parasitaphelenchus* sp.) of the smaller European elm bark beetle exist as saprophytes in the debris of the insect's galleries in the plant. Protozoan ectosymbiotes frequently are quite restricted in their pathogenesis in plants. They commonly are limited to latex or resin ducts or other localized plant tissues into which their symbiotic insect (e.g., plant bug) host inoculated them. However, some cause extensive vascular disease (e.g., phloem necrosis of coffee).

Present information thus suggests that the degree of independence which the phytopathogenic ectosymbiotes of various insects demonstrate in the genesis of plant disease varies widely. However, the details of most phytopathogenic activities of this type remain poorly defined. It seems highly probable that the dependence of these microbes on their symbiotic insect host, once in the plant, far exceeds that which is currently thought to exist. Insects obviously are vectors and inoculators, if not perpetual caretakers, of many more species of phytopathogenic microbes than is now recognized. Plant pathologists and entomologists have been somewhat negligent in their consideration of insects as host symbiotes of phytopathogenic microbes. This is especially true for fungi, bacteria, nematodes, and protozoans.

## B. Endosymbiotes of Phytopathogenic Insects

Many phytopathogenic insects are symbiotic hosts for phytopathogenic microbes that spend part of their life cycle as endosymbiotes of the insect and the rest of their life as ectosymbiotes of the same insect and as pathogens of plants. The previously mentioned nematode, *Rhadina-phelenchus cocophilus*, which causes red-ring disease of coconut palms, is a prime example of such a pathogen. Nematodes invade the hemocoel of the palm weevil, *Rhynchophorus palmarum*, and remain sexually immature endosymbiotes while in the insect. When the adult weevil invades palms, the nemas exit from the insect, mature to sexually reproducing adult nematodes, and contribute as pathogens to the genesis of red-ring disease of these plants. Many viruses and mycoplasms also are excellent examples of phytopathogenic agents that are both ecto- and endosymbiotes of phytopathogenic insects (e.g., several species of leafhoppers). Thus, it seems clear that phytopathogenic agents show the complete gamut of symbiotic interrelations with phytopathogenic insects. These agents range from strict ectosymbiotes through combined ecto- and endosymbiotes to strict endosymbiotes. The claim that a strict endo-symbiotic microbe in a phytopathogenic insect qualifies as a plant pathogen may seem unwarranted. However, the view is expressed mainly to remind us of the vastly important roles that endosymbiotic microbes play in enabling phytopathogenic insects to feed, grow, metamorphose, and reproduce. Because the majority of insects that attack plants would not be reproducing organisms without their microbial endosymbiotes, such symbiotes obviously contribute to pathogenesis in such plants.

## C. Supraspecies as Phytopathogenic Entities

Modern biology increasingly recognizes "life" as a continuum whose interdependent components include described species; however, the self-sufficiency of many described species is still vastly overrated. Our brief consideration of insects as components of phytopathogenic symbiotic complexes surely illustrates this point. Future understanding of pathogenesis in plants should benefit greatly from further research consideration of complexes of synergistic species as causal entities. The term *supraspecies* (i.e., above species) is proposed as an appropriate designation for such multispecies entities. Our current and future technological capabilities in research do, or will, allow effective analyses of the etiologies of diseases involving such supraspecies entities.

## IV. SUMMARY

This chapter represents a brief analysis of selected insects as inducers of plant disease based on the literature and my research in the field over a 25-year period. It has sought to demonstrate how insects contribute to long-term deleterious effects (i.e., disease) in plants. Useful extensions and expansions of some of the pioneering views and research, especially of Leach (1940) and Carter (1973), also have been attempted. It is hoped that the facts and judgments presented here will stimulate thought and research in this fascinating, but underinvestigated, realm of biology.

The view that this topic addresses the most common major "void" in contemporary considerations of plant disease situations has been presented. Many erroneous conclusions have been made in entomology and plant pathology because of faulty experimental designs for studies in this area. Entomologists and plant pathologists must collaborate if more meaningful data are to be collected in the future.

Some persons may be significantly disturbed by the semantic aspects of considering insects as pathogens of plants. They are urged to think more of the real dynamic nature of biology and to consider the evolution of thinking that has occurred with regard to plant-infesting nematodes.

The term "phytoallactin" has been proposed to encompass all insect-derived compounds that induce disease in plants. Thus, phytotoxins and phytohormones are types of phytoallactins. Theories on the molecular bases of insect-induced phytotoxemias have been discussed. Some aspects of each major theory seem likely to become fact as our knowledge expands.

Growth and differentiation disorders in plants induced by insect feeding or ovipositional activities have been examined. Insect-altered quantities of phytohormones have been considered as causes of both galls and tumors in plants. Indoleacetic acid probably is a major insect-derived growth regulatory chemical involved in the induction of growth and differentiation disorders in plants. The subject of cecidogenesis (gall formation) as induced by insect secretions is essentially "virgin" territory. Combinations of *in vitro* plant cell and tissue culture with aseptic rearing of cecidogenic insects on such tissues should provide promising technological means for definitive analyses of the insect-derived chemical inducers of galls. The inherently programmed behavior of the inducing insect is fundamentally involved in the overall process of cecidogenesis.

The concept of many phytopathogenic insects being the "host" component in numerous phytopathogenic symbiotic complexes of species, especially involving microbes, has been discussed. This discussion led to the suggestion that the future frontiers in phytopathological research will largely involve the consideration of complexes of species as causal entities. The term "supraspecies" has been proposed as an appropriate designation for such multispecies complexes.

In conclusion, this chapter seeks to demonstrate that entomologists and plant pathologists researching with plants will greatly increase the validity of their experimental data by cooperating with each other. If such scientists wish to make significant progress on lingering major problems in crop protection, they have no choice but to proceed as interdisciplinary teams. Let us put the skeletons of our classical disciplines into the closet and break scientifically into the bright lights of greater understanding that is inevitable if we truly work together symbiotically.

### References

Anders, F. (1961). Untersuchungen Uber das cecidogene Prinzip der Reblaus (*Viteus vitifolii* Shimer) III. Biochemische Untersuchungen über das galleninduzierende Agenus. *Biol. Zentralbl.* **80**, 199–233.

Baker, K. F., Snyder, W. C., and Holland, A. H. (1946). *Lygus* bug injury of lima bean in California. *Phytopathology* **36**, 493–503.

Beardsley, J. (1959). On the taxonomy of pineapple mealybugs of Hawaii, with a description of a previously unnamed species (Homoptera: Pseudococcidae). *Proc. Hawaii. Entomol. Soc.* **17**, 29–37.

Brown, E. S. (1958). Injury to cacao by *Amblypelta* Stål (Hemiptera, Coreidae) with a summary of food-plants of species of this genus. *Bull. Entomol. Res.* **49**, 543–554.

Carle, P., Carde, J. P., and Bonlay, C. (1970). Feeding behavior of *Matsucoccus feytaudi* Duc. Histological and histochemical characterization of the tissue disorganization caused in the host plant (*Pinus pinaster* Ait. var *mesogeenis*). *Ann. Sci. For.* **27**, 89.

Carpenter, I. P. (1928). Study of the life history and spotting habits of *Eutettix chenopodii* (Homoptera, Cicadellidae). *Univ. Kans. Sci. Bull.* **18**, 457–483.

Carter, W. (1933). The spotting of pineapple leaves caused by *Pseudococcus brevipes*, the pineapple mealybug. *Phytopathology* **23**, 243–259.

Carter, W. (1939). Injuries to plants caused by insect toxins. *Bot. Rev.* **5**, 273–326.

Carter, W. (1942). The geographical distribution of mealybug wilt with notes on some other insect pests of pineapple. *J. Econ. Entomol.* **35**, 10–15.

Carter, W. (1949). Insect notes from South America with special reference to *Pseudococcus brevipes* and mealybug wilt. *J. Econ. Entomol.* **42**, 761–766.

Carter, W. (1973). "Insects in Relation to Plant Disease." Wiley, New York.

Deemsteegt, V. D., Webger, A. J., and Graham, C. L. (1969). A chlorotic leaf spot of rice; insect induction. *Phytopathology* **59**, 1556.

Fenton, F. A. (1925). Notes on the biology of the leafhopper *Eutettix strobi* Fitch. *Proc. Iowa Acad. Sci.* **31**, 437–440.

Flock, R. A., and Wallace, J. M. (1955). Transmission of fig mosaic by the eriophyid mite *Aceria ficus*. *Phytopathology* **45**, 52–54.

Gibbs, D. G., and Leston, D. (1970). Insect phenology in a forest cocoa farm locality in West Africa. *J. Appl. Ecol.* **7**, 519–548.

Gilmer, R. M., and McEwen, F. L. (1958). Chlorotic fleck, an eriophyid mite injury of Myrobalan plum. *J. Econ. Entomol.* **51**, 335–337.

Goodchild, A. J. P. (1952). Digestive system of the West African cacao capsid bugs (Hemiptera: Miridae). *Proc. Zool. Soc. London* **122**, 543–572.

Heitefuss, R., and Williams, P. H., eds. (1976). "Physiological Plant Pathology." Springer-Verlag, Berlin and New York.

Hughes, J. H. (1943). The alfalfa plant bug *Adelphocoris lineolatus* Goeze and other Miridae (Hemiptera) in relation to alfalfa-seed production in Minnesota. *Minn. Univ. Agric. Exp. Stn., Tech. Bull.* No. 161.

Jensen, D. D. (1954). The effect of aphid toxins on *Cymbidium* orchid flowers. *Phytopathology* **44**, 493–494.

Kloft, W. (1960). Wechselwirkungen zwischen pflanzensaugenden Insekten und den von ihnen besogenen Pflanzengeweben. *Z. Angew. Entomol.* **45**, 337–381; **46**, 42–70.

Knorr, L. C., and Denmark, H. A. (1970). Injury to citrus by the mite *Brevipalpus phoenicis. J. Econ. Entomol.* **63**, 1966.

Komarek, J. (1946). The physiological damage upon the ash tree made by the scale insect *Lecanium coryli* L. *Vestn. Cesk. Spol. Zool.* **10**, 156–165.

Leach, J. G. (1940). "Insect Transmission of Plant Diseases." McGraw-Hill, New York.

Leach, R., and Smee, C. (1933). Gnarled stem canker of tea caused by the capsid bug (*Helopeltis bergrothi* Reut.). *Ann. Appl. Biol.* **20**, 691–706.

Maramorosch, K. (1977). "The Atlas of Insect and Plant Viruses." Academic Press, New York.

Miles, P. W. (1968). Insect secretions in plants. *Annu. Rev. Phytopathol.* **6**, 137–164.

Miles, P. W. (1972). The saliva of Hemiptera. *Adv. Insect Physiol.* **9**, 183–255.

Norris, D. M. (1975). Chemical interdependencies among *Xyleborus* spp. ambrosia beetles and their symbiotic microbes. *Mater. Org., Beih.* **3**, 479–488.

Schäller, G. (1968). Biochemische Analyse des Aphidenspeichels und seine Bedeutung für die Gallenbildung. *Zool. Jahrb. Physiol.* **74**, 54–87.

Schmitz, G. (1958). *Helopeltis* du contonnier en Africa centrale. *Publ. Inst. Natl. etude Agron. Congo, Ser. Sci.* No. 71.

Smith, F. F., and Poos, F. W. (1931). The feeding habits of some leafhoppers of the genus *Empoasca. J. Agric. Res.* **43**, 267–286.

Stace-Smith, R. (1954). Chlorotic spotting of black raspberry induced by the feeding of *Amphorophora rubitoxica* Knowlton. *Can. Entomol.* **86**, 232–235.

Strobel, G. A. (1974). Phytotoxins produced by plant parasites. *Annu. Rev. Plant Physiol.* **25**, 541–566.

Strong, F. E. (1970). Physiology of injury caused by *Lygus hesperus. J. Econ. Entomol.* **63**, 808–814.

Vindane, C. (1967). Internal and external symptomology from plant sucking insects on vitis (Trans.). *Ann. Accad. Agric. Torino* **109**, 117–136.

Wadley, F. M. (1929). Observations on the injury caused by *Toxoptera graminum* Rond. (Homoptera: Aphididae). *Proc. Entomol. Soc. Wash.* **31**, 130–134.

Way, M. J. (1953). Studies on *Theraptus sp.* (Coreidae); the cause of the gumming disease of coconuts in East Africa. *Bull. Entomol. Res.* **44**, 657–667.

Wilks, J. M. (1964). The spined stink bug; a cause of cottony spot in pear in British Columbia. *Can. Entomol.* **96**, 1198–1201.

Wilson, N. S., and Cochran, L. C. (1952). Yellow spot, and eriophyid mite injury on peach. *Phytopathology* **42**, 443–447.

# Chapter 14

# How Viruses and Viroids Induce Disease

MILTON ZAITLIN

*It is to be hoped that symptomatology will develop beyond the practice of simply recording the visible effects of infection; the subject should be more than descriptive and it is high time that attempts were made to get explanations in biochemical terms for the many and varied changes viruses cause in plants.*

Bawden (1964)

## I. INTRODUCTION

In reality, this chapter would be better entitled, "How *do* viruses and viroids induce disease"? because unfortunately we have not advanced significantly in our understanding of the basis for plant viral—and now viroid—symptomatology since Sir F. C. Bawden's admonition. With the information currently available we cannot give a definitive exposition of the underlying biochemical determinants of disease. We can describe symptoms, both external and internal, we can make assessments of bio-

257

chemical changes accompanying disease, but we really do not know how symptoms (i.e., disease) arise.

Plant virologists are not alone in their ignorance of the underlying basis for disease. In a recent text, our animal virus colleagues concluded that "despite the impressive expansion in our knowledge of the molecular biology of viral multiplication over the last decade, we still are not at all certain about how viruses kill cells—or indeed, at another level, how they cause disease" (Fenner *et al.*, 1974).

I do not want to give the impression, however, that this subject is not worthy of consideration at this time. Some recent findings in molecular biology and virology have given us some clues that now allow us to speculate with more clarity as to what virus- or viroid-engendered constituent(s) might interact with the host to induce disease and possibly which cellular processes are affected first.

## II. DEFINITION OF DISEASE

For the purposes of our consideration, it is important to settle on a definition of disease. In Chapter 3, Volume III, Bateman (1978) examined the question in detail, citing the opinions of a number of plant pathologists. Most of their definitions give considerable weight to a disturbed cellular activity. I prefer, as far as virus diseases are concerned, to agree with the concept of Horsfall and Cowling (1977) that "disease is a malfunctioning process." Cellular malfunction is a very important consideration for plant virology (and "viroidology," if I may coin the term) because there are a number of examples in which these agents replicate within plants but induce no overt symptoms or obvious diminution of plant growth or yield. One such example is the case of infection by potato virus S. This "disease" was first detected serologically, thus demonstrating only that the virus had replicated in certain potato clones (Rozendaal and van Slogteren, 1958). Are these plants diseased? I would argue that they are not, and thus I am equating disease with symptomatology. There is a fine line here because, given more subtle or sensitive measuring capabilities, one might be able to detect a "malfunction" that had escaped detection earlier. It is certainly possible that plants with infections that at first glance appear to be completely benign may be shown, on closer examination, to be less vigorous than noninfected plants.

Plants may also harbor and replicate "foreign" RNAs without apparent harm to themselves. For example, with the technique of polyacrylamide gel electrophoresis, double-stranded RNAs may be detected in diseased fungi, but some otherwise healthy fungi may also have them (Sanderlin and Ghabrial, 1978). I would guess that a similar phenomenon will be observed in higher plants. Moreover, with the electrophoretic technique, "new" viroids have recently been found in several species of apparently healthy plants (Owens *et al.,* 1978). Clearly, then, cells can support "foreign" nucleic acids and can allow for their replication. Are such organisms diseased? Once again, I submit that they are not; that is not to say, however, that foreign nucleic acids that are harmless in one organism when introduced into another species might not induce disease. Indeed, they often do.

## III. FACTORS INFLUENCING SYMPTOM EXPRESSION

It is not my intention to describe the range of symptoms induced in plants by viruses. They are manifold and have been well catalogued by Holmes (1964), who suggested that we see such a diversity because "viruses and their innumerable strains appear able to upset in almost every conceivable way the processes by which plants grow and maintain themselves." Symptoms may reflect a direct response to virus replication in the infected tissue, or they may be a secondary response of a tissue to a perturbation elsewhere. For example, wilting and death of leaves can result from vascular plugging in the stem.

Symptom expression is a dynamic phenomenon. Symptoms induced by a given virus may differ in different plant species and/or varieties, and within any one host, symptoms may be influenced by the duration of the infection and by a range of environmental factors, such as temperature, light quantity and quality, and nutrition. With a *given* virus in a *given* host, environmental factors that diminish or enhance symptoms generally affect virus replication in the same positive or negative manner.

The developmental status of a cell at the time of infection normally has a profound influence on its symptoms. Infections initiated on a mature leaf frequently engender few symptoms. On the other hand, when the same virus enters the dividing, developing cells in the apex, it may induce drastic aberrations in chloroplasts, resulting in severe disease (Matthews, 1970).

It is interesting that plant viruses induce no visible symptoms in mesophyll protoplasts infected *in vitro.* This may reflect, in part, a requirement for the cells to be dividing or developing at the time of infection in order to show chloroplast abnormalities. One striking case is the hypersensitive local-lesion response of certain virus–host combinations. In leaf tissues, several cultivars of tobacco develop necrotic local lesions in response to tobacco mosaic virus (TMV) infection, but mesophyll cell protoplasts isolated from these tissues show no symptoms and support virus replication very well (Otsuki et al., 1972). One possible explanation is that the interaction of a number of closely associated cells is required for hypersensitivity since only tissue culture fragments above a certain size exhibit this response upon infection by TMV (Beachy, personal communication).

The presence of a second virus can also influence symptomatology in both a negative and a positive way (Bennett, 1953; Ross, 1974). For example, the presence of one virus or strain may inhibit the expression of symptoms by another; this phenomenon is known as cross-protection and has recently been shown for viroids as well (Niblett et al., 1978). On the positive side, viruses often cause synergistic effects; two viruses replicating in one host may greatly enhance total symptom expression, the sum being greater than that induced by the two individually. The timing of the infection by each of the viruses influences the degree of their interaction (Ross, 1974).

## IV. EARLIER SPECULATIONS ON DISEASE INDUCTION

The authors of the principal texts on plant virology (Bawden, 1964; Matthews, 1970; Gibbs and Harrison, 1976) either do not speculate on mechanisms by which viruses induce disease, or they conclude that we know too little to speculate with much authority. There is a consensus, however, that something beyond the stresses of virus replication per se is involved. Matthews (1970) described in detail much of the research that documents virus-induced biochemical aberrations but concluded that "these findings are probably secondary consequences of virus infection, not essential for virus replication."

The first comprehensive consideration of this subject was that of Diener (1963) in a review article on the physiology of virus-infected plants. Although he did not address the mechanism of induction of disease directly, he did emphasize how some of the immediate effects of virus disease might be brought about by growth regulatory imbalances and by amino acid accumulation in infected tissue. In the latter case, he suggested that the virus-induced accumulation of free amino acids in leaves may be responsible for certain symptoms, although this postulate has been challenged (reviewed in Goodman et al., 1967, p. 165). Diener's conclusion follows from the observation that some viral symptoms resemble the physiological disorder "frenching," which can be induced by adding certain amino acids to tobacco plants in axenic culture.

Hirai and Wildman (1967) noted the remarkable similarity between the symptoms produced in tobacco plants by actinomycin D and by TMV. Actinomycin D, which is known to inhibit DNA-dependent RNA synthesis, induces mosaic patterns and vein-clearing symptoms in leaf tissues to which the drug is translocated. These authors suggested that this similarity is a reflection of the fact that TMV symptoms arise because of an interference with transcription of host DNA to messenger RNA and that the difference in symptoms seen among strains is a reflection of the degree of this interference.

Speculations concerning the cause of symptoms associated with animal virus infections have taken a different direction from that of plant viruses because most, but certainly not all, animal viral infections result in a "cytopathic effect"; i.e., the cell is killed. During the senescence that precedes death, there is a generalized, nonspecific shutdown of host protein and nucleic acid synthesis, making it easy to visualize why cells die. Obviously, the cytopathic effect is not the only direct cause of disease, because, if it were, all viruses that induce cell death in a given tissue would show the same symptoms, which they do not. A central role in the cytopathic effect is seen for double-stranded RNA, a very potent inhibitor of *in vitro* protein synthesis and a stimulator of the synthesis of the interferon involved in the host defense against virus-induced disease (Carter and DeClercq, 1974; Carrasco, 1977). Shutdown phenomena are not characteristic of virus-induced diseases in plants. As a rule, infected plant cells do not die, and there is no consistent trend toward viral inhibition of protein synthesis in the host. On the contrary, in a recent study with plant protoplasts, TMV-directed protein synthesis was shown to take place in addition to apparently unaffected synthesis of host proteins (Siegel *et al.*, 1978). There is some evidence, however, for viral interference with host RNA synthesis, particularly chloroplastic rRNA (Hirai and Wildman, 1969; Pring, 1971). A study involving two TMV strains that elicit markedly different symptoms indicated that breakdown of chloroplast rRNA was greatest in the plants infected by the strain that resulted in the most chlorotic symptoms (Fraser, 1969).

Cell death does occur in some plant virus–host combinations, resulting in necrotic local lesions, but these are specialized situations and are not very representative of the general pattern of plant virus-induced disease. It is discouraging that, although the local lesion is the most extensively studied single phenomenon in plant virology, we still do not understand how it is initiated by the infecting virus!

Kado and Knight (1966) localized the capacity of TMV RNA to induce local lesions on hypersensitive hosts to a region of the RNA about 25% of the distance from its 3' end. [In their paper they concluded that it was 25% from the 5' end, but recent technical reassessments have shown that the RNA ends were misidentified (Wilson et al., 1976)]. Kado and Knight termed RNA in this region the "local-lesion gene," although they recognized that it possibly overlapped or was concomitant with the coat protein gene. It is evident from their work that nucleotide sequences in this region of the RNA are responsible for the interaction with some component(s) of the host, resulting in hypersensitivity, but I find it difficult to accept the proposition that there is a specific local-lesion "gene" (and presumably a gene product) responsible for this response, just as there is also probably no "mosaic" gene or "enation" gene, etc. TMV RNA is large enough to contain only a relatively small number of genes, all of which are probably involved in potentiating viral replication. Those genes, or portions of them, or their translation products might also function in disease, but only in an indirect or secondary manner.

## V. PRODUCTS OF VIRUS AND VIROID INFECTION THAT MAY INDUCE DISEASE

Our knowledge of virus and viroid composition and replication has advanced to the point where we are able to identify the principal constituents and direct products of replication, some of which must be responsible for the initial event(s) that result in disease. I am stressing the necessity to identify such an initial interaction because it is crucial in engendering disease. Just as we can describe macroscopic symptoms, we can describe disease at the cellular and subcellular levels by making measurements of the altered quantities or activities of host constituents or enzymes. In most instances, we are undoubtedly describing the consequences of disease and not its causes. Obviously, when comparisons are made between a diseased plant and its unaffected counterpart, we will observe differences; the plant is sick! Early literature on plant virology records many such measurements (reviewed in Goodman et al., 1967; Matthews, 1970), but studies of this kind are not as popular now, which most likely reflects the realization that such measurements probably will not lead us to the cause of disease. It is not possible to reject the possibility, however, that some of these biochemical changes represent the next step after the initial event and result in aberrations that ultimately cascade into the disease syndrome.

However, accepting the conclusion that a virus or viroid constituent or product induces the initial event resulting in disease, let us examine the potential candidates.

## A. Viruses

The direct products of virus replication include the following: (1) the virus particles (virions) themselves, (2) the virus coat protein(s), (3) other proteins, which are direct viral messenger RNA translational products, (4) the viral RNA (or DNA in those few viruses that contain it), (5) the complementary strand of the nucleic acid, and (6) double-stranded forms consisting of the viral RNA and its complement. Depending on the virus, the viral RNAs may be either single- or double-stranded. In many cases the total viral genome is divided among several RNA molecules, which may be encapsidated in separate nucleoprotein particles. Furthermore, in infected host cells, small RNAs may be found which are specific pieces of the full-sized "parental" viral RNAs; some of these RNAs serve as monocistronic messenger RNAs for the viral-coded polypeptides (Beachy and Zaitlin, 1977). As would be expected, DNA plant viruses also generate RNA molecules in the infected cell (Howell and Hull, 1978).

In many cases, virions also contain trace amounts of metals; certain ones may contain lipids, carbohydrates, polyamines, transfer RNAs, and even host proteins.

## B. Viroids

In contrast to viruses, viroids appear to have only three possible products that could be the initiators of disease, namely, (1) the small viroid RNA itself, which may be either linear or circular, (2) its recently discovered complementary strand (Grill and Semancik, 1978), and possibly (3) a double-stranded form consisting of the viroid hydrogen-bonded to its complement. Comprehensive investigations have not uncovered any coat protein or translational product of viroid RNAs; they apparently do not contain genetic information that codes for protein (reviewed in Diener and Hadidi, 1977).

## VI. WHICH PRODUCTS ARE INVOLVED?

Specific information as to the possible involvement of all the virus or viroid products in disease is not available, but at least three groups can be eliminated as having a primary role in disease on largely circumstantial grounds.

## A. Virus Replication

It is generally conceded that the symptoms induced by a virus are not due to the direct effect of the stress put on the host by being forced to synthesize extra protein and nucleic acid, although it is possible that in specialized cases such stress might affect a plant adversely. There is, however, certainly no general correlation between the amount of virus nucleoprotein synthesized and symptom intensity. For example, both TMV and barley yellow dwarf virus produce severe diseases, but yields of extractable virus are typically 5–10 mg and 50–150 ng per gram of tissue, respectively. This represents about a 100,000-fold difference between the two! I wish to emphasize, however, that for a specific virus–host interaction, the amount of virus replication may affect symptom intensity. Manipulations of the environment that affect symptom development also may bring about a concomitant response in the amount of virus produced. This is certainly not unexpected because, if disease is brought about by some viral or viroid product, more of that product is likely to result in more disease.

A further suggestion that symptom expression is separable from virus replication may be gleaned from recent studies of Tomlinson *et al.* (1976), in which carbendazim, a breakdown product of the fungicide Benomyl, when applied to the roots of plants inhibited symptom production by two viruses. Replication was not affected for the one virus (TMV) in which it was studied. Infectivity measurements showed that the chemical had no effect on the amount of TMV per unit of tissue, although more recently Fraser and Whenham (1978), using a very different methodology, found a marked suppressive effect of carbendazim on TMV RNA synthesis, clearly in conflict with the finding of Tomlinson *et al.* (1976).

## B. Viral Coat Protein

In all probability, coat protein is unlikely to be a primary determinant of symptom production. The most salient examples come from studies with strains and mutants of TMV. For instance, the Holmes masked (symptomless) strain and the common strain of the virus have identical coat proteins but produce very different symptoms. Other examples may be seen with chemically induced TMV mutants that have very minor differences in their coat protein sequences but nevertheless induce very different symptoms (Hennig and Wittmann, 1972).

Another example of the probable minimal role of coat protein in disease can be seen in studies with multicomponent viruses. In these viruses, the total viral genome is divided among several RNA molecules, and in some cases it is possible to make "pseudorecombinant" virus infections using combinations of RNAs taken from different viruses. For instance, infection by a pseudorecombinant obtained by mixing two large RNAs from tomato aspermy virus with a small, coat-protein-encoding RNA from cucumber mosaic virus results in the replication of virions encapsidated with cucumber mosaic virus coat protein. The symptoms, however, are those of tomato aspermy virus (Habili and Francki, 1974). Thus, in this system, neither coat protein nor the RNA that encodes it play a role in symptomatology.

Coat protein is very significant to symptomatology in one restricted situation, however, and that is when no functional coat is produced. Defective mutants of several plant viruses, in which the coat protein is nonfunctional and is unable to encapsidate the RNA, all bring about characteristic symptoms that are different from those of wild-type virus. Because the viral RNA is unprotected, it spreads only from cell to cell and does not move rapidly throughout the plant, as do functional virions. Thus, no true systemic symptoms develop, and somewhat unique primary symptoms are observed (Siegel and Zaitlin, 1965). Surprisingly, the specific type of defect in the coat protein does influence the color and severity of the primary symptoms on the inoculated leaves. Defective coat proteins that have lost the capacity for reversible aggregation cause more severe symptoms than defective proteins that have retained that property (Jockusch and Jockusch, 1968).

## C. Other Viral Constituents

The metals, lipids, carbohydrates, polyamines, and transfer RNAs found in some virions are all of host origin. Thus, it is difficult to visualize a role for them as primary determinants of disease. Furthermore, except for metals, which are probably constituents of most viruses, these substances are found only in certain viruses, which precludes them from having a general role.

## VII. THE LESSONS FROM VIROIDS AND SATELLITE RNAs

As discussed earlier, viroid infection is initiated by a very small RNA molecule. Thus, disease must be initiated either by this molecule, its RNA complement, or possibly a double-stranded molecule formed by the hydrogen-bonded pairing of the viroid to its complementary RNA. Nevertheless, whichever of these molecules is involved, this represents the simplest disease-inducing system known, and the interaction of this molecule with its host must initiate disease!

Some recent studies by Elizabeth Dickson and her colleagues (1979) show how subtle disease initiation can be. It has been known for a number of years that the viroid that induces the spindle tuber disease of potato has a number of strains that induce symptoms ranging from very mild to very severe. Furthermore, prior infection of a plant with a mild strain can protect a plant from disease produced by subsequent infection by a severe strain (Fig. 1), although both viroids probably replicate in the host tissue (Niblett et al., 1978). Dickson et al. employed the technique of RNA fingerprinting to characterize the sequence and complexity of the viroid RNA molecules. They found that of the 359 nucleotides comprising potato spindle tuber viroid (Gross et al., 1978), nucleotide substitutions involving a very small number of them (perhaps even one!) constituted the only difference between mild and severe strain RNAs (Fig. 2), but these few changes had a profound effect on symptomatology (Fig. 1). This is a meaningful lesson, which probably can be extrapolated to help us decide how viral symptoms are induced;

Fig. 1. Symptoms produced on tomato by mild (PM) and severe (PS) strains of potato spindle tuber viroid and the disease protection afforded by sequential inoculation with PM followed by PS. The inoculation procedures are as follows: (A) PS only; (B) buffer and 14 days later with PS; (C) PM only; (D) PM and 14 days later with PS; no symptoms of PS were evident; and (E) buffer only. Plants were photographed 52 days after the first inoculation. (Courtesy of C. L. Niblett.)

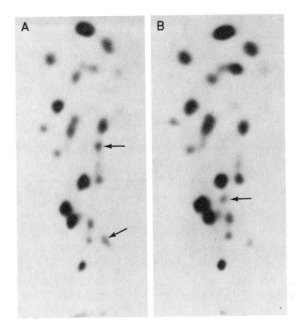

**Fig. 2.** RNA fingerprints of two strains of potato spindle tuber viroid: (A) Strain PS, causing severe disease on tomato (Fig. 1A); and (B) strain PM, causing mild disease on tomato (Fig. 1C). Purified viroids were labeled *in vitro* with [125]I and digested into fragments of various sizes by ribonuclease $T_1$, which cleaves RNA between all guanilic acid residues and adjacent nucleotides. The fragments were subjected to electrophoresis (the horizontal axis of the fingerprint), which sorts them on the basis of their charge, and then by homochromatography in the vertical direction, which separates them according to size. The larger (and more informative) fragments are found in the lower part of the fingerprint. The autoradiograms of the fingerprints of the two strains are very similar, showing that their nucleotide sequences are similar, but they do show small differences. The arrows point to the spots on the fingerprints that were consistently different in comparisons of several isolates of mild and severe strains. These experiments are detailed in Dickson *et al.* (1979). (Courtesy of Elizabeth Dickson.)

namely, *symptoms can be altered drastically as a result of a small number of changes in the sequence of an RNA molecule.*

Another striking example of the disease-modulating potential of small RNA molecules comes from the work of J. M. Kaper and his colleagues involving a small (100,000 molecular weight) satellite RNA, which sometimes replicates in tissues infected with cucumber mosaic virus (CMV). This RNA, which they have termed CARNA 5, is completely dependent on CMV for its replication and encapsidation. What is interesting in the context of our consideration here is the relation of CARNA

5 to disease. CMV infections in tomato and tobacco cause recognizable disease but, when CARNA 5 is also replicating, the disease in tomato becomes more severe, whereas in tobacco it becomes milder (Kaper and Waterworth, 1977)! Clearly, this is a case in which a small RNA can modulate disease, although it apparently has no role in viral replication per se.

## VIII. CURRENT SPECULATIONS ON DISEASE INDUCTION

On the basis of the arguments presented in Section VII, it is likely that some form of viral or viroid RNA is responsible for the interaction with the host to signal disease. RNA molecules possess the specificity required to guarantee a specific interaction with host DNA; as few as 10 to 20 hydrogen-bonded base pairs yield a stable RNA–DNA duplex. RNA molecules have been considered as potential regulators of biological processes (Britten and Davidson, 1969; Dickson and Robertson, 1976). Reanney (1975) has proposed that there is a general regulatory role for viral RNA in eukaryotes. Such a role has also been predicted for plant viral RNAs (Hirai and Wildman, 1967; Ross, 1974) and for viroids (Diener, 1971; Semancik and Weathers, 1972).

Both Reanney and others (Dickson and Robertson, 1976; Dickson, 1979) provide examples of how, in biological systems, small RNA molecules can be generated in host cells by selective transcription, mRNA processing, etc., in order to be available to act as regulators. Reanney argues that "small sections (often less than 5%) of eukaryote viral RNAs constitute aberrant examples of regulatory signals which occur in normal systems." RNA regulatory molecules should, he predicts, be highly structured, possessing a high degree of self-complementarity in order to be safe from nuclease degradation before they have an opportunity to pair with their target DNA sequences. Viroids, which are highly base paired, and segments of viral RNAs have these properties, as do the double-stranded RNAs generated during replication.

## IX. CONCLUSIONS

Little is known at present about how viruses and viroids induce disease. Our profound ignorance of these fundamental aspects of disease processes stands in distinct contrast to our rapidly growing understanding of the mechanism by which viruses and viroids replicate in their hosts. Both of these subjects are important to plant pathology. However, progress is being made only in one of them. Why is this so? In my opinion, it is mainly because we have had sound conceptual theories and reliable methods with which to make progress in studying virus replication but not disease induction. It has not been possible to identify the site(s) in the cell that viruses and viroids perturb first or to characterize the nature of the perturbation.

In this brief chapter I have given some evidence that points to virus- or viroid-engendered RNA molecules as the agents that interact with host DNA. How they interact and what functions they perturb are matters for speculation; unfortunately, we have little hard evidence to guide us at this time. Dickson (1979) has suggested that they could serve as primers for aberrant DNA or RNA synthesis. I believe that technology becoming available will enable us to investigate the nature of the virus–host interaction and then to examine how specific cellular or organelle functions are affected. The primary candidate for examination, in my opinion, is the chloroplast, because it is usually subject to considerable insult during both virus- and viroid-induced disease.

The unraveling of the mysteries of disease induction will provide plant scientists with new vistas for research. It should be an exciting experience!

## References

Bateman, D. F. (1978). The dynamic nature of disease. In "Plant Disease: An Advanced Treatise" (J. G. Horsfall and E. B. Cowling, eds.), Vol. 3, Ch. 3. Academic Press, New York.

Bawden, F. C. (1964). "Plant Viruses and Virus Diseases," 4th Ed., p. 26. Ronald Press, New York.

Beachy, R. N., and Zaitlin, M. (1977). Characterization and in vitro translation of the RNAs from less-than-full-length, virus-related, nucleoprotein rods present in tobacco mosaic virus preparations. Virology 81, 160–169.

Bennett, C. W. (1953). Interactions between viruses and virus strains. Adv. Virus Res. 1, 40–67.

Britten, R. J., and Davidson, E. H. (1969). Gene regulation for higher cells: A theory. Science 165, 349–357.

Carrasco, L. (1977). The inhibition of cell function after viral infection. A proposed general mechanism. FEBS Lett. 76, 11–15.

Carter, W. A., and DeClercq, E. (1974). Viral infection and host defense. Science 186, 1172–1178.

Dickson, E. (1979). Viroids: Infectious RNA in plants. In "Nucleic Acids in Plants" (T. C. Hall and J. W. Davies, eds.). Chem. Rubber Press, Gainesville, Florida (in press).

Dickson, E., and Robertson, H. D. (1976). Potential regulatory roles for RNA in cellular development. Cancer Res. 36, 3387–3393.

Dickson, E., Robertson, H. D., Niblett, C. L., Horst, R. K., and Zaitlin, M. (1979). Minor differences between nucleotide sequences of mild and severe strains of potato spindle tuber viroid. Nature (London) 277, 60–62.

Diener, T. O. (1963). Physiology of virus-infected plants. *Annu. Rev. Phytopathol.* 1, 197–218.

Diener, T. O. (1971). Potato spindle tuber "virus" IV. A replicating, low molecular weight RNA. *Virology* 45, 411–428.

Diener, T. O., and Hadidi, A. (1977). Viroids. *Compr. Virol.* 11, 285–337.

Fenner, F., McAuslan, B. R., Mims, C. A., Sambrook, J., and White, D. O. (1974). "The Biology of Animal Viruses," 2nd Ed., p. 339. Academic Press, New York.

Fraser, R. S. S. (1969). Effects of two TMV strains on the synthesis and stability of chloroplast ribosomal RNA in tobacco leaves. *Mol. Gen. Genet.* 106, 73–79.

Fraser, R. S. S., and Whenham, R. J. (1978). Inhibition of the multiplication of tobacco mosaic virus by methyl benzimidazol-2-yl carbamate. *J. Gen. Virol.* 39, 191–194.

Gibbs, A., and Harrison, B. (1976). "Plant Virology The Principles." Arnold, London.

Goodman, R. N., Király, Z., and Zaitlin, M. (1967). "The Biochemistry and Physiology of Infectious Plant Disease." Van Nostrand, Princeton, New Jersey.

Grill, L. K., and Semancik, J. S. (1978). RNA sequences complementary to citrus exocortis viroid in nucleic acid preparations from infected *Gynura aurantiaca. Proc. Natl. Acad. Sci. U.S.A.* 75, 896–900.

Gross, H. J., Domdey, H., Lossow, C., Jank, P., Raba, M., Alberty, H., and Sänger, H. L. (1978). Nucleotide sequence and secondary structure of potato spindle tuber viroid. *Nature (London)* 273, 203–208.

Habili, N., and Francki, R. I. B. (1974). Comparative studies on tomato aspermy and cucumber mosaic viruses. III. Further studies on relationship and construction of a virus from parts of the two viral genomes. *Virology* 61, 443–449.

Hennig, B., and Wittmann, H. G. (1972). Tobacco mosaic virus: Mutants and strains. *In* "Principles and Techniques in Plant Virology" (C. I. Kado and H. O. Agrawal, eds.), pp. 546–594. Van Nostrand-Reinhold, New York.

Hirai, A., and Wildman, S. G. (1967). Similarity in symptoms produced in tobacco plants by actinomycin D and TMV. *Virology* 31, 721–722.

Hirai, A., and Wildman, S. G. (1969). Effect of TMV multiplication on RNA and protein synthesis in tobacco chloroplasts. *Virology* 38, 73–82.

Holmes, F. O. (1964). Symptomatology of viral diseases in plants. *In* "Plant Virology" (M. K. Corbett and H. D. Sisler eds.), pp. 17–38. Univ. of Florida Press, Gainesville.

Horsfall, J. G., and Cowling, E. B. (1977). Prologue: How disease is managed. *In* "Plant Disease: An Advanced Treatise" (J. G. Horsfall and E. B. Cowling, eds.), Vol. I, Ch. 1. Academic Press, New York.

Howell, S. H., and Hull, R. (1978). Replication of cauliflower mosaic virus and transcription of its genome in turnip leaf protoplasts. *Virology* 86, 468–481.

Jockusch, H., and Jockusch, B. (1968). Early cell death caused by TMV mutants with defective coat proteins. *Mol. Gen. Genet.* 102, 204–209.

Kado, C. I., and Knight, C. A. (1966). Location of a local lesion gene in tobacco mosaic virus RNA. *Proc. Natl. Acad. Sci. U.S.A.* 55, 1276–1283.

Kaper, J. M., and Waterworth, H. E. (1977). Cucumber mosaic virus associated RNA 5: Causal agent for tomato necrosis. *Science* 196, 429–431.

Matthews, R. E. F. (1970). "Plant Virology." Academic Press, New York.

Niblett, C. L., Dickson, E., Fernow, K. H., Horst, R. K., and Zaitlin, M. (1978). Cross protection among four viroids. *Virology* 91, 198–203.

Otsuki, Y., Shimomura, T., and Takebe, I. (1972). Tobacco mosaic virus multiplication and expression of the N gene in necrotic responding tobacco varieties. *Virology* **50**, 45–50.

Owens, R. A., Smith, D. R., and Diener, T. O. (1978). Measurement of viroid sequence homology by hybridization with complementary DNA prepared *in vitro*. *Virology* **89**, 388–394.

Pring, D. R. (1971). Viral and host RNA synthesis in BSMV-infected barley. *Virology* **44**, 54–66.

Reanney, D. C. (1975). A regulatory role for viral RNA in eukaryotes. *J. Theor. Biol.* **49**, 461–492.

Ross, A. F. (1974). Interaction of viruses in the host. *Int. Symp. Virus Dis. Ornamental Plants, 3rd; Tech. Commun. Int. Soc. Hortic. Sci.* No. 36, pp. 247–260.

Rozendaal, A., and van Slogteren, D. H. M. (1958). A potato virus identified with potato virus M and its relationship to potato virus S. *Proc. Conf. Potato Virus Dis., 3rd, Lisse-Wageningen, 1957* pp. 30–36.

Sanderlin, R. S., and Ghabrial, S. A. (1978). Physicochemical properties of two distinct types of virus-like particles from *Helminthosporium victoriae*. *Virology* **87**, 142–151.

Semancik, J. S., and Weathers, L. G. (1972). Exocortis disease: Evidence for a new species of "infectious" low molecular weight RNA in plants. *Nature (London), New Biol.* **237**, 242–244.

Siegel, A., and Zaitlin, M. (1965). Defective plant viruses. *Perspect. Virol.* **4**, 113–126.

Siegel, A., Hari, V., and Kolacz, K. (1978). The effect of tobacco mosaic virus infection on host and virus-specific protein synthesis in protoplasts. *Virology* **85**, 494–503.

Tomlinson, J. A., Faithfull, E. M., and Ward, C. M. (1976). Chemical suppression of the symptoms of two virus diseases. *Ann. Appl. Biol.* **84**, 31–41.

Wilson, T. M. A., Perham, R. N., Finch, J. T., and Butler, P. J. G. (1976). Polarity of the RNA in the tobacco mosaic virus particle and the direction of protein stripping in sodium dodecyl sulfate. *FEBS Lett.* **64**, 285–289.

# Chapter 15

# How Air Pollutants Induce Disease

EVA J. PELL

## I. INTRODUCTION

How air pollutants induce disease is a complex issue. Just as there are many types of biotic pathogens with various modes of activity, there are also many types of air pollutants—gas, liquid, and particulate—each with different physicochemical potentials. Rather than examine pollutants individually, I will focus on certain aspects of plant response that might provide some unifying concept applicable to diverse air pollutants. The literature reveals only scattered answers to the basic question of how air pollutants induce disease. At the conclusion of this chapter I will consider why this question has yet to be answered and what should be done to make future efforts more successful.

## II. QUESTIONS BEING POSED BY AIR POLLUTION PHYTOTOXICOLOGISTS

Identification of an air pollution problem on plants generally begins with the observation of a symptom that cannot be attributed to a biotic pathogen. The appearance and distribution of the symptom contribute to initial suspicions concerning its origin. Identification of a toxic sub-

PLANT DISEASE, VOL. IV

stance as the causal agent is often followed by a detailed characterization of these symptoms, often at macroscopic and microscopic levels of description.

Responses of foliage to air pollutants are diverse and may include chlorosis, necrosis, or pigment production. Often one pollutant will cause several symptoms. For example, acute exposure of plants to sulfur dioxide ($SO_2$) will induce a necrotic symptom, and chronic exposures will elicit chlorosis (van Haut and Stratmann, 1970). On some occasions foliage injured by ozone ($O_3$) exhibits a reddish necrotic stipple; on others, it exhibits a whitish fleck (Brandt and Heck, 1967). The reasons for this variation in host response are obscure. Plant organs other than leaves occasionally are affected by air pollutants. For example, hydrogen fluoride (HF) induces soft-suture disease of peach (Benson, 1959), ethylene prevents opening of flowers (Abeles, 1973; Rhoads et al., 1973), and $O_3$ alters pollen germination of petunias (Feder, 1968).

After the causal agent has been identified and the plant response has been characterized, susceptibility and tolerance of different genera, species, and cultivars are determined. The diagnostician can identify a field problem on the basis of suspicious symptoms appearing on a susceptible species. If an air pollutant poses a serious economic threat, losses can be reduced by selection of tolerant cultivars. Some researchers have contrasted physiological responses of genetically different plant material to determine mechanisms of action of air pollutants, but the potential of this tool has not been exploited fully.

In common with pathologists studying biotic pathogens, air pollution researchers have spent many years documenting and defining the environmental variables that influence disease development. Light, temperature, humidity, soil moisture and quality, and nutrition are only a few of the identified variables that alter plant response to air pollutants (Heck, 1968; Dunning and Heck, 1977; Leone and Brennan, 1972). The results of these studies have helped to explain why symptoms occur on vegetation at some times and not at others. Similarly, there has been interest in interactions between biotic pathogens and abiotic stress. The ability of viruses, fungi, and bacteria to influence the sensitivity of plants to air pollutants has been demonstrated in laboratory and field experiments (Heagle, 1973). Many of these investigations, although initiated for practical reasons, are providing clues that could lead to an understanding of the mechanisms of action of air pollutants and the mechanisms of tolerance in some plants.

Some researchers have investigated the impact of air pollutants on yield and quality of crops. Reductions in yield associated with air pollution injury to foliage probably can be attributed to inhibition or alteration of physiological processes, including photosynthesis and nitrogen fixation. Similarly, lowered sugar content of grape crops whose leaves have been injured by photochemical oxidants are reflective of slower rates of photosynthesis (Thompson *et al.*, 1969). Qualitative changes, including decreases in β-carotene concentrations in alfalfa foliage (Thompson and Kats, 1976) and increases in ascorbic acid concentrations in corn, are reflective of alterations in intermediary metabolism (Pippen *et al.*, 1975).

## III. INTERACTIONS OF AIR POLLUTANTS WITH THE PLANT

The observations thus far considered are the result of physiological changes initiated by air pollutants. In this chapter the charge is to address the question of how these effects are induced. The literature is replete with papers that describe the impact of air pollutants on a single cellular component, but rarely, if ever, have these findings provided the true explanation for altered plant response. Rather than encyclopedically review these papers, I propose to examine a few specific aspects of plant response and to see where some of the results may offer encouraging hypotheses to explain a number of the observed effects of air pollutants.

The plant is comprised of many organs: leaves, stems, roots, fruits, flowers, seeds, etc. This discussion will be confined to effects on leaves, because they are the primary receptors of many of the air pollutants and because much of the literature concerns the impact of these substances on foliage. I shall look at the leaf, at the organ, cellular and subcellular levels, in order to explore possible explanations for the changes that these compartments sustain. Interactions between air pollutants and stomata, cell surfaces, plasma membranes, chloroplasts, and mitochondria will be examined. These structures represent sites along the path of an air pollutant into the cell. At each juncture the reactions between a toxic substance and the host will contribute to the determination of the future of the plant.

### A. Air Pollutant Response at the Leaf Surface—Stomatal Interactions

Most phytotoxic air pollutants injure leaves, which produce the carbohydrates that sustain the rest of the plant. Leaves have a very large surface area and, most importantly, have a very large number of portals for ingress of air pollutants, namely, stomata. The impact of two classes of air pollutants on foliage will be evaluated: (1) gases and (2) liquids—specifically, acidic precipitation.

Many characteristics of foliage are of interest. Consider the impact of air pollutants on stomatal aperture. Stomata account for most transfer of gases into and out of plant foliage; ingress of gases is related to stomatal number, size, and degree of opening. Many researchers discuss the response of guard cells to air pollutants. They usually address two questions: (1) Does stomatal aperture control whether a plant expresses foliar injury? (2) If stomatal aperture is altered by an air pollutant, by what mechanism is this change accomplished?

Engle and Gabelman (1966) reported that guard cells of an onion cultivar tolerant to $O_3$ collapsed more rapidly in response to the gas than did guard cells of a susceptible cultivar. This mechanism permitted less ingress of gases into the tolerant cultivar and more into the susceptible one, thus explaining tolerance by exclusion. No other plant system has been found that demonstrates this mechanism of tolerance (Dugger *et al.*, 1962; Ting and Dugger, 1968; Turner *et al.*, 1972). Stomatal closure in response to $O_3$ has been well documented (Hill and Littlefield, 1969; Turner *et al.*, 1972; Vargo *et al.*, 1978). The length of time required for stomatal closure varies with the species and $O_3$ concentration and duration (Hill and Littlefield, 1969). The mechanism of stomatal closure in response to $O_3$ is not understood. Histological observations revealed that occasional guard cells are injured (Hill *et al.*, 1961), but the frequency of this injury would not seem to account for observed reduction in stomatal function. Ozone inhibits photosynthesis (Botkin *et al.*, 1972; Hill and Littlefield, 1969; Pell and Brennan, 1973), and this inhibition may be sufficient to account for the elevated concentration of starch and increased turgor of guard cells.

The explanation for guard cell response may not be found in these cells themselves. Regulation of guard cells has traditionally been thought to rest with their starch–sugar status and more recently with regulation of flux of potassium ions (Zelitch, 1969). Hormonal control of stomatal aperture is another possibility. There are many reports that abscisic acid (ABA) induces stomatal closure (Mansfield, 1976; Milborrow, 1974). Adedipe *et al.* (1973) reported that foliar application of ABA reduces $O_3$ injury. Whether $O_3$ might stimulate increases in concentration of ABA and thus trigger stomatal closure is open to speculation. Since wilting of wheat leaves can lead to a 40-fold increase in ABA (Wright and Hiron, 1969) there is reason to believe that a stress factor, such as $O_3$ or another air pollutant, might cause a similar elevation. Furthermore, the increase in ABA occurs rapidly; wilting bean leaves exhibit a $3\mu g/kg$ fresh weight increase in ABA in 7 minutes (Milborrow, 1974). Since it takes more than 7 minutes for $O_3$ to induce stomatal closure, it seems plausible that the gas could be altering guard cell behavior indirectly by elevating ABA levels. Ozone is known to alter hormonal status both directly and indirectly; induction of premature senescence by the oxidant has been documented numerous times (Rich, 1964). Abscisic acid levels will increase as tissue senesces normally (Milborrow, 1974), and $O_3$ may be turning on a mechanism that permits elevation of this hormone.

A microorganism–$O_3$ interaction studied in our laboratory provides additional cause for speculation on the role of $O_3$ in altering the behavior of guard cells. Stomata of unifoliolate leaves of soybean plants exhibiting severe bud blight caused by the tobacco ring spot virus (TRSV) closed more rapidly in response to $O_3$ than the pores of non-infected plants (Vargo et al., 1978). Because TRSV attacks meristematic tissue, there is unquestionable hormone imbalance. It is not known whether ABA concentrations are higher in TRSV-infected than in non-infected plants. Since ABA can be synthesized and transported in the plant (Milborrow, 1974), it is tempting to speculate that virus-infected plants may contain more ABA and that $O_3$ stress may merely direct movement of hormone to foliar sites. The validity of this hypothesis awaits laboratory evaluation.

Rist (1977) and others (Unsworth et al., 1972) have reported that conductance of foliage, as determined by a diffusive porometer, decreases when plants are exposed to $SO_2$. Biscoe et al. (1973) reported that broad bean plants exposed to a range of $SO_2$ concentrations exhibited elevated conductance levels with exposure to the gas. Biscoe et al. (1973) suggested that rather than affecting guard cells directly, $SO_2$ is absorbed by subsidiary cells that in turn respond with decreases in turgor; subsequently, guard cells collapse in the open position, permitting greater conductance and greater $SO_2$ uptake. Whether these and similar results reported for corn differ from those of pinto bean by virtue of genetic differences is not clear. Specific comparisons of $SO_2$ susceptibility for corn, broad bean, and pinto bean have not been reported, but it is doubtful that the explanation relates to genetic complement. Broad bean and pinto bean are both apparently quite susceptible to $SO_2$ (Guderian, 1977). It is more likely that the lower concentrations used by Rist (1977) were nearly twice as great as the upper limits utilized by Biscoe et al. (1973). Hence, stomatal response may well vary with gas, concentration, and environmental conditions.

In our laboratory, plants exposed to peroxyacetyl nitrate (PAN) showed lower leaf conductance during the day and higher conductance at night when compared with healthy leaves (Metzler and Pell, 1977). Since a guard cell malfunction was suspected, a histological evaluation of guard cell integrity was made, and no correlation was found between altered function and observable structure of these cells. Adjacent epidermal cells were injured, and this damage probably resulted in altered guard cell function. The lack of normal fluctuation of guard cells probably accounted for the desiccation and death of mesophyll cells that followed.

There are numerous other examples of gaseous pollutant effects on stomata, but it suffices to say that these pores are not altered in the same way by all gases. Acid precipitation may preferentially affect regions surrounding stomata, at least in pinto bean, sunflower, and tradescantia (Evans et al., 1977; Evans and Raynor, 1976). Whether this preferential injury results from depressions on the cuticle surface, which collect precipitation, or whether it is due to the sensitivity of guard cells has not yet been determined. Ninety-five percent of all acid-induced lesions of bean and sunflower occur adjacent to trichomes or stomata. Foliage becomes more sensitive as leaves expand, at which time the number of trichomes and stomata per unit area of leaf is reduced. It is therefore possible that some other portion of the leaf surface is interacting with the acid.

Abundance of stomata has been thought to influence the susceptibility or tolerance of various plant species to air pollutants. Plants growing in polluted areas may have fewer stomata than plants in nonpolluted areas (Sharma, 1976; Sharma and Tyree, 1973). Gesalman and Davis (1978) reported no correlation between $O_3$ susceptibility in azalea cultivators and stomatal frequency. Before stomatal frequency can be used to select for air pollutant-tolerant plant material, more convincing data concerning the role of this characteristic must be obtained.

The role of stomata in determining response of foliage to air pollutants is clearly complex. A susceptible plant may protect itself by closing its stomata. However, a tolerant plant may remain tolerant even when stomata are open. It will therefore be necessary to examine the interior of the leaf to determine how air pollutants induce disease.

## B. Air Pollutant Interactions at the Cell Surface

When an air pollutant enters the foliage it usually affects only certain types of cells. If, for example, mesophyll tissue is injured, reduction in primary productivity will follow. If epidermal cells collapse, desiccation is a primary response. Thus, plant response can be explained in part by the type of cell that is injured. If we knew the mechanisms by which air pollutants injure specific types of cells, it might be possible to explain their differential susceptibility as well. There are at least four factors that could determine why some cells are injured by certain pollutants and others are not: (1) mode of entry, (2) preferential susceptibility of a cell type based on morphological configuration and/or biochemical composition of the cell, (3) chemical reactivity of the pollutants, and (4) ability of a cell to repair itself.

The types of cells injured by acid precipitation are undoubtedly influenced by mode of entry. The first cells that come into contact with acid are the epidermal cells; after they collapse, the mesophyll cells below become vulnerable to injury (Evans et al., 1977; Evans and Raynor, 1976). An interesting question arises. Does mesophyll tissue collapse because of direct association with acid or through desiccation that results from erosion of upper or lower layers of cells? Some acid does directly enter the mesophyll layer via the stomata.

Differential cellular effects are most interesting when gases are considered. Most toxic gases apparently penetrate the leaf through stomata. While a small percentage of gas moves through the cuticle, there is no evidence that it is significant. Why then does PAN injure epidermal cells (Glater, 1970; Metzler and Pell, 1977) and $O_3$ and $SO_2$ injure mesophyll cells (Hill et al., 1961; Ledbetter et al., 1959; Pell and Weissberger, 1976; Solberg and Adams, 1956)? The former two gases both have high oxidizing potentials, and yet they injure different cell types. Sulfur dioxide is a reducing or oxidizing agent, and yet both $SO_2$ and $O_3$ injure parenchyma cells.

Sulfur dioxide injures both palisade and spongy mesophyll cells to a similar extent (Solberg and Adams, 1956). Researchers report that $O_3$ preferentially injures palisade parenchyma (Hill et al., 1961). From research conducted in our laboratory it would appear that both types of cells may be equally susceptible to $O_3$ injury (Pell and Weissberger, 1976). At any rate, no one has ever explained why $O_3$ or $SO_2$ preferentially injures parenchyma cells, although there are many descriptions of symptom expression at the light and electron microscopic levels. If the relation of cells to each other is examined, some clues as to the mechanism of differential cells susceptibility are provided. Mesophyll cells are joined at cell wall interfaces. Cells also are bounded by gas spaces. The interface between mesophyll cells and gas spaces is the site of exchange of carbon dioxide, water, oxygen, and probably any other gas gaining access to the interior of leaves (Ledbetter and Porter, 1970).

The walls at these interfaces contain no lignin or "encrusting" substance and are less dense than other portions of the wall of a given cell (Ledbetter and Porter, 1970). These unlignified surfaces may provide avenues for penetration of toxic gases into the cell. I propose that cells which are injured are those bounded by numerous gas spaces and lacking lignin, whereas cells which are not injured by $O_3$ or $SO_2$ are those that are lignified to some degree. While there is a little lignin in mesophytic foliar cells (Martin and Juniper, 1970), there are no quantitative comparisons of lignin concentration in different types of foliar cells. Lignin is subject to attack by $O_3$ or $SO_2$ *in vitro* (Sarkanen and Ludwig, 1971). It is plausible that this polymer of phenylpropane would also react with these toxic gases *in vivo*. Hence, lignins could be acting as a "sink" for $SO_2$ and $O_3$, or, to put it another way, cell walls high in lignin might be providing a tolerance mechanism for their protoplasts.

Several lines of evidence are consistent with this hypothesis. Young plant tissue, which is usually tolerant to $SO_2$ and $O_3$, has far fewer intercellular spaces than mature tissue; hence, unlignified portals of entry for gases may not be as available. (Glater *et al.*, 1962; Uhring, 1978). Petunia plants treated with benomyl or the growth retardent daminozide, both of which prevent $O_3$ injury, also have less internal gas space (Uhring, 1978). Epidermal cell walls, which are usually tolerant to $SO_2$ and $O_3$ injury, are thicker than parenchyma cells, presumably due to lignification (Martin and Juniper, 1970). If tolerance of epidermal cells to the gases is support for the hypothesis discussed above, the following exception should be cited. Krause and Weidensaul (1978) reported that $O_3$ can injure epidermal and guard cells of geranium. Geranium could be unique in its response to $O_3$; it is also possible that scanning electron microscopy, utilized in this study, revealed heretofore unobserved phenomena. If the latter explanation were true, this hypothesis would be significantly weakened.

While the hypothesis is intriguing, there is little information contrasting lignin content of walls of different cell types or ages, or between cells from foliage of a variety of species. Not only would such a contrast be illuminating within a genus, but a contrast between parenchyma cells of cultivars exhibiting differential susceptibility would be of interest. While absence of lignin may provide a mechanism of ingress for a toxic gas into the cell, it cannot always account for subsequent injury. Once within the cell wall, an air pollutant will face the next challenge of injuring the cell or being detoxified. Hence, while $O_3$ and $SO_2$ may enter or be excluded from the cell by a similar mechanism, lists of species susceptibility for the two gases are not the same.

Unlike $O_3$ and $SO_2$, PAN does injure epidermal cells (Glater, 1970). Peroxyacetyl nitrate is among the most reactive of all the prominent air pollutants (Stephens, 1969). Hence, it may simply react with the first cells it contacts. Such a gas could conceivably overcome a mechanism that affords protection to somewhat less reactive gases.

Leaf geometry may provide another explanation for differential cellular susceptibility. It has been suggested that cell surface-to-volume ratios may dictate which cells "see" the most pollutant and hence sustain the greatest degree of injury (Evans and Ting, 1974). To test these alternate hypotheses, researchers might isolate cells from species exhibiting differential susceptibility to a given gas and expose them *in vitro*. A number of physically limiting factors could be removed in this manner, and cell response to a gas could be measured based on inherent characteristics of the cell. Rhoads and Brennan (1978) exposed isolated mesophyll cells of Bel-B and Bel-W tobacco, tolerant and susceptible cultivars, respectively, to $O_3$, and found a similar repression of $O_2$ evolution in both. Obviously, $O_2$ evolution is only one of many parameters that could have been measured including cell death and membrane permeability. Furthermore, it is only one pair of myriad comparisons among cell types that might be worthy of consideration. However, it is this kind of comparison that is necessary to determine whether cells are inherently different in susceptibility or are simply reflecting their location within the leaf.

## C. Air Pollutant Interactions with Subcellular Compartments

It seems apparent that cellular properties influence air pollutant susceptibility. If a toxic substance penetrates the cell wall, the next potential sites of action are the cell membranes and the subcellular compartments within the cell. Disruption of membrane integrity will lead to collapse of protoplasts. If chloroplasts are injured, primary productivity will be impaired. If mitochondria are damaged respiration will be altered. Explanations for how air pollutants alter cellular compartments are necessary to finally determine how air pollutants induce disease.

The subcellular components of the cell could be considered morphologically, physiologically, or biochemically. Since there is considerable debate concerning the presence of structure without function, we should look at these parameters as the interactors they are.

### 1. Passage through the Cell Membrane

Responses leading to change in a specific physiological function can result from direct impact on an organelle or indirect effects of changes in other parts of the cell. For example, many changes observed in $O_3$-injured cells can result from direct effects on cell membrane peremeability. Ozone may react directly with the plasmalemma and other internal cell membranes (Christensen and Giese, 1954; Mudd et al., 1971). Proteins or lipids, both susceptible to oxidation, comprise the bulk of these surfaces. There is considerable literature pertaining to the impact of $O_3$ on these membrane components. Substrate movement and ion and water transport, all membrane functions, are altered by $O_3$ (Evans and Ting, 1973; Spotts et al., 1975a; Ting et al., 1974). Changes in concentration of ions or degree of hydration of a cell or changes in location of crucial cellular constituents will have an impact on many cellular functions.

Ultrastructural evidence supporting membranes as a target is somewhat less prominent. Stain deposition along membrane surfaces of injured cells has been observed by many researchers. The significance of these deposits is not clear. Although transmitting electron microscopy (EM) is an excellent tool, it has limitations. Viewing the plasmalemma in cross section provides little opportunity for observing aberrations. Cunningham and Swanson (1977) examined freeze-fractured plasma, nuclear, and chloroplast membranes of pinto bean and observed a tenfold increase in the deposition of intramembranous protein particles after an $O_3$ exposure. This transformation occurs at a time prior to observation of changes in thin sections observed with conventional transmitting EM. It is intriguing to speculate on the explanation for the response. Are these particles responsible for altered permeability that occurs prior to detection of morphological changes, or are these cells destined to be injured? During many $O_3$ exposures a relatively small number of cells are injured. Hence, changes observed in membranes may reflect the susceptible response of injured cells or tolerance or repair in the larger number of cells for which injury cannot be observed.

## 2. Encounter with an Organelle—A Chloroplast or Mitochondrion

Once an air pollutant has penetrated the cell membrane it can interact with a variety of subcellular components. The impact of air pollutants on chloroplasts and mitochondria has received much consideration.

Conventional transmitting EM investigations revealed changes in structure of the chloroplast in response to $O_3$, PAN, $SO_2$, $NO_2$, HF, and HCl (Pell and Weissberger, 1976; Swanson and Thomson, 1973; Thomson et al., 1965, 1966, 1974; Wellburn et al., 1972; Wei and Miller, 1972; A. Endress, personal communication). Changes in size and shape of the organelle have been reported; altered stromal density and swelling of granal compartments have also been observed. Several researchers have noticed the presence of large crystals in the chloroplast stroma of plants exposed to $O_3$, PAN, $NO_2$, and HCl (Thomson et al., 1965, 1966, 1974; Wellburn et al., 1972; A. Endress, personal communication).

Endress (personal communication) reported that crystal formation is transitory and disappears with time after exposure to HCl. Similar crystals have been noted in other abiotically stressed plants, most notably in stroma of water-stressed plants (Wrischer, 1973; Gunning et al., 1968). Gunning et al. (1968) suggested that these structures are crystals of the enzyme ribulose-1,5-diphosphate carboxylase, the most prominent protein found in fraction I of the stroma. These crystals are not always observed; neither Swanson and Thomson (1973) nor Pell and Weissberger (1976) observed them in $O_3$-treated tobacco or soybean foliage, respectively.

Several explanations could account for the lack of consistency in these observations. Crystals may be formed during a water efflux period which often accompanies exposure to air pollution. Since this efflux may reverse itself with time, so could crystal formation. Crystal formation is reversible in water-stressed plants. Furthermore, crystals are found more readily in younger than older leaves when both are subject to water stress (Wrischer, 1973). It is possible that different physiological age of foliage in various air pollutant experiments could account for the presence or absence of crystal formation. Finally, crystal formation may be genetically determined. Gunning et al. (1968) induced crystal formation in bean but not in corn. The authors suggested that inherently lower concentrations of fraction I protein in the stroma of corn than in bean may make crystal formation less likely in bean than in corn.

The obvious question that follows is, What does alteration of chloroplasts do to their function? Photosynthetic performance is one of the first effects that is of concern to plant physiologists. There is ample evidence that $O_3$, $SO_2$, and HF all inhibit photosynthesis (Bennett and Hill, 1974; Hill and Littlefield, 1969). This phenomenon has been demonstrated with intact foliage, leaf discs, isolated cells, and both intact and fragmented isolated chloroplasts. It has been shown that air pollutants alter photosynthetic rates of alga, angiosperm, conifers, and lichens.

In the case of $SO_2$, the chloroplast may be a primary site of action. Sulfur dioxide acts mainly as $HSO_3^-$ or $SO_3^{2-}$. In this form Ziegler (1975) suggests that "the most characteristic action of sulfite . . . is found in those enzymes which metabolize $CO_2$." Sulfite inhibits ribulose diphosphate carboxylase, phosphoenolpyruvate carboxlase, and malic enzyme noncompetively with regard to their respective substrates. Inhibition of these enzymes is competitive when bicarbonate and sulfite ions are compared. Lichens are extremely sensitive to $SO_2$, while Chlorella and spinach are less so (Ziegler, 1977). When the sulfite sensitivity of ribulose diphosphate carboxylase extracted from Pseudevernia furfuraceae, a lichen, was compared with the activity of enzyme from Chlorella vulgaris or spinach, it was found to be equivalent in all three species. Therefore, while susceptibility may be explained by vulnerability of this enzyme, comparative tolerance could not be explained by a difference in enzymatic response of the other two species to sulfite. Thus, some mechanism of avoidance may be responsible for tolerance. It is also possible that enzymes in an *in vitro* state will respond differently from those *in vivo*.

Many aspects of photosynthesis are altered by air pollutants. Inhibition of photosynthesis inevitably leads to less plant productivity. The relationship between altered productivity and inhibition of photosynthesis is an area currently receiving a great deal of attention but is beyond the scope of this chapter.

Because respiration involves uptake of $O_2$ it is understandable that the impact of air pollutants on this function has received considerable attention. Ozone and PAN induce swelling of mitochondria of pinto bean, while HF induces a similar effect in soybean (Thomson *et al.*, 1965, 1966; Wei and Miller, 1972). The physiological function of respiration is altered by many pollutants, but the trends are less clear. Ozone, $SO_2$, and HF all can initiate an increase or decrease in respiratory rate (Lee, 1967; Macdowall, 1965; Pell and Brennan, 1973; Chang, 1975; Ziegler, 1975). While both responses are symptomatic of cellular alteration, the responses may be different based on the time sequence at which the measurements are made. For example, tissue severely injured by HF decreased in rate of respiration, while respiration rates increased in the same tissue sampled prior to the appearance of visual injury (Yu and Miller, 1967). Obviously, when tissue becomes severely injured, cells die and mitochondria no longer function. Earlier in symptom development, mitochondria may swell prior to cell destruction. Such speculation is supported by EM studies. Swelling of mitochondria leads to uncoupling of phosphorylation, which is reflected in increased rates of respiration (Racker, 1965). A somewhat different example can be derived from the $O_3$ literature. It has been shown that respiration can decrease early in symptom development and increase later. The early decrease may be related to the water-soaking symptom often associated with initial $O_3$ injury. Oxygen may penetrate such cells at a slower rate, thus reducing their rate of reaction with the gas (Ripaldi *et al.*, 1972). Increases observed later may result from swelling of that organelle. Hence, two different sets of conditions may influence respiratory response.

The interest in photosynthesis and respiration has stimulated curiosity concerning effects on bioenergetics of the cell. Both reduction and stimulation in phosphorylation have been reported in mitochondria exposed to $O_3$ *in vivo* and *in vitro* (Lee, 1967; Macdowall, 1965). No change in photosphosphorylation was reported when isolated chloroplasts were exposed to $O_3$ (Coulson and Heath, 1974). Reports of increases and decreases in adenine nucleotides have been reported for $O_3$ and HF using different methods of detection and different pollutant exposure conditions (Pell, 1974; Pell and Brennan, 1973; Tomlinson and Rich, 1968; Chang, 1968; McCune et al., 1970; McNulty and Lords, 1960). It is not surprising that the literature is not consistent. The bioenergetic status of cells is delicately balanced between organelles. In order to more clearly understand the meaning of changes such as those described above, it will be necessary to expose intact plants to air pollutants and analyze either nucleotide status and/or phosphorylative capacity of all pertinent cell fractions, including chloroplasts and mitochondria, after an *in situ* exposure. Current methods employing homogenized tissue from exposed plants or organelles that are exposed *in vitro* will not determine changes in energy balance that might alter cell or leaf behavior. While this criticism was developed with regard to data on bioenergetics, it should not be regarded as unique to this function but applicable to many applied biochemical questions. Alterations in the status of sterols (Spotts et al., 1975b; Tomlinson and Rich, 1971), amino acids (Tingey, 1974; Tomlinson and Rich, 1967), fatty acids (Tomlinson and Rich, 1969), etc., determined on the basis of foliar analysis, tell us little about the sites of alteration.

Three major cellular effects of air pollutants have been treated briefly. These areas do not represent an exhaustive list of impacts. Since the main objective of this chapter is to consider modes of action, I leave further exploration of the literature to the reader.

## IV. QUESTIONS AIR POLLUTION
## PHYTOTOXICOLOGISTS HAVE ANSWERED

When a pollutant comes into contact with a plant, it encounters a series of obstacles. At each roadblock the pollutant has the potential to penetrate further or to be detoxified. At the stomata the pollutant will either gain ingress or be excluded. Once within the leaf, at the cell surface, the pollutant can react and be rendered nontoxic, react and form a toxic secondary product, or penetrate without reacting and encounter substrates within the cell. As air pollutants pass through the cell membrane, they may either disrupt it and the protoplast it envelops, or nondestructively enter the cell. Once within the protoplast, the toxic substance could react with a broad range of organelles and pathways, any one of which could lead to malfunction and/or cell death. Determining the site of injury induced by a given pollutant and relating it to altered physiological function is being accomplished in many laboratories.

Answers are more difficult to come by in explaining the basic mechanism of action. A variety of experimental approaches have been utilized to secure evidence concerning these questions. Many researchers have considered the kinetic relationship between changes in several physiological parameters and pollutant exposure. The timing of a physiological change and visible symptom expression often will be compared in order to determine whether a given physiological change is a primary or secondary response to a pollutant. While sequences of changes have been established, they do not indicate which change(s) are responsible for all subsequent effects. Researchers are limited by the number of measurements that can be made during a single exposure. Consider photosynthesis as an example. Any attempt to determine whether some aspect of this function is the target of attack by a pollutant should consider all the enzymes involved in the light or dark reactions, electron transport reactions, membranes of the grana, stromal density, $CO_2$ fixation, etc.

*In vitro* systems have been utilized to determine how a pollutant alters a given physiological system. Isolated chloroplasts (Coulson and Heath, 1974), mitochondria (Lee, 1968), red blood cells (Goldstein *et al.*, 1974), cell suspensions (Rhoads and Brennan, 1978), etc., have been exposed to pollutants, and specific answers have been obtained for these systems. In air pollution research as well as in physiological studies in general, *in vitro* studies tell us potential effects but cannot predict actual *in situ* responses.

On the basis of knowledge of the chemistry of the pollutants, some speculation about mechanisms of action can be justified. Knowledge of the chemistry of $SO_2$ in solution has led to the theory that $HSO_3^-$ would compete with $HCO_3^-$ ($CO_2$ in solution) (Ziegler, 1975). Hence, enzymes that normally act as substrates for $CO_2$ would be logical sites of action for $SO_2$. This has indeed turned out to be a mechanism of $SO_2$ toxicity. Similarly, when HF goes into solution, the fluoride anion behaves similarly to other anions, moving through the plant by passive diffusion (Chang, 1975). As an anion, it is postulated that fluoride could bind with cations. Since many cations are cofactors for enzymes, it is not surprising to discover that fluoride is an enzyme inhibitor. Which specific enzyme(s) are involved in fluoride toxicity is open to question, however.

Ozone is a molecule with an extremely high oxidizing potential. It has been postulated that $O_3$ would oxidize the first molecules contacted provided that they had a complementary chemical potential. For this reason the plasmalemma has been postulated as the site of action of this oxidant (Rich, 1964). There is considerable ultrastructural, biochemical, and physiological evidence to support this hypothesis. The precise molecules that are first oxidized by $O_3$ have not been defined. It is possible that they are not always the same. Membrane conformations change with environmental conditions, plant age, etc., and components and their ratios vary with species.

To date, no crucial experiments have been reported to identify the phytotoxic mechanisms of action of any air pollutant. It is even difficult to piece the literature together. Comparison of investigations is hampered by varying experimental conditions, biological conditions, and knowledge of the physiology of the healthy plant. Unfortunately, exposure facilities and plant species utilized vary tremendously among laboratories. In some ways this is desirable, since a mechanism should be demonstrated for many species. However, it is sometimes difficult to extrapolate results from one series of experiments to another. Last, applied physiologists are dependent on the data provided by basic physiologists, and often such information is not available.

## V. CONCLUSIONS

Looking toward the research goals of the future prompts me to wonder whether previous emphasis has been directed toward susceptibility rather than tolerance. Mechanisms of susceptibility and tolerance are in many ways two sides of the same coin. However, as already noted, tolerance does not only result because a particular site is resistant in a given species. Exclusion and alternate sites of degradation may account for some mechanisms of tolerance. When we look at the numbers of tolerant species and the numerous conditions during which tolerance is apparent, we find that tolerance may be the rule and susceptibility the exception. The practical applications of tolerance are obvious. Inclusion of characteristics of tolerant material in breeding programs ultimately would resolve many agricultural problems in air pollution.

In this chapter I have attempted to introduce some of the approaches physiologists have taken to air pollution research. Occasional personal speculation as to why specific observations were made seemed necessary, but these guesses should not be construed as generalizations. The diversity of plant species attacked by air pollutants is so large that we cannot expect uniform responses. Examples of plant responses were drawn from a limited band of pollutants that were selected because of their importance. No slight was intended toward particulates, including heavy metals, or toward other gaseous air pollutants. It is hoped that similar approaches will prove useful with other pollutant systems. Only a handful of physiological responses were considered, although many more descriptions of plant reactions to pollutants have been documented. My only hope is that my modest effort to weave a few thoughts together will add to the width of the fabric already on the loom.

## References

Abeles, F. B. (1973). "Ethylene in Plant Biology." Academic Press, New York.

Adedipe, N. O., Khatamian, H., and Ormrod, D. P. (1973). Stomatal regulation of ozone phytotoxicity in tomato. Z. *Pflanzenphysiol.* **68**, 323–328.

Bennett, J. H., and Hill, A. C. (1974). Acute inhibition of apparent photosynthesis by phytotoxic air pollutants. *In* "Air Pollution Effects on Plant Growth" (M. Dugger, ed.), Am. Chem. Soc. Symp. Ser. **3**, 115–127.

Benson, N. R. (1959). Fluoride injury or soft suture and splitting of peaches. *Proc. Am. Soc. Hortic. Sci.* **74**, 184–198.

Biscoe, P. V., Unsworth, M. H., and Pinckney, H. R. (1973). The effects of low concentrations of sulphur dioxide on stomatal behaviour in *Vicia faba. New Phytol.* **72**, 1299–1306.

Botkin, D. B., Smith, W. H., Carlson, R. W., and Smith, T. L. (1972). Effects of ozone on white pine saplings: Variation in inhibition and recovery of net photosynthesis. *Environ. Pollut.* **3**, 273–289.

Brandt, C. S., and Heck, W. W. (1967). Effects of air pollutants on vegetation. *In* "Air Pollution" (A. Stern, ed.), Vol. 1, pp. 401–443. Academic Press, Inc., New York.

Chang, C. W. (1968). Effect of fluoride on nucleotides and ribonucleic acid in germinating corn seedling roots. *Plant Physiol.* **43**, 669–674.

Chang, C. W. (1975). Fluorides. *In* "Responses of Plants to Air Pollution" (J. B. Mudd and T. T. Kozlowski, eds.), pp. 57–95. Academic Press, New York.

Christensen, E., and Giese, A. C. (1954). Changes in absorption spectra of nucleic acids and their derivatives following exposure of ozone to ultra violet radiations. *Arch. Biochem. Biophys.* **5**, 208–216.

Coulson, C., and Heath, R. L. (1974). Inhibition of the photosynthetic capacity of isolated chloroplasts by ozone. *Plant Physiol.* **53**, 32–38.

Cunningham, W. P., and Swanson, E. S. (1977). Ozone induced changes in bean leaf cellular membrane structure. A freeze-fracture electron microscopic study. *Plant Physiol.* **59**, Suppl. 124. (Abstr.)

Dugger, W. M., Jr., Taylor, O. C., Cardiff, E., and Thompson, C. R. (1962). Stomatal action in plants as related to damage from photochemical oxidants. *Plant Physiol.* **37**, 487–491.

Dunning, J. A., and Heck, W. W. (1977). Response of bean and tobacco to ozone: Effect of light intensity, temperature, and relative humidity. *J. Air Pollut. Control Assoc.* **27**, 882–886.

Engle, R. L., and Gableman, W. H. (1966). Inheritance and mechanism for resistance to ozone damage in onion, *Allium cepa* L. *Proc. Am. Soc. Hortic. Sci.* **89**, 423–430.

Evans, L. S., and Raynor, G. S. (1976). "Acid Rain Research Program, Annual Progress Report." Brookhaven Natl. Lab., Upton, New York.

Evans, L. S., and Ting, I. P. (1973). Ozone-induced membrane permeability changes. *Am. J. Bot.* **60**, 155–162.

Evans, L. S., and Ting, I. P. (1974). Ozone sensitivity of leaves: Relationship to leaf water content, gas transfer resistance, and anatomical characteristics. *Am. J. Bot.* **61**, 592–597.

Evans, L. S., Gmur, N. F., and DaCosta, F. (1977). Leaf surface and histological perturbations of leaves of *Phaseolus vulgaris* and *Helianthus annuus* after exposure to simulated acid rain. *Am. J. Bot.* **64**, 903–913.

Feder, W. A. (1968). Reduction in tobacco pollen germination and tube elongation, induced by low levels of ozone. *Science* **160**, 1122.

Gesalman, C. M., and Davis, D. D. (1978). Ozone susceptibility of ten azalea cultivars as related to stomatal frequency or conductance. *J. Am. Soc. Hortic. Sci.* **103**, 489–491.

Glater, R. B. (1970). "Smog and Plant Structure in Los Angeles County," Rep. No. 70-17. U.C.L.A. Sch. Eng. Appl. Sci., Los Angeles, California.

Glater, R. B., Solberg, R. A., and Scott, F. M. (1962). A developmental study of leaves of *Nicotiana glutinosa* as related to their smog-sensitivity. *Am. J. Bot.* **49**, 954–970.

Goldstein, B. D., Lai, L. Y., and Cuzzi-Spada, R. (1974). Potentiation of complement-dependent membrane damage by ozone. *Arch. Environ. Health* **28**, 40–42.

Guderian, R. (1977). "Air Pollution. Phytotoxicity of Acidic Gases and its Significance in Air Pollution Control." Springer-Verlag, Berlin and New York.

Gunning, B. E. S., Steer, M. W., and Cochrane, M. P. (1968). Occurrence, molecular structure, and induced formation of the "stromacentre" in plastids. *J. Cell Sci.* **3**, 445–446.

Heagle, A. S. (1973). Interactions between air pollutants and plant parasites. *Annu. Rev. Phytopathol.* **11**, 365–388.

Heck, W. W. (1968). Factors influencing expression of oxidant damage to plants. *Annu. Rev. Phytopathol.* **6**, 165–188.

Hill, A. C., and Littlefield, N. (1969). Ozone. Effect of apparent photosynthesis, rate of transpiration, and stomatal closure in plants. *Environ. Sci. Technol.* **3**, 52–56.

Hill, A. C., Pack, M. R., Treshow, M., Downs, R. S., and Transtrum, L. G. (1961). Plant injury induced by ozone. *Phytopathology* **51**, 356–363.

Krause, C. R., and Weidensaul, T. C. (1978). Ultrastructural effects of ozone on the host–parasite relationship of *Botrytis cinerea* and *Pelargonium hortorum*. *Phytopathology* **68**, 301–307.

Ledbetter, M. C., and Porter, K. R. (1970). "Introduction to the Fine Structure of Plants Cells." Springer-Verlag, Berlin and New York.

Ledbetter, M. C., Zimmerman, P. W., and Hitchock, A. E. (1959). The histological effects of ozone on plant foliage. *Contrib. Boyce Thompson Inst.* **20**, 225–282.

Lee, T. T. (1967). Inhibition of oxidative phosphorylation and respiration by ozone in tobacco mitochondria. *Plant Physiol.* **42**, 691–696.

Lee, T. T. (1968). Effect of ozone on swelling of tobacco mitochondria. *Plant Physiol.* **43**, 133–139.

Leone, I. A., and Brennan, E. (1972). Modification of sulfur dioxide injury to tobacco and tomato by varying nitrogen and sulfur nutrition. *J. Air Pollut. Control Assoc.* **22**, 544–547.

McCune, D. C., Weinstein, L. H., and Mancini, J. F. (1970). Effects of hydrogen fluoride on the acid-soluble nucleotide metabolism of plants. *Contrib. Boyce Thompson Inst.* **24**, 213–226.

Macdowall, F. D. (1965). Stages of ozone damage to respiration of tobacco leaves. *Can. J. Bot.* **43**, 419–427.

McNulty, I. B., and Lords, J. L. (1960). Possible explanation of fluoride-induced respiration in *Chlorella pyrenoidosa. Science* **132**, 1553–1554.

Mansfield, T. A. (1976). Stomatal behaviour. Chemical control of stomatal movements. *Philos. Trans. R. Soc. London, Ser. B* **273**, 541–550.

Martin, J. T., and Juniper, B. E. (1970). "The Cuticles of Plants." Arnold, London.

Metzler, J. T., and Pell, E. J. (1977). The influence of peroxyacetyl nitrate on stomatal conductance of primary bean leaves as it relates to macroscopic and microscopic symptom expression. *Proc. Am. Phytopathol. Soc.* **4**, 193. (Abstr.)

Milborrow, B. V. (1974). The chemistry and physiology of abscisic acid. *Annu. Rev. Plant Physiol.* **25**, 259–307.

Mudd, J. B., McManus, T. T., Ongun, A., and McCullogh, T. E. (1971). Inhibition of glycolipid biosynthesis in chloroplasts by ozone and sulfhydryl reagents. *Plant Physiol.* **48**, 335–339.

Pell, E. J. (1974). The impact of ozone on the bioenergetics of plant systems. *In* "Air Pollution Effects on Plant Growth" (M. Dugger, ed.), *Am. Chem. Soc. Symp. Ser.* **3**, 106–113.

Pell, E. J., and Brennan, E. (1973). Changes in respiration, photosynthesis, adenosine 5'-triphosphate, and total adenylate content of ozonated pinto bean foliage as they relate to symptom expression. *Plant Physiol.* **51**, 378–381.

Pell, E. J., and Weissberger, W. C. (1976). Histopathological characterization of ozone injury to soybean foliage. *Phytopathology* **66**, 856–861.

Pippen, E. L., Potter, A. L., Randall, V. G., Ng, K. C., Reuter, F. W., and Morgan, A. I. (1975). Effect of ozone fumigation on crop composition. *J. Food Sci.* **40**, 672–676.

Racker, E. (1965). "Mechanisms in Bioenergetics." Academic Press, New York.

Rhoads, A. F., and Brennan, E. (1978). The effect of ozone on chloroplast lamellae and isolated mesophyll cells of sensitive and resistant tobacco cultivars. *Phytopathology* **68**, 883–886.

Rhoads, A. F., Troiano, J., and Brennan, E. (1973). Ethylene gas as a cause of injury to easter lilies. *Plant Dis. Rep.* **57**, 1023–1024.

Rich, S. (1964). Ozone damage to plants. *Annu. Rev. Phytopathol.* **2**, 253–266.

Ripaldi, C. P., Brennan, E., and Leone, I. A. (1972). Inhibition of foliar respiration in ozone-exposed pinto bean plants. *Phytopathology* **62**, 499. (Abstr.)

Rist, D. L. (1977). The influence of exposure temperature and humidity on the response of pinto bean to $SO_2$, in relation to stomatal conductance and $SO_2$ uptake. M. S. Thesis, Dep. Plant Pathol. Cent. Air Environ. Studies, Pennsylvania State Univ., Univ. Park.

Sarkanen, K. V., and Ludwig, C., eds. (1971). "Lignins." Wiley (Interscience), New York.

Sharma, G. K. (1976). Cuticular features as indicators of environmental pollution. *Proc. Int. Symp. Acid Precip. For. Ecosyst., 1st; U.S. For. Serv. Gen. Tech. Rep. NE* NE-23.

Sharma, G. K., and Tyree, J. (1973). Geographic leaf cuticular and gross morphological variations in *Liquidambar styraciflua* L. and their possible relationship to environmental pollution. *Bot. Gaz.* **134**, 179–184.

Solberg, R. A., and Adams, D. F. (1956). Histological responses of some plant leaves to hydrogen fluoride and sulfur dioxide. *Am. J. Bot.* **43**, 755–760.

Spotts, R. A., Lukezic, F. L., and Hamilton, R. H. (1975a). The effects of benzimidazole on some membrane properties of ozonated pinto bean. *Phytopathology* **65**, 39–45.

Spotts, R. A., Lukezic, F. L., and Lacasse, N. L. (1975b). The effect of benzimidazole, cholesterol, and a steroid inhibitor on leaf sterols and ozone resistance of bean. *Phytopathology* **65**, 45–49.

Stephens, E. R. (1969). Chemistry of atmospheric oxidants. *J. Air Pollut. Control Assoc.* **19**, 181–185.

Swanson, E. S., and Thomson, W. W. (1973). The effect of ozone on leaf cell membranes. *Can. J. Bot.* **51**, 1213–1219.

Thompson, C. R., and Kats, G. (1976). Effect of photochemical air pollution on two varieties of alfalfa. *Environ. Sci. Technol.* **10**, 1237–1241.

Thompson, C. R., Hensel, E., and Kats, G. (1969). Effects of photochemical air pollutants on Zinfandel grapes. *Hortiscience* **4**, 222–224.

Thomson, W. W., Dugger, W. M., Jr., and Palmer, R. L. (1965). Effects of peroxyacetyl nitrate on ultrastructure of chloroplasts. *Bot. Gaz.* **126**, 66–72.

Thomson, W. W., Dugger, W. M., Jr., and Palmer, R. L. (1966). Effects of ozone on the fine structure of the palisade parenchyma cells of bean leaves. *Can. J. Bot.* **44**, 1677–1682.

Thomson, W. W., Nagahashi, J., and Platt, K. (1974). Further observation on the effects of ozone on the ultrastructure of leaf tissue. *In* "Air Pollution Effects on Plant Growth" (M. Dugger, ed.), *Am. Chem. Soc. Symp. Ser.* **3**, 83–93.

Ting, I. P., and Dugger, W. M., Jr. (1968). Factors affecting ozone sensitivity and susceptibility of cotton plants. *J. Air Pollut. Control Assoc.* **18**, 810–813.

Ting, I. P., Perchorowicz, J., and Evans, L. (1974). Effect of ozone on plant cell membrane permeability. *In* "Air Pollution Effects on Plant Growth" (M. Dugger, ed.), *Am. Chem. Soc. Symp. Ser.* **3**, 8–21.

Tingey, D. T. (1974). Ozone induced alterations in the metabolite pools and enzyme activities of plants. *In* "Air Pollution Effects on Plant Growth" (M. Dugger, ed.), *Am. Chem. Soc. Symp. Ser.* **3**, 40–57.

Tomlinson, H., and Rich, S. (1967). Metabolic changes in free amino acids of bean leaves exposed to ozone. *Phytopathology* **57**, 972–974.

Tomlinson, H., and Rich, S. (1968). The ozone resistance of leaves as related to their sulfhydryl and ATP content. *Phytopathology* **58**, 808–810.

Tomlinson, H., and Rich, S. (1969). Relating lipid content and fatty acid synthesis to ozone injury of tobacco leaves. *Phytopathology* **59**, 1284–1286.

Tomlinson, H., and Rich, S. (1971). Effects of ozone on sterols and sterol derivatives in bean leaves. *Phytopathology* **61**, 1404–1405.

Turner, N. C., Rich, S., and Tomlinson, H. (1972). Stomatal conductance, fleck injury and growth of tobacco cultivars varying in ozone tolerance. *Phytopathology* **62**, 63–67.

Uhring, J. (1978). Leaf anatomy of petunia in relation to pollution damage. *J. Am. Soc. Hortic. Sci.* **103**, 23–27.

Unsworth, M. H., Biscoe, P. V., and Pinckney, H. R. (1972). Stomatal responses to sulphur dioxide. *Nature (London)* **239**, 458–459.

van Haut, D., and Stratmann, H. (1970). "Color-Plate Atlas of the Effects of Sulfur Dioxide on Plants." Verlag W. Girardet, Essen.

Vargo, R. H., Pell, E. J., and Smith, S. H. (1978). Induced resistance to ozone injury of soybean by tobacco ringspot virus. *Phytopathology* **68**, 715–719.

Wei, L. L., and Miller, G. W. (1972). Effects of HF on the fine structure of mesophyll cells from *Glycine max* Merr. *Fluoride* **5**, 67–73.

Wellburn, A. R., Majernik, O., and Wellburn, F. A. M. (1972). Effects of SO$_2$ and NO$_2$ polluted air upon the ultrastructure of chloroplasts. *Environ. Pollut.* **3,** 37–49.

Wright, S. T. C., and Hiron, R. W. P. (1969). (+)–Abscisic acid, the growth inhibitor induced in detached wheat leaves by a period of wilting. *Nature (London)* **224,** 719–720.

Wrischer, M. (1973). Protein crystalloids in stroma of bean protoplasts. *Protoplasma* **77,** 141–150.

Yu, M. H., and Miller, G. W. (1967). Effect of fluoride on the respiration of leaves from higher plants. *Plant Cell Physiol.* **8,** 483–493.

Zelitch, I. (1969). Stomatal control. *Annu. Rev. Plant Physiol.* **20,** 329–350.

Ziegler, I. (1975). The effect of SO$_2$ pollution on plant metabolism. *Residue Rev.* **56,** 79–105.

Ziegler, I. (1977). Sulfite action on ribulosediphosphate carboxylase in the lichen *Pseudovernia furfuracea. Oceologia* **29,** 63–66.

Chapter 16

# How Parasitic Seed Plants
# Induce Disease in Other Plants

DONALD M. KNUTSON

## I. INTRODUCTION

During the first century A.D., Pliny the Elder wrote his remarkable book "Natural History." In Book XVI he included some observations on the diversity of plants that live on other plants:

> For some varieties of plants cannot grow in the earth, and take root in trees, because they have no abode of their own and consequently live in that of others: instances of this are mistletoe and the plant in Syria called cadytas, which twines itself around not only trees, but even brambles, and likewise in the district about Tempe in Thessaly the plant called polypodium, and also the dolichos and the serphyllum.

With this auspicious beginning in the first century of the previous millennium it would appear to be an easy task in the late 1970s to compose a brief essay on how these fascinating plants induce disease in other plants. But appearances can be deceiving. Most of the past 19 centuries were spent in the dark. More than 1000 years after Pliny,

PLANT DISEASE, VOL. IV
ISBN-0-12-356404-2

Albertus Magnus also observed mistletoe. He conceived the idea that mistletoes not only were rooted in the tissues of trees but derived their nourishment from the trees as well. His brilliant idea was right on the mark. But it was not accepted for yet another 600 years. Thus our knowledge about parasitic seed plants and the diseases they cause has developed almost wholly within the past 100 years. To be sure, we have learned a great deal about the taxonomy and general biology of mistletoes and other parasitic seed plants, but our knowledge about how they induce disease remains fragmentary and filled with conjecture.

My purpose in this essay is to evaluate what we know and to synthesize some new ideas about mechanisms of pathogenesis and parasitism. I will focus primarily on those parasitic angiosperms that cause serious disease, with special emphasis on the dwarf mistletoes, the most advanced of the aerial parasites. But first, a few words about the distribution of parasitic seed plants and the diseases they induce.

## II. AN OVERVIEW

The 3000 parasitic angiosperms in 15 plant families (Table I) all are dicotyledonous angiosperms, with the exception of a single report of a parasitic gymnosperm *Podocarpus ustus* (De Laubenfels, 1959).

### A. Losses

Although Pliny mentioned that mistletoes have an ability to slowly kill trees, it remained for European pathologists of the nineteenth century to draw professional attention to damage by *Viscum, Cuscuta, Lathraea, Pedicularis*, and other parasites on horticultural tree crops and herbaceous food plants (Hartig, 1882; Prillieux, 1895). James Weir (1916), an American pathologist, was the first to quantify the growth loss in coniferous forest trees due to the dwarf mistletoes (*Arceuthobium* sp.).

Certainly many parasitic plants are serious pathogens. The following examples are presented to indicate some of the damage done to food and fiber crops by these remarkable plants.

Heald (1926) spoke of *Cuscuta* (dodder) so severe that entire fields of clover were ruined. Mijatovic and Stojanovic (1973) cite a case in Yugoslavia in which *Orobanche* (broomrape) reduced sunflower seed yields by 33%. Gharib (1973) states that in Iraq broomrape is so destructive that entire fields of tomatoes and tobacco are left uncropped. The loss of 40% to 70% of sorghum to *Striga* (witchweed) has been reported (Kasasian, 1973). Fiber crops are similarly affected. The dwarf

**TABLE I**

Families of Vascular Plants with Parasitic Genera [a,b]

| | No. of parasitic species | Parasites of | |
| --- | --- | --- | --- |
| | | Root | Top |
| Magnoliidae | | | |
| Magnoliales | | | |
| Lauraceae | 20 | | Hemi |
| Rosidae | | | |
| Santalales | | | |
| Balanophoraceae * | 100 | Holo | |
| Eremolepidaceae | 10 | | Hemi |
| Loranthaceae * | 700 | Hemi | Hemi |
| Misodendraceae * | 10 | | Hemi |
| Olacaceae | 20 | Hemi | |
| Santalaceae * | 400 | Hemi | Hemi |
| Viscaceae * | 400 | | Hemi |
| Rafflesiales | | | |
| Hydnoraceae * | 20 | Holo | |
| Rafflesiaceae * | 50 | Holo | Holo |
| Polygalales | | | |
| Krameriaceae * | 20 | Hemi | |
| Asteridae | | | |
| Polemoniales | | | |
| Cuscutaceae * | 200 | | Hemi |
| Lennoaceae * | 10 | Holo | |
| Scrophulariales | | | |
| Scrophulariaceae ° | 600 | Hemi | |
| Orobanchaceae ° | 200 | Holo | |

[a] From Atsatt (1973), Barlow and Wiens (1971), Cronquist (1968), and Kuijt (1969).

[b] In families marked with an asterisk, all species are parasitic. Most top parasites are chlorophyllous hemiparasites. Root parasites are designated as to whether the majority of species are holo- or hemiparasites.

mistletoes cause an estimated annual loss of 20 million cubic meters of wood fiber in North America and in parts of Europe and Asia (Hawksworth, 1973). The leafy mistletoes, also, are very damaging. *Viscum album* is a serious pathogen in horticultural trees (Gäumann, 1950) and coniferous forests in Europe, most notably in France where 25% reductions in growth have been reported (Plagnat, 1950). Australian eucalyptus forests, plagued with *Amyema* sp. sustain growth losses of 50% or more (Greenham and Brown, 1957).

## B. Distribution and Host Range

Most genera in this diverse group of plants are tropical or subtropical; but some of the most serious disease-inducing genera have adapted to the northern and southern temperature regions (Table II). *Viscum album* in Sweden and *Arceuthobium americanum* in Canada extend to nearly 60° north latitude. *Myzodendron* parasitizes antarctic beech (*Nothofagus*) forests in South America as far south as Cape Horn. Their ranges also vary greatly. *Arceuthobium oxycedri,* for example, is found from the Mediterranean area of Europe and Africa to Asia Minor and India, a 6000-mile range. *Areuthobium verticilliflorum,* on the other hand, is known only in a small area west of Durango, Mexico. Similar diversity exists in the degree of host specificity. *Dendrophthoe falcata* parasitizes more than 350 hosts in India whereas *Arceuthobium minutissimum* is found only on *Pinus wallichiana* in the Himalayas.

The tremendous diversity of forms among the parasitic plants (Teryokhin, 1977) prompted Kuijt (1969) to propose that parasitism arose at least eight different times in unrelated groups of dicotyledons as an interaction of the roots of two different species. Possibly, root nodules were ancestral to haustoria (Atstatt, 1973). From whatever origin, parasitism affords several distinct survival advantages, the primary one being a source of minerals in the mineral-poor tropical soils, the original home of most parasitic plants. By escaping into the less competitive niche afforded by parasitism, the formerly autotrophic plants also found a source of elaborated carbon and nitrogenous compounds, a reliable supply of water, and, for aerial parasites, a better chance to capture light energy in dense tree canopies.

Central to survival of parasitic plants is the hazardous necessity of finding a proper host and establishing an organic union. Strategies have evolved for insuring contact with a proper host. In addition to broad host ranges, various genera have such survival traits as fantastically high seed production (*Striga*), seed germination primarily near host roots (*Striga, Orobanche*), vectoring by birds associated with host trees (*Phoradendron, Viscum*), and explosive fruits that cast seeds 15 meters from the source (*Arceuthobium*). Once at the host surface, other mechanisms prepare the seed for host penetration. The sticky seeds of dwarf mistletoe, for example, dry on the host bark in such a fashion that when the seed germinates, the radicle is aimed toward the host tissue. Positive thigmatropism and negative phototropism also insure growth of the

## TABLE II
### Important Disease-Inducing Genera, Their Hosts, and Regions of Serious Crop Losses [a]

*Root parasites*
A. Broomrapes (Orobanchaceae)
   1. *Aeginetia*—maize, rice, sugarcane (India, Pakistan, Ceylon, S.E. Asia, Phillipines)
   2. *Christisonia*—sugarcane (Phillipines)
   3. *Orobanche*—legumes, tobacco, tomato, cabbage, flax, hemp, grapes, watermelon, cucurbits, mint, sunflower, clover, eggplant (Mediterranean region, Europe, New Zealand, S. Africa)
B. Parasitic figworts (Scrophulariaceae)
   1. *Alectra (Melasma)*—cowpeas, soybeans, peanuts, legumes, sugarcane (S. Africa, Rhodesia, W. Indies)
   2. *Rhamphicarpa*—maize, cowpeas, rice, sorghum (Madagascar, E. Africa)
   3. *Striga* (witchweed)—maize, sorghum, sugarcane, tobacco, grasses (S. Africa, S.E. United States)
*Stem or leaf parasites*
A. Cuscutaceae
   *Cuscuta*—alfalfa, clover, sunflower, potato, sugar beets, tobacco, bamboo, asters (Europe, United States, India, Russia, Ceylon, Puerto Rico, China, Australia)
B. Lauraceae
   *Cassythia*—evergreen shrubs and ornamentals, orange trees (India) (Hawaii, E. Indies, Puerto Rico, India)
C. Viscaceae
   1. *Arceuthobium*—conifers (United States, Canada, Mexico, China, India, Pakistan, Kenya, Mediterranean area, Middle East)
   2. *Dendrophthora*—rubber, mango, avocado, cacao (South America)
   3. *Korthalsella*—acacia, eucalyptus (Hawaii, Australia)
   4. *Notothixos*—eucalyptus (Australia)
   5. *Phoradendron*—coffee, avocado, teak, various forest trees (Bolivia, Central America, Mexico, United States, W. Indies)
   6. *Viscum*—rubber, conifers, fruit trees, deciduous trees (Europe, Asia, Africa)
D. Loranthaceae
   1. *Amyema*—eucalyptus (Australia)
   2. *Elytranthe*—rubber, cashew (Malaya)
   3. *Loranthus* (no longer a valid genus; members have been distributed among other genera)—rosewood, acacia, teak, mango, citrus, guava, pomegranate, jujube, tea, rubber, tapioca, casuarina, eucalyptus, cacao (India, Phillipines, Malaya, Indonesia, Africa)
   4. *Phthirusa*—rubber (Brazil)
   5. *Psittacanthus*—citrus trees (Mexico)

[a] From Plagnat (1950), Steward (1958), Gill and Hawksworth (1961), King (1966), Kuijt (1969), Hawksworth and Wiens (1972), Hepper (1973), and Lubenov (1973).

parasite embryo toward the host. Only after the success of these mechanisms can the parasitic plant begin its imperative function of establishing organic union.

## III. THE INVADING PARASITE

The angiosperm parasite, like all other parasites discussed in this volume, must invade the host after establishment on or near the host surface.

### A. The Holdfast

The parasite first produces a contact organ called a holdfast. In fungi this is called an appressorium. The holdfast and the nutrient bridge between the host and parasite performs the three functions of (1) attachment to the host, (2) penetration into living host tissue, and (3) establishment of nutrient-flow pathways. With the exception of the unique stem connections of *Cuscuta* and *Cassythia*, these organs are highly modified roots.

In many genera, development begins, as in *Castilleja*, with an enlargement of root cortex cells into a spherical structure, often with numerous root hairs associated (Dobbins and Kuijt, 1973). The cells of this structure differentiate into three principle tissues: a zone of pericycle derivatives (xylem vessel elements), a central core of thick-walled parenchyma cells that become the intrusive organ, and an outer area of cortical parenchyma cells. These are thinned-walled cells, laid down at the host interface, which appear to discharge their contents (Dobbins and Kuijt, 1973). This is probably the source of the cutinlike substance reported by Thoday (1951) that fastens the holdfast to the host.

### B. Penetration to the Xylem

From the holdfast, the parasite penetrates the xylem of the host by means of a wedge-shaped "intrusive organ" (infection peg in fungi), which penetrates the suberized dermal tissue of the host by enzymatic and mechanical action (Thoday, 1951). The intrusive organ is composed of its own xylem elements and a central core of thick-walled parenchyma with dense cytoplasm and prominent nuclei (Piehl, 1963).

Penetration seems to be rapid (Rogers and Nelson, 1962). Some intrusive organs grow right through the host cells, as in *Cuscuta*, but more commonly, as in the dwarf mistletoes, the host cells are crushed in advance of the parasite as it occupies the space. This establishes the endophyte, or endophytic system, within the host and is followed by the rapid union of parasite xylem with host xylem, or even development of new host xylem. Saghir *et al.* (1973) report that *Orobanche*, infecting tomato, seems able to modify host cortical cells into xylem elements that unite immediately with the stele of the parasite. The parasite then ramifies through the tissues of the host and forms the endophyte, also called the endophytic system. Mistletoes parasitize phloem tissue and in addition produce growth structures called sinkers, which become associated with both the phloem and xylem portion of the wood rays. When it comes in contact with the host cambium, the dwarf mistletoe stimulates production of ray initials so that there are more rays per unit area in infected wood. As the tree grows and lays down annual rings, the sinkers become incorporated into the central ray tissue. Often the rays are greatly enlarged by proliferation of ray initials and parasitic tissue, and several infected rays often fuse into one (Srivastava and Esau, 1961a,b). Moreover, mistletoe initials, present in the plane of the host cambium, produce derivative cells in both the host phloem and xylem, greatly increasing the amount of contact surface between parasite and host tissue. The invariable xylem association in secondary host tissue is a dominant feature of all parasitic genera, even in those not typically pathogenic, such as *Phoradendron* (Calvin, 1967). If parasitism arose originally in mineral-poor regions, this invariable xylary union is evidence of an early need for xylem-transported materials, especially nitrogen. In some genera, such as in *Arceuthobium* the main body of the parasite is in phloem tissue, suggesting an evolution in the direction of exclusive phloem parasitism.

## IV. DISEASE INDUCED BY THE PARASITE

When the parasitic angiosperm sets up a "condominium" with the host, the collective response is in agreement, in most details, with that of other parasite systems previously described by Wood (1967).

A tremendous burst of metabolic activity occurs. The contact cells of *Phthirusa pyrifolia* (Loranthaceae), for example, contain numerous Golgi bodies, ribosomes, and an extensive endoplasmic reticulum, all suggesting the mobilization of energy-producing mechanisms.

The new host tissue being formed at the infection site respires and transpires more and, after a few days, photosynthesizes less than normal tissues. Also, the infection site becomes a sink for organic compounds, such as sugars, and ions, such as phosphorus, potassium, and sulfur.

As the parasite penetrates the tissue, it often stimulates an increase in the size and number of host cells, which can be seen macroscopically as a swelling at the site of invasion. This suggests that hormones are involved.

## A. Cytokinins

The compound most likely to be involved in this process is cytokinin. Some evidence for this is the conversion by *Orobanche* of cortical host tissue to xylem elements (Saghir *et al.*, 1973). This is strikingly similar to the induction of xylem vessel elements by cytokinins when added to carrot phloem tissue (Mizuno *et al.*, 1974).

Several observations of dwarf mistletoes also suggest cytokinin involvement. When ponderosa pine seedlings are inoculated with dwarf mistletoe, the earliest evidence of infection is often the activation of lateral buds that are normally inhibited by auxin. Presumably auxin inhibition is reversed by cytokinins (Hall, 1973).

The suppression of chlorophyll senescence has been observed in the green island effect produced by *Cuscuta* (Yarwood, 1967). This probably functions in forest trees where retention of green needles is greatest on twigs infected with dwarf mistletoe (Weir, 1916). Dwarf mistletoe also stimulates the formation of witches'-brooms, as do certain bacteria and fungi. These microbe-induced brooms can be duplicated by the application of cytokinins (Thimann and Sachs, 1966).

In most plants, cytokinins are produced primarily by roots, move through the xylem, and modify localized growth patterns (Leopold and Kriedmann, 1975). Indoleacetic acid (IAA), on the other hand, is produced in buds, moves primarily through phloem tissues, is functional at a considerable distance from the source, and inhibits root growth. Although endophytic systems of the parasites are not true roots, they perform the root functions of anchoring and uptake. Other characteristics of parasitism by angiosperms that probably involve cytokinins are mobilization of nutrients and uptake of low molecular weight metabolites. Although they could be involved in the actual uptake of molecules, the cytokinins more likely influence synthesis of the proteins that function in uptake and accumulation (Skoog and Armstrong, 1970).

Regulation of plant growth is a subtle process that typically involves more than one growth substance. These complexities will be better understood as more is learned about parasitic plants. At this point in the infection process, the stage is set for the development of a normal source–sink relationship between the exporting organ of the host and the receiving organ of the parasite.

## B. The Nutrient Sink

Of the organic nutrients, soluble sugars make up the bulk of the nutrient stream. Whether an organ acts as a source or sink at a given time depends on its concentration of these sugars (Epstein, 1972) and the level of metabolic activity. A flowering stalk of a host plant, for example, can compete successfully for metabolites against established root parasites. Although the exact mechanism of carbohydrate movement is unknown (Smith et al., 1969), Wardlaw (1974) proposed that a message produced in the sink may cause hormones to direct and control the movement of sugars along the source–sink pathway.

The movement of sucrose and inorganic ions (sulfur, phosphorus) is well established for Orobanche (Saghir et al., 1973), Striga (Rogers and Nelson, 1962; Okonkwo, 1966), Cuscuta (Ciferri and Poma, 1963), Odontites (Govier et al., 1967), and dwarf mistletoe (Hull and Leonard, 1964).

Solutes move only from host to parasite. In dwarf mistletoes the endophytic system is a strong sink for photosynthate both from the needles of the infected branch and from the main body of the tree. The interesting dynamics of simultaneous movement of photosynthate to two sinks or from two sources are discussed by Epstein (1972). Labeled photosynthate moves from treated foliage to the infection site and can also move through the dwarf mistletoe infection and into the tree, although often at a reduced rate (Hull and Leonard, 1964).

The relation of nutrient accumulation to disease can readily be seen by comparing the dwarf and leafy mistletoes. In contrast to the dwarf mistletoes, which act as powerful nutrient sinks and cause a very serious disease, the leafy mistletoes (Phoradendron) do not attract carbon-labeled metabolites from their hosts and are thus not serious pathogens. The leafy mistletoe also exports carbohydrates from its leaves and feeds its own endophytic system; the dwarf mistletoe does not (Leonard and Hull, 1965).

Sucrose is the most common carbohydrate transported in host plants, but it is glucose and fructose that accumulate in the tissue of the parasites Orobanche (Whitney, 1973) and Odontites (Govier et al., 1967). In sugarcane infected with Aegineta (Orobanchaceae) there was a threefold decrease in sucrose and a comparable increase in glucose and other reducing sugars (Lee and Goseco, 1932).

The increase in monosaccharides is correlated with a buildup of invertase (sucrase), the enzyme that splits sucrose into fructose and glucose. The conversion to monosaccharides raises the osmotic pressure, which promotes the movement of water and minerals to the infection site. Whitney (1972) reported that monosaccharides in *Orobanche* on bean are responsible for 17–30% of the osmotic pressure in the parasite but less than 5% in the host. Whether or not sugars effectively increase osmotic pressure, certainly it is true that the energy needed for growth of the parasite is derived largely from the breakdown of the monosaccharides.

Inorganic phosphorus, potassium, and nitrogen all accumulate at infection sites to levels substantially higher than in adjacent host tissue in *Orobanche, Cuscuta, Viscum, Loranthus* (Nicoloff, 1923; Gill and Hawksworth, 1961), and *Arceuthobium* (McDowell, 1964). Magnesium levels are variable, while calcium appears to be actively inhibited from moving to infection sites.

The accumulation of carbohydrates and inorganic ions at the infection site occurs early in the infection process—before sporulation in fungi (Wood, 1967) and before aerial shoot production in dwarf mistletoes (Hull and Leonard, 1964). Because of the tremendous changes in host metabolism associated with early infection, one is forced to conclude that disease induction begins at the moment parasitism is established. Initial disease symptoms—host cell division and nutrient mobilization—are a host response to invasion by agents, whether animate or inanimate. Subsequent pathogenicity depends largely on the degree of dependency on the host and the number of infections per plant.

Although there is clear evidence of nutrient transport to the infection site, it is far less certain how these compounds enter the parasitic tissue. Soluble sugars normally move through phloem sieve elements, yet there generally is no phloem tissue in the haustoria. The xylem vessels of host and parasite are connected and share what often appears to be a common cell wall. Thus, although they share a common apoplast, the living protoplasm, or symplast, of each is separate from the other. In some host–parasite combinations the two separate organisms may fuse via plasmodesmata (Hoch, 1977), but none have been reported for parasitic angiosperms, although half-plasmodesmata have been reported in dwarf mistletoe infections (Tainter, 1971).

In the absence of plasmic continuity there must be another scheme for nutrient movement. We must look to the concept of transfer cells for the solution to this enigma. Transfer cells are specialized cells in plants that function in the short-range transport of solutes. They can develop in various plant tissues, such as epidermis, companion cells, phloem, and xylem parenchyma. By means that are not understood, transfer cells develop specialized, thickened secondary wall deposits. The living plasma membrane then follows the contour of the wall ingrowths (Pate and Gunning, 1972). Ingrowth formation of wall coincides with the onset of intensive solute transport and seems actually to be a response to the presence of solutes (Pate and Gunning, 1972). Cells with such wall ingrowths have been reported in giant cells of nematode-infected tissue (Dropkin, 1969), as well as in *Orobanche, Cuscuta* (Pate and Gunning, 1972), and *Arceuthobium* (Alosi, 1978).

In coniferious hosts infected with *Arceuthobium*, there are areas in the sinkers where host ray cells and sinker parenchyma cells are closely associated, in fact, their cell walls appear fused. Areas of wall thinning between these two, similar to primary pit fields, can be seen, although there are neither actual pits nor plasmodesmatal connections in these areas. Generally, the parasite cell wall displays more thinning than the host walls. As with normal transfer cells, the inner wall of parasite wall has irregular contours, with the plasmalemma closely associated with the wall ingrowth, which provides high plasmalemma surface area and localized increase in apoplast volume (Alosi, 1979).

The most likely explanation for movement of solutes into the parasite is that these host cells become "leaky," releasing nutrients into the common cell wall that are actively taken up by the symplast of the parasite (Alosi, 1979). Although there is no known mode of function of transfer cells (Pate and Gunning, 1972), there is a body of evidence showing that sugars do move through cell walls (Giaguinta, 1976; Hawker, 1965).

## C. Respiration, Transpiration, and Photosynthesis

As with fungus-induced diseases, plants infected by parasitic angiosperms exhibit high rates of respiration, transpiration, and, at least initially, photosynthesis. There are, of course, many variations in the general pattern. For example, petunia infected with *Orobanche* has elevated root respiration, but not shoot respiration (Singh and Krishnan, 1971). Spruce branches infected with dwarf mistletoe have higher rates of both dark respiration and photosynthesis than normal branches. According to Wareing *et al.* (1968), both reactions often accompany carbohydrate removal and hormone imbalance as occurs with infection by parasitic plants. Chlorophyllous aerial shoots of *Arceuthobium* also consume twice as much oxygen as their coniferous hosts, with the highest respiration occurring in spring (Miller and Tocher, 1975).

Higher transpiration rates for parasitic plants than for hosts have been reported for *Amyema* (Wood, 1924), *Striga* (Saunders, 1933), *Orobanche* (Rakhimov, 1967), *Phoradendron* (Harris *et al.*, 1930), and *Arceuthobium* (Gill and Hawksworth, 1961). This corresponds roughly to a higher osmotic pressure for the parasite cells (Korstian, 1924). This imbalance in transpiration between host and parasite maintains a steep water potential gradient, which keeps water flowing to the parasite

The subsequent water stress of the host contributes significantly to the diseased state. The dwarf mistletoe, for example, can maintain a lower water potential than the host tree, even when the tree is under strong drought stress (Mark and Reid, 1971). It is not surprising that these parasites can colonize trees in areas of very low rainfall or that the pathogenicity is very high in dry forest areas of the western United States.

Whitney (1972) determined that the osmotic pressure of *Orobanche* is higher than that of broad bean (*Vicia faba*) roots but lower than that of the tops. This suggests that water is drawn from host roots but not from host leaves or stems. A different relationship must hold for *Striga*, which causes pronounced wilting of parasitized corn (Hattingh, 1954). The effect of parasitism by *Orobanche* on broad bean is reduced at high soil moisture (Kasasian, 1973).

Although photosynthetic rates may increase initially as a response to infection (Clark and Bonga, 1970), there follows the inevitable senescence of chlorophyll and decline of host vigor. Petunia infected with *Orobanche* has significantly less chlorophyll than uninfected plants (Singh and Krishnan, 1971). Petunia also has less root protein and less root biomass, a characteristic also noted for pine seedlings infected with dwarf mistletoe (Knutson and Toevs, 1973).

In addition to affecting host photosynthesis, many parasitic angiosperms are themselves chlorophyll-bearing plants, a unique feature among plant pathogens.

Parasites with a high rate of photosynthesis are generally nonpathogenic. The rate of photosynthesis of *Phoradendron* mistletoe, an innocuous hemiparasite, is of the same magnitude as that of the host, whereas the rate for serious plant pathogens such as *Cuscuta* and dwarf mistletoe is about 10% that of the host (MacLeod, 1961; Hull and Leonard, 1964; Miller and Tocher, 1975). The importance of photosynthesis as a source of fixed $CO_2$ in these pathogens can be questioned since *Striga* flowers and produces viable seed in total darkness (Rogers and Nelson, 1962). Similarly, aerial shoots of dwarf mistletoe appeared completely normal after 9 months of dark treatment (Hull and Leonard, 1964).

A more important role of chloroplasts in hemisparasites may involve lipid production, since choloroplasts are a major site of lipid synthesis (Yang and Stumpf, 1965). Lipids have been reported in the chloroplasts of all tissues of dwarf mistletoe, including the endophytic system (Tainter, 1971), as well as the plastids of *Cuscuta* (Dörr, 1972) and *Orobanche* (Dodge and Lawes, 1974).

In dwarf mistletoes, lipids and fatty acids are also present in great abundance in the cytoplasm of parenchyma cells of aerial shoots and the endophytic system (Tainter, 1971).

Although it is tempting to consider the lipids as a trapping agent that prevents host utilization of these nutrients, it is not clear that this actually occurs.

The oil droplets accumulate early in the infection process, at the time of highest metabolic activity of the parasite and maximal movement of sucrose from the host to the infection site. However, even in very old mistletoe tissue, the oil droplets persist without evidence of degradation (Tainter, 1971). If these droplets are neutral lipids (Tainter, 1971), probably triglycerides, they could readily be hydrolyzed under either acidic or alkaline conditions to form free glycerol and fatty acids and be converted through the glyoxalate cycle to sucrose. This evidently does not happen; apparently the need for converting the lipids back to sucrose does not arise.

Smith *et al.* (1969) suggested the interesting possibility that sugar alcohols (arabitol, mannitol) might act as nutrient traps in parasitic angiosperms much as they do in fungi. They reviewed work showing that mannitol and dulcitol are abundant in many genera in six families of parasitic angiosperms, particularly Scrophulariaceae and Orobanchaceae. Frequently, the polyols present in the parasites were absent from host tissue. Smith *et al.* (1969) recognized that their hypothesis was weakened somewhat by the fact that polyols were not present in *Arceuthobium* and were frequently found in various autotrophic families. It can be generally said that there is, in parasitic angiosperms, an accumulation of organic compounds—starches (Gill and Hawksworth, 1961), sugar alcohols (Smith *et al.*, 1969), and phenolic compounds (Khanna *et al.*, 1968). These materials are probably of host origin and are formed and maintained by considerable expenditure of energy. Presumably, they contribute to specific functions of the parasites, but in ways not yet known.

## V. A NITROGEN STARVATION HYPOTHESIS OF PATHOGENICITY

These energy-utilizing reactions increase the rate of glycolysis and thus the continuous requirement for glucose and inorganic phosphate from the host.

This loss of carbohydrates and other nutrients, plus the subsequently higher rates of respiration and other metabolic functions constitute one component of disease induction. Disease, however, is induced not only by this loss of metabolites but, more importantly, by the effect parasitism has on the regulatory mechanism in the plant which governs synthesis of organic nitrogen compounds. Subsequent imbalances in the circulation of organic compounds through the plant induce the common symptoms of disease. Plants under any stress, such as drought or infection, become initially richer in available amino acids such as proline, phenylalanine, and glutamic acid (Savnier *et al.* 1968; Kemble and MacPherson, 1954).

Substantial amounts of protein and other nitrogenous compounds are required for growth of the often-large vegetative body of parasite plants, as well as for the large seed crops they produce. The nitrogen for this protein synthesis must all come from the host plant.

The high level of both bound and free amino acids in three mistletoe genera, *Amyema, Arceuthobium,* and *Phoradendron* (Greenham and Leonard, 1965), and the presence of aspartate and glutamate transaminase enzymes in *Cuscuta* extracts (MacLeod, 1963) attest to a high level of protein synthesis. Since these enzymes can produce several amino acids from corresponding $\alpha$-keto acids and glutamic acids, their presence suggests that the parasite does not need to obtain all its amino acids from the host (MacLeod, 1963). Generally, the amino acids found in the parasite resemble the host pattern, but not exactly. For example, certain mistletoes contain hydroxyproline and asparagine, whereas their hosts do not (Greenham and Leonard, 1965).

MacLeod (1963) reported the presence of glutamate dehydrogenase from *Cuscuta,* thus providing a link between nitrogen and carbohydrate metabolism.

Nitrogen and sugar metabolism are linked in normal plants. Parasitic plants upset this linkage in two ways, and in so doing activate the primary mechanism whereby they induce disease.

Most nitrogen moves upward through the xylem in organic form, primarily as amino acids and amides. Aspartic acid, asparagine, and glutamine have been reported for a wide range of plants. The remainder consists of small amounts of alanine, leucine, methionine, serine, threonine, and valine (Bollard, 1960).

These organic nitrogenous compounds are synthesized in plant roots from sucrose transported down through phloem tissue and from inorganic nitrogen (nitrate) absorbed from the soil solution. The conversion is rapid: 60% of all sugar transported to roots is transformed to amino acids and organic acids within 3 hours. About one-half of the labeled carbon eventually returns to the upper plant through the xylem (Kursanov, 1957). Any deficiency of nitrogen causes a sharp reduction in sugar transport to the roots, which reduces synthesis of organic nitrogen, thus further contributing to functional nitrogen deficiency.

Here, then, we see the strategic value of the early association of parasite and host xylary cells. Because the xylem elements of the parasites are in union with those of the hosts, they have direct access to the host-produced organic nitrogen compounds. Diversion of xylem sap to the infection site is promoted by the high transpiration rates and high osmotic concentration due to the accumulation of carbohydrates at the infection site. The subsequent reduction in nitrogen available to the host reduces sucrose demand by the roots. There is also a simultaneous demand for sugar at the infection site. Both contribute to reduced sugar flow to roots, reduced production of amino nitrogen, and further symptoms of nitrogen deficiency. This positive feedback loop between carbohydrate and nitrogen metabolism is a likely explanation for the dramatic disease symptoms caused by parasitic angiosperms.

The testing of this hypothesis and the investigation of the many subtle effects of parasitic plants on root exudates and on colonization by mycorrhizal fungi or soilborne pathogenic microorganisms await the efforts of plant pathologists. Worldwide research on parasitic angiosperms is accelerating, perhaps fulfilling a 15-year-old prophecy: "The field of phanerogamic parasitism of both plant roots and tops should soon gain as much distinction in its relationship to phytopathology as have bacteriology, virology, hematology, and mycology" (Wellman, 1964).

## VI. SUMMARY

The basic quandry in this chapter was to resolve a seeming contradiction: General thought holds to the notion that parasitic plants rob hosts primarily of carbohydrates, especially sugars. Sugars move through phloem tissue normally, yet the overwhelming association of conduction tissue between parasitic plants and hosts are xylem connections, established quickly by the parasites. How can this be explained? The basic point of this chapter is that parasitic plants induce disease by first penetrating the host and producing hormones (in the case of parasitic angiosperms, cytokinins seem especially important) and hormone imbalances. These cause host plants to "speed up" and we can measure higher respiration, sometimes greater photosynthesis, increased cell division, etc. Such host responses are common with most pathogens, whether or not they are animate.

Accompanying these events is a source–sink relationship, with a migration of sugars and inorganic ions (P, K, S, Mg) to the infection sites. The loss of these nutrients to the host constitutes one aspect of diseasedness: the host is simply deprived of some necessary foodstuffs. A more insidious aspect, to my mind, is a pathogen-induced deregulation of a sucrose–amino acid cycle. Sugar normally moves to the roots through the phloem, combines (in part) with inorganic nitrogen, and is exported topside through the xylem as amino acids and amides. These two transport systems are linked: A reduction in one results in a reduction in the other. Parasitic plants accumulate sugar and reduce this transport to the roots. Reduced root sugar causes reduced production and export of organic nitrogen. The xylem connection between parasites and hosts ensures a bountiful flow of amino acids and amides to the parasite and reduces the amount of nitrogen available for the host. This slight nitrogen deficiency actually causes an additional reduction in the transport of sugar, which is separate from the sugar that accumulates at infection sites. The parasites, therefore, cause a vicious cycle, less sugar → less nitrogen → less sugar, which results in the chlorosis, growth reduction, and yield loss so commonly observed in parasitized plants.

One aspect of parasitism of trees by mistletoes and other parasitic angiosperms goes beyond disease and suggests an ecosysytem function. One frequently sees 500-year-old Douglas fir trees that have massive crowns totally infected by dwarf mistletoe, but yet have lived in equilibrium for probably a century. Juniper trees, infected with *Phoradendron* often have virtually no green foliage except mistletoe foliage, yet they live on, actually laying down annual growth rings. From such observations I am forced to think of these tree mistletoe organisms as a separate organization with a dynamism and physiology distinct from that of either the host or the parasite. Feibleman's Theory of Integrative Levels states: "For an organization at any given level, its mechanism lies at the level below and its purpose at the level above" (Feibleman, 1954). If, then, the "organization" is the tree–mistletoe system and if its mechanisms are at the level of physiological interdependence, what, then, is the purpose, which Feibleman states lies at the level above, and what is that level?

I propose that the level is the ecological level and the "purpose" is to provide ecosystem level stability at the cost of individual tree damage.

Mistletoe induces early spring bud break of Douglas-fir at a season when food high in nitrogen is scarce for ungulates and insects. Infected Douglas-fir tissue often supports a congregation of specialized aphids, feeding on nitrogen-rich tissue as solemnly as a full house of congressman debating a pay raise for themselves. And ants attend the aphids and surely birds eat the ants.

Other ecosystems functions include nesting by specific birds in mistletoe brooms, birds that also act as vectors. Fallen brooms of seral tree species, such as ponderosa pine and Douglas-fir, are a source of fuel for flash fires that tend to perpetuate the seral stage in the forest succession (B. Tinnin, personal communication).

Thus, although parasite flowering plants are very serious pathogens of both food and fiber crops, they are also functional parts of natural ecosystems, providing food and shelter for many creatures as one part of the complex mechanism that maintains stability of natural biological communities.

### References

Alosi, M. C. (1979). Morphological and cytological studies on *Arceuthobium* (*Viscaceae*) in relationship to host phloem, with studies on the healthy phloem in *Pinus sabiniana* (Pinaceae). Ph.D Thesis, Portland State Univ., Portland, Oregon.

Atsatt, P. R. (1973). Parasitic flowering plants: How did they evolve? *Am. Nat.* **107**, 502–510.

Barlow, B. A., and Wiens, D. (1971). The cytogeography of the Loranthaceous mistletoes. *Taxon* **20**, 291–312.

Bollard, E. G. (1960). Transport in the xylem. *Annu. Rev. Plant Physiol.* **11**, 141–166.

Calvin, C. L. (1967). Anatomy of the endophytic system of the mistletoe, *Phoradendron flavescens*. *Bot. Gaz.* **128**(2), 117–137.

Ciferri, O., and Poma, G. (1963). Fixation of carbon dioxide by *Cuscuta epithymum*. *Life Sci.* **3**, 158–162.

Clark, J., and Bonga, J. M. (1970). Photosynthesis and respiration in black spruce (*Picea mariana*) parasitized by eastern dwarf mistletoe (*Arceuthobium pusillum*). *Can. J. Bot.* **48**, 2029–2031.

Cronquist, A. (1968). "The Evolution and Classification of Flowering Plants." Houghton, Boston, Massachusetts.

De Laubenfels, D. J. (1959). Parasitic conifer found in New Caledonia. *Science* **130**, 97.

Dobbins, D. R., and Kuijt, J. (1973). Studies on the haustorium of *Castilleja* (Scrophulariaceae). I. The upper haustorium. II. The endophyte. *Can. J. Bot.* **51**, 917–931.

Dodge, J. D., and Lawes, G. B. (1974). Plastid ultrastructure in some parasitic and semi-parasitic plants. *Cytobiologie* **9**, 1–9.

Dörr, I. (1972). Der Anschlub der *Cuscuta*-Hyphen an die Siebrohren ihrer Wirtspflanzen. *Protoplasma* **75**, 167–184.

Dropkin, V. H. (1969). Cellular responses of plants to nematode infections. *Annu. Rev. Phytopathol.* **7**, 101–122.

Epstein, E. (1972). "Mineral Nutrition of Plants: Principles and Perspectives." Wiley, New York.

Feibleman, J. K. (1954). Theory of Integrative levels. *Br. J. Philos. Sci.* **5**, 59–66.

Gäumann, E. (1950). "Principles of Plant Infection. A Text-Book of General Plant Pathology for Biologists, Agriculturists, Foresters, and Plant Breeders. Crosby Lockwood, London and Hafner, New York.

Gharib, M. S. (1973). Biological and economical aspects of the broomrape (*Oro-*

*banche* spp.) in northern Iraq. *Proc. Eur. Weed Res. Counc. Symp. Parasit. Weeds,* pp. 44–47.

Giaquinta, R. (1976). Evidence of phloem loading from the apoplast. *Plant Physiol.* **57,** 872–875.

Gill, L. S., and Hawksworth, F. G. (1961). The mistletoes: A literature review. *U.S. Dep. Agric., Tech. Bull.* No. 1242.

Govier, R. N., Nelson, M. D., and Pate, J. S. (1967). Hemiparasitic nutrition in angiosperms. I. The transfer of organic compounds from host to *Odontites verna* (Bell.) Dum. (Scrophulariaceae). *New Phytol.* **66,** 285–297.

Greenham, C. G., and Brown, A. G. (1957). The control of mistletoe by trunk injection. *J. Aust. Inst. Agric. Sci.* **23,** 308–318.

Greenham, C. G., and Leonard, O. A. (1965). The amino acids of some mistletoes and their hosts. *Am. J. Bot.* **52**(1), 41–47.

Hall, R. H. (1973). Cytokinins as a probe of developmental processes. *Annu. Rev. Plant Physiol.* **24,** 415–444.

Harris, J. A., Harrison, G. J., and Pasco, T. A. (1930). Osmotic concentration and water relations in the mistletoes, with special reference to the occurrence of *Phoradendron californicum* on *Covillea tridentata. Ecology* **11,** 687–702.

Hartig, R. (1882). "Lehrbuch der Baumkrankheiten." Springer-Verlag, Berlin.

Hattingh, I. D. (1954). Control of witchweed-*Striga lutea. Farming S. Afr.* **29,** 316–318.

Hawker, J. S. (1965). The sugar content of cell walls and intercellular spaces in sugar cane stems and its relation to sugar transport. *Aust. J. Biol. Sci.* **18,** 959–969.

Hawksworth, F. G. (1973). Dwarf mistletoes (*Arceuthobium*) of coniferous forests of the world. *Proc. Eur. Weed Res. Counc. Symp. Parasit. Weeds,* pp. 231–235.

Hawksworth, F. G., and Wiens, D. (1972). Biology and classification of dwarf mistletoes (*Arceuthobium*). *U.S. Dep. Agric., Agric. Handb.* No. 401.

Heald, F. D. (1926). "Manual of Plant Diseases." McGraw-Hill, New York.

Hepper, F. N. (1973). Problems in naming *Orobanche* and *Striga. Proc. Eur. Weed Res. Counc. Symp. Parasit. Weeds,* pp. 9–17.

Hoch, H. C. (1977). Mycoparasitic relationships: *Gonatobotrys simplex* parasitic on *Alternaria tenuis. Phytopathology* **67,** 309–314.

Hull, R. J., and Leonard, O. A. (1964). Physiological aspects of parasitism in mistletoes (*Arceuthobium* and *Phoradendron*). I. The carbohydrate nutrition of mistletoe. II. The photosynthetic capacity of mistletoe. *Plant Physiol.* **39,** 996–1017.

Kasasian, L. (1973). Miscellaneous observations on the biology of *Orobanche crenata* and *O. aegyptica. Proc. Eur. Weed Res. Counc. Symp. Parasit. Weeds,* pp. 68–75.

Kemble, A. R., and MacPherson, H. T. (1954). Liberation of amino acids in perennial rye grass during wilting. *Biochem. J.* **58,** 46–49.

Khanna, S. K., Viswanathan, P. N., Tewari, C. P., Krishnan, P. S., and Sanwal, G. G. (1968). Biochemical aspects of parasitism by the Angiosperm parasites: Phenolics in parasites and hosts. *Physiol. Plant.* **21,** 949–959.

King, L. J. (1966). "Weeds of the World." Wiley (Interscience), New York.

Knutson, D. M., and Toevs, W. J. (1973). Dwarf mistletoe reduces root growth of ponderosa pine seedlings. *For. Sci.* **18**(4), 323–324.

Korstian, C. F. (1924). Density of cell sap in relation to environmental conditions in the Wasatch mountains of Utah. *J. Agric. Res.* **28,** 845–907.

Kuijt, J. (1969). "The Biology of Parasitic Flowering Plants." Univ. of California Press, Berkeley.

Kusanov, A. L. (1957). The root sytem as an organ of metabolism. *Radioisot. Sci. Res., Proc. UNESCO Int. Conf., Paris* **4**, 494–509.

Lee, A., and F. Goseco. (1932). Studies of the sugar cane root parasite *Aeginetia indica. Proc. Int. Soc. Suger-Cane Technol. Congr., San Juan* **4**, 1–12.

Leonard, O. A., and Hull, R. J. (1965). Translocation relationships in and between mistletoes and their hosts. *Hilgardia* **37**, 115–153.

Leopold, A. C., and Kriedmann, P. E. (1975). "Plant Growth and Development," 2nd Ed. McGraw-Hill, New York.

Lubenov, J. (1973). Sur le probleme pose par les mauvaises herbes parasites en Bulgarie. *Proc. Eur. Weed Res. Counc. Symp. Parasit. Weeds*, pp. 18–27.

McDowell, L. L. (1964). Physiological relationships between dwarf mistletoe and ponderosa pine. (Ph.D. Thesis, Oregon State Univ., Corvallis.) *Diss. Abstr.* **25**(1), 53.

MacLeod, D. G. (1961). Photosynthesis in *Cuscuta. Experientia* **17**, 542–543.

MacLeod, D. G. (1963). The parasitism of *Cuscuta. New Phytol.* **62**, 257–263.

Mark, W. R., and Reid, C. P. P. (1971). Lodgepole pine-dwarf mistletoe xylem water potentials. *For. Sci.* **17**, 470–471.

Mijatovic, K., and Stojanovic, D. (1973). Distribution of *Orobanche* spp. on the agricultural crops in Yugoslavia. *Proc. Eur. Weed Res. Counc. Symp. Parasit. Weeds*, pp. 28–34.

Miller, J. R., and Tocher, R. D. (1975). Photosynthesis and respiration of *Arceuthobium tsugense* (Lorathaceae). *Am. J. Bot.* **62**, 765–769.

Mizuno, K., Komamine, A., and Shimokoriyama, M. (1974). Isolation of substances inducing vessel element formation in cultured carrot root slices. *Proc. Int. Conf. Plant Growth Substances, 8th, Tokyo*, ??.

Nicoloff, T. H. (1923). Contribution à la physiologie de la nutrition des parasites végétaux supérieurs. *Rev. Gen. Bot.* **35**, 545–554, 593–601.

Okonkwo, S. N. C. (1966). Studies on *Striga senegalensis*. II. Translocation of [14]C-labelled photosynthate, urea-[14]C and sulphur[35] between host and parasite. *Am. J. Bot.* **53**, 142–148.

Pate, J. S., and Gunning, B. E. S. (1972). Transfer cells. *Annu. Rev. Plant Physiol.* **23**, 173–196.

Piehl, M. A. (1963). Mode of attachment, haustorium structure, and hosts of *Pedicularis canadensis. Am. J. Bot.* **50**, 978–985.

Plagnat, F. (1950). Le gui du sapin. *Ann. Ec. Nat. Eaux For.* **12**, 155–231.

Prillieux, E. (1895). "Maladies des plantes agricoles et des Arbres fruitiers et forestiers." Librairie de Firmin. Didot et Cie, Paris.

Rakhimov, U. K. (1967). Transpiration and diffusion pressure deficit of broom rape and the plant host. *Sov. Plant Physiol.* **14**, 631–632.

Rogers, W. E., and Nelson, R. R. (1962). Penetration and nutrition of *Striga asiatica. Phytopathology* **52**, 1064–1070.

Saghir, A. R., Foy, C. L., Hameed, K. M., Drake, C. R., and Tolin, S. A. (1973). Studies on the biology and control of *Orobanche ramosa* L. *Proc. Eur. Weeds Res. Counc. Symp. Parasit. Weeds*, pp. 106–116.

Saunders, A. R. (1933). Studies on phanerogamic parasitism with particular reference to *Striga lutea* (Lour.). *Union S. Afr., Dep. Agric. Tech. Serv., Sci. Bull. No.* 128.

Savnier, R. E., Hull, N. M., and Ehrenreich, J. H. (1968). Aspects of drought tolerance in Creosote bush (*Larrea divaricata*). *Plant Physiol.* **43**, 401–404.

Singh, P., and Krishnan, P. S. (1971). Effect or root parasitism by *Orobanche* on the respiration and chlorophyll content of petunia. *Phytochemistry* **10**, 315–318.

Skoog, F., and Armstrong, D. J. (1970). Cytokinins. *Annu. Rev. Plant Physiol.* **21**, 359–384.

Smith, D., Muscatine, L., and Lewis, D. (1969). Carbohydrate movement from autotrophs to heterotrophs in parasitic and mutualistic symbiosis. *Biol. Rev. Cambridge Philos. Soc.* **44**, 17–90.

Srivastava, L. M., and Esau, K. (1961a). Relation of dwarf mistletoe (*Archeuthbium*) to the xylem tissue of conifers. I. Anatomy of parasite sinkers and their connection with host xylem. *Am. J. Bot.* **48**, 159–167.

Srivastava, L. M., and Esau, K. (1961b). II. Effect of the parasite on the xylem anatomy of the host. *Am. J. Bot.* **48**, 209–215.

Steward, A. N. (1958). "Manual of Vascular Plants of the Lower Yangtze Valley, China." International Academic Print. Co., Tokyo.

Tainter, F. H. (1971). The ultrastructure of *Arceuthobium pusillum*. *Can. J. Bot.* **49**, 1615–1622.

Teryokhin, E. S. (1977). Origin and evolution of the basic types and modes of parasitism in flowering plants (in Russian). *Bot. Zh.* (*Leningrad*) **62**(6), 777–792. [Engl. transl., *Curr. Adv. Ecol. Sci.* **3**(9), (1977).]

Thimann, K. V., and Sachs, T. (1966). The role of cytokinins in the fasciation disease caused by Corynebacterium fascians. *Am. J. Bot.* **53**, 731–739.

Thoday, D. (1951). The haustorial system of *Viscum album*. *J. Exp. Bot.* **2**, 1–19.

Wardlaw, I. F. (1974). Phloem transport: Physical, chemical or impossible. *Ann. Rev. Plant Physiol.* **25**, 515–539.

Wareing, P. F., Khalifa, M. M., and Treharne, K. J. (1968). Rate-limiting processes in photosynthesis at saturating light intensities. *Nature* (*London*) **220**, 453–457.

Weir, J. R. (1916). Mistletoe injury to conifers in the Northwest. *U.S. Dep. Agric., Bull.* No. 360.

Wellman, F. L. (1964). Parasitism among neotropical phanerogams. *Annu. Rev. Phytopathol.* **2**, 43–56.

Whitney, P. J. (1972). The carbohydrate and water balance of beans (*Vicia faba*) attacked by broomrape (*Orobanche crenata*). *Ann. Appl. Biol.* **70**, 59–66.

Whitney, P. J. (1973). Transport across the region of fusion between bean (*Vicia faba*) and broomrape (*Orobanche crenata*). *Proc. Eur. Weed Res. Counc. Symp. Parasit. Weeds*, pp. 154–166.

Wood, J. G. (1924). The relations between distribution, structure, and transpiration of arid South Australia plants. *R. Soc. So. Austral. Trans.* **48**, 226–235.

Wood, R. K. S. (1967). "Physiological Plant Pathology." Blackwell, Oxford.

Yang, S. F., and Stumpf, P. K. (1965). Fat metabolism in higher plants. XXI. Biosynthesis of fatty acids by avocado mesocarp enzyme systems. *Biochim. Biophys. Acta* **98**, 19–26.

Yarwood, C. E. (1967). Response to parasites. *Ann. Rev. Plant Physiol.* **18**, 419–438.

Chapter 17

# Allelopathy

RICHARD F. FISHER

## I. INTRODUCTION AND TERMINOLOGY

Allelopathy is a beautifully euphonious word. It comes from the Greek, *allēlōn*, "of each other," and *pathos*, "to suffer"—"the suffering of each other."

The purpose of this chapter is to consider that body of knowledge which concerns the production by one plant of chemicals that induce

PLANT DISEASE, VOL. IV
Copyright © 1979 by Academic Press, Inc.
All rights of reproduction in any form reserved
ISBN-0-12-356404-2

suffering in another. This is allelopathy. This is chemical pathogenesis. Most investigators consider this to be a one-way street, from plant A to plant B, but we must not overlook the possibility of plant B fighting back with its own chemicals.

Allelopathy, as used here, includes the case of plants that are pathogenic to plants of other species but are not parasitic on them. Knutson discusses the latter in this volume, Chapter 16. In Chapter 18 Hoestra describes the competition between plants of the same species.

Allelopathy, as used here, must not be confused with physical competition, such as crowding or shading. Muller (1969) suggested that the total influence of one plant on another should be termed "interference"; thus, interference includes both physical and chemical effects.

The term *allelopathy* was coined by Molisch (1937) to refer to both detrimental and beneficial biochemical interactions among all classes of plants, including microorganisms. The root word, *pathy,* however, implies detrimental interactions, and in general the use of the term has been restricted to detrimental effects. In this chapter, I shall also use the term in its restricted sense. Although allelopathic interactions occur between all types of plants, I shall limit the present discussion chiefly to those between seed plants in nature or in horticulture or agriculture. The chemicals involved are sometimes called kolines (Rice, 1974).

## II. HISTORICAL BACKGROUND

The study of allelopathy has had a cloudy past. In the early nineteenth century De Candolle (1832) reported that root excretions of several plants were injurious to common crop plants. He was a "prophet crying in the wilderness," however, because for a hundred years after his report only an occasional lonely soul, such as Way (1847), dared to depart from the dogma that plants compete only physically for "lebensraum," water, light, and nutrients.

If De Candolle's thory had been accepted by others, the study of allelopathy would have advanced much faster. Instead, the work of von Leibig and later of Lawes and Gilbert showed that mineral nutrition could answer many of the questions raised by the soil sickness dilemma. Plant physiology launched into the essential element era, and allelopathy and De Candolle were forgotten.

In the early 1900s Schreiner and associates, working on the problem of fatigued soils, rediscovered allelopathy (Schreiner and Lathrop, 1911). They found that they could recover substances from crop plants and from previously cropped soil that were deleterious to many common crop plants. Their work suggested that crop rotations combat more than simply nutritional, insect, and biotic disease problems. Again, the work hardly caused a stir in the scientific community. Chemical fertilizers were becoming commonplace, and they seemed to hold the answer to the problem of fatigued soil. Studies of allelopathy again disappeared from the scene.

In the middle of the twentieth century, new interest in crop residues and allelopathy arose. Bonner found that the residue of guayule, *Parthenium argentatum*, produces *trans*-cinnamic acid, which is toxic to young guayule plants. He found also that the cinnamic acid is slowly decomposed in the soil so that the effect dissipates with time (Bonner and Galston, 1944; Bonner, 1946). These findings led McCalla and associates to consider allelopathy as a potential cause of the stubble mulch problem (McCalla and Haskins, 1964). In the arid Great Plains the abandonment of moldboard plowing in favor of planting the new crop right through the stubble or residue of the previous crop had long been advocated. McCalla and Patrick (Patrick *et al.*, 1964) found that, during decomposition, substances that inhibit germination or growth of the new crop are produced from the residues of the previous crops. As in Bonner's work, these effects disappear with time. After more than 100 years, De Candolle's theories had become well documented in the agronomic arena.

In ecology a similar pattern developed. After Way's work in the middle of the nineteenth century, little was done until Cowles (1911) suggested that biochemical interactions among plants might be important in determining the course and rate of succession. Ecologists who were interested in succession were ill equipped to follow Cowles' lead, and the plant physiologists who could were simply not interested. It was not until the late 1950s that workers trained in physiology but interested in succession began to unlock the chemical secrets of the dynamics of plant communities.

Muller in California and Rice in Oklahoma, along with many others, began to look closely at allelopathy as an important phenomenon in succession. This work, more than anything else, has led to renewed interest in allelopathy and, in fact, to this chapter. One can see that, although the concept of allelopathy has been around for a long time, the study of allelopathy is nonetheless a young field.

This chapter raises more questions than it answers, but perhaps it will stimulate interest in what will become an increasingly active, interesting, and controversial area of study in the future.

## III. ALLELOPATHIC INTERACTIONS IN
## NATURAL ECOSYSTEMS

### A. Phytoplankton Succession

The causes of plant succession have been the subject of considerable research in the past three or four decades. The rapid buildup of phytoplankton during a "bloom" in a body of water, the succession of species during the course of the bloom, and the rapid disappearance of the bloom have all been puzzling problems for limnologists and phycologists. The work of T. R. Rice (1954) and others has shown in culture that some algae produce autotoxins that limit their own populations. Other workers (Proctor, 1957; Lefevre, 1952) have shown that algae produce substances in culture that are algastatic, algacidal, or algadynamic. The evidence from culture studies indicates that allelopathic mechanisms may control algal populations, or at least play a role in their control. Although there is no direct evidence for allelopathic control of phytoplankton blooms or pulses in natural bodies of water, this is a promising area for future research.

### B. Terrestrial Succession

The evidence that allelopathy is an important factor in terrestrial succession is more complete. Rice (1974) and his students have shown that the rate and course of secondary succession on abandoned fields in the central Great Plains is controlled largely by allelopathic responses. The rapid disappearance of the pioneer weed stage is due to autotoxicity and the production of kolines by many of the pioneer plants. Johnson grass (*Sorghum halepense*), sunflower (*Helianthus annus*), crabgrass (*Digitaria sanguinalis*), ragweed (*Ambrosia psilostachya*), and *Euphorbia supina* all produce kolines that are effective against each other as well as other pioneer-stage plants. Triple awn grass (*Aristida oligantha*), however, is not affected by these allelopathic chemicals; therefore, it quickly dominates the second stage of succession. After the rapid developments early in succession the rate slows markedly, and many decades may elapse before the true climax plants occupy the area.

### C. Inhibition of Nitrogen Fixation and Nitrification

Rice (1974) believes that the slowing of succession in the second stage is due largely to a very slow rate of nitrogen accrual in the soil. Nitrogen accretion is accomplished mainly by biological fixation of atmospheric nitrogen. Rice and co-workers have found that second-stage grasses produce chemicals that inhibit nitrogen fixation by both free-living and symbiotic nitrogen-fixing bacteria. Through this activity, second-stage plants that require little nitrogen can delay the invasion of climax plants that require much nitrogen.

The slowing in the second stage of succession need not be controlled by nitrogen availability alone, however. In the eastern deciduous forest both woody and nonwoody second stage-plants produce allelopathic compounds that help delay the invasion of the climax species (Horsley, 1977a,b; Fisher et al., 1978).

In the sclerophyllous chaparral area of California another successional cycle is mediated by allelopathic activity (Muller et al., 1968). In chaparral several species of shrubs exercise dominance over the herbaceous plants. Fire often occurs in this type of vegetation, and the shrubs are generally consumed by the fire. In the following rainy season the burned areas are invaded by many herbaceous plants. As the shrubs sprout again or new seedlings develop, allelopathic compounds are released that again inhibit the herbaceous plants. This occurs first as small rings around the shrubs, but as the shrub canopy closes, the herbs disappear until another fire destroys the shrub overstory.

Another potential role of allelopathy in succession is the inhibition of nitrification by climax vegetation. Rice (1974) presents an intricate argument for this role. Presumably, climax vegetation inhibits the formation of the easily leached nitrate ion in favor of the more stable ammonium ion. However, the soil–plant system is very complex, and there are other explanations for much of the data Rice uses to support this argument. The idea that successional stages control the form of available soil nitrogen is an intriguing one, and it should receive a good deal more attention.

## D. Prevention of Seed Decay

As they germinate in the soil, seeds are subject to attack by many microorganisms. Plants have evolved chemical defenses against these hazards. Most seeds produce one or more allelopathic chemicals that inhibit one or more types of soil microorganisms (Nickell, 1960). In fact, many plant reproductive parts may have such defenses. The chestnut blight fungus (Endothia parasitica) attacks shoot, but not root, tissue. This indicates that roots also produce some protective chemicals that shoots do not. Mycorrhizal roots of pine have a built-in antibiotic system that helps retard pathogenic attack. These sorts of mechanisms, although they never seem to prevent disease, serve a vital role in host–pathogen interactions.

## IV. ALLELOPATHIC INTERACTIONS IN
## MANAGED ECOSYSTEMS

### A. Allelopathy in Agriculture

The importance and influence of allelopathy in agriculture has been changing rapidly in recent years. The study of allelopathic interactions and the development of modern agriculture have been closely related. The problem of soil sickness had important allelopathic origins but was essentially solved by fertilization. The problem of stubble mulch or inhibition by crop residue, a clear case of allelopathy, was solved when normal moldboard plowing in the autumn was resumed.

In general, allelopathic problems have been solved inadvertently by the intensive cultivation, fertilization, and weed control practices of modern agriculture. What will happen when and if energy contraints make these techniques too expensive to use? The problem of stubble mulch is a preview of this potential event. When minimal tillage is used, allelopathic interactions prevent certain crop rotations. If the use of fertilizers and weed control chemicals is curtailed in the future, a much better understanding of the role of allelopathy in agriculture will be required in order to maintain crop yields.

### B. Allelopathy in Horticulture

The problems involving orchard replanting of apples, peaches, and other members of the family Roseaceae as discussed by Hoestra in this volume, Chapter 18, can often be traced to allelopathic interactions. Peaches produce amygdalin, which is hydrolyzed in the soil to produce HCN. Other members of the Roseaceae produce amygdalin or other allelochemicals. In many cases these plant products are toxic only when they are acted on by microorganisms (Patrick et al., 1964). When the large amount of root material from an old orchard tree is decomposed, a sufficient quantity of allelopathic chemical is produced to inhibit plants in the area, even new plants of the same species.

Many plants that do not commonly grow together are often put together in ornamental plantings. There is almost nothing in the literature about allelopathic interactions among such plants, but gardeners and nurserymen know many of the interactions well. Solonaceae, Cruciferae, and Roseaceae are well known for their allelopathic or soil-poisoning effects. Other trees and shrubs are thought to do poorly in certain combinations. With the large folklore on this subject it seems that much more research should be done in the area.

Certain types of bark mulch and bark or other organic waste used as soil amendments have strong allelopathic effects. Of course, nutritional problems often occur after the addition of large quantities of organic material to the soil. However, even when large amounts of fertilizers are used to combat this problem, fresh oak bark and green pine straw have inhibitory effects on nursery plants. Of the bark mulches used Douglas-fir, pine, and hardwood barks have the most inhibitory effect on annuals and on ground-cover plants, such as *Pachysandra* and *Cotoneaster*.

Many ground-cover plants, such as *Hypericum* and those in the family Ericaceae, have strong allelopathic effects on some annual and perennial plants. No doubt much of their success as ground covers is due to their allelopathic activity. This is another area where observation has led to many hypothetical interactions that must be corroborated by research findings.

### C. Allelopathy in Forestry

Forestry offers many potential instances of allelopathy, but few have been confirmed by research. It appears that old field weeds are allelopathic toward black cherry (*Prunus serotina*) and maple (*Acer saccharum*). Bracken fern (*Pteridium aquillinum*) has also been shown to be detrimental to the establishment of Douglas-fir (*Pseudotsuga menziesii*), and *Festuca* and *Muhlenbergia* grasses are allelopathically active against ponderosa pine (*Pinus ponderosa*). Since most trees are species of the second or a later successional stage, there are many chances for potential allelopathic interactions.

Research conducted at the University of Toronto showed that foliage extracts of bearberry (*Arctostaphylos uva-ursi*) and sheep laurel (*Kalmia angustifolium*) were only slightly inhibitory to the germination and early growth of jack pine (*Pinus banksiana*). The same extracts were progressively more inhibitory to red pine (*P. resinosa*), white pine (*P. strobus*), white spruce (*Picea glauca*), and balsam fir (*Abies balsamea*). Since this is approximately a successional sequence for these species, it indicates one reason why jack and red pine are the most successful pioneers in the group (Fisher, unpublished observations).

As in agriculture, it appears that foresters have learned to control many allelopathic interactions without knowing that they exist. Intensive site preparation, fertilization, weed control, and even the planting of seedlings rather than seed all reduce the allelopathic effects that the new trees must endure. When and if forests must be established without these intensive silvicultural treatments, a more thorough knowledge of allelopathy will be necessary. Obviously, nature combats allelopathic effects, or there would be no forests; therefore, it seems likely that many ways to deal with allelopathic problems can be found.

## V. SOURCE AND RELEASE OF
## ALLELOPATHIC CHEMICALS

### A. Plant Parts That Contain Allelopathic Chemicals

As mentioned earlier, many seeds contain inhibitors of microorganisms and seed germination. This source of allelochemicals is very specific, and, although quite important to the seed, the chemicals do not have a long-term effect in the ecosystem. Fruits, on the other hand, contain inhibitors that not only are important during seed germination, but also play other important roles in the ecosystem. One of the best examples is the fruit husk of walnut (*Juglans*). This husk contains juglone, a chemical with wide phytotoxic properties. Some flowers undoubtedly also contain toxins, but we know little about the chemicals or the frequency of occurrence.

Stems, leaves, and roots have often been assayed for inhibitors. Stems have not received the same attention as leaves and roots, but they undoubtedly contain some inhibitors. Essentially all classes of inhibitors have been isolated from leaves. Roots generally have been found to contain fewer toxins at lower concentrations than leaves. Since chemicals can be localized within plants, it is common to find inhibitors in only one plant part; in some cases, however, the inhibitor occurs throughout the plant. Extensive lists of inhibitory chemicals and the plant parts that contain them are available in Ohman and Kommendahl (1960), Harborne (1964), Bonner (1965), and Rice (1974).

### B. Release of Allelopathic Chemicals into the Environment

Allelopathic chemicals are released into the environment through volatilization, leaching, root exudation, and the decomposition of plant parts. Volatilization is mainly a phenomenon of plants in arid and semiarid environments. A large number of taxonomically diverse genera have been shown to release inhibitors in this way: *Ambrosia, Artemisia, Eucalyptus, Heteromeles, Lepechinia, Prunus, Salvia,* and *Umbillularia* (Muller, 1966; Rice, 1974).

Volatilized inhibitors are generally terpenoid essential oils of the monoterpene or sesquiterpene group. They are often released into the atmosphere from special glands on the stems or leaves. They may be absorbed through the cuticle of neighboring plants directly from the air or as condensate in dew. They may also be adsorbed on dry soil, released into soil solution in wet soil, and taken up by the plant (Muller and del Moral, 1966).

Many organic chemicals, often in large quantities, can be leached from foliage by rain water or fog drip. The variety of chemicals that are leachable seems endless. Free sugars, sugar alcohols, amino sugars, pectic substances, amino acids, many organic acids, gibberellins, terpenoids, alkaloids, and many phenolic substances have been shown to be leached. Tukey and his students have done much of the work on leaching, and Tukey (1970) has reviewed the phenomenon and the factors that control it.

Root exudates have often been mentioned as potential sources of allelopathic chemicals, but little careful work has been done on them. Horsley (1977a) demonstrated that root washings of *Solidago* and *Aster* inhibit black cherry (*Prunus serotina*), but Fisher *et al.* (1978) found that root exudates of *Solidago* and *Aster* are not as effective against sugar maple (*Acer saccharum*) as are compounds released upon decomposition. A good deal more $^{14}C$ tagging of allelopathic chemicals, such as the work of Linder *et al.* (1957), must be carried out to assess the true potential of root exudates as sources of allelopathic chemicals.

Following the death of a plant, the integrity of the tissues and cells is lost, and the ease with which many chemicals are leached increases markedly. In living tissue many compounds are compartmented so that they do not move freely in the plant and do not come together. The release of HCN from the cyanogenic glycoside, amygdalin, is a case in point. The roots of many *Prunus* species contain both amygdalin and emulsin, the enzyme responsible for its breakdown to HCN and benzaldehyde. While the root is alive these chemicals are separated, but, after death and loss of cell integrity, the amygdalin and emulsin react and HCN is released (Patrick, 1955).

Some plants do not produce toxins themselves but provide microorganisms with the raw materials for the production of inhibitors (Norstadt and McCalla, 1968). Often it is difficult to separate chemicals produced by higher plants and released during decomposition from chemicals produced by the microorganisms responsible for decomposition. Regardless of the exact source, it appears that the highest concentrations of allelopathic chemicals are released into the environment during plant decomposition.

## C. Factors Affecting the Quantities of Allelopathic Chemicals Produced

There are considerable data on the effect of environmental conditions on the amount of phenolic acids, terpenes, and alkaloids in crop plants such as tobacco. Since many inhibitors are in these chemical groups, it seems reasonable that they would respond similarly. Increased intensity of ionizing radiation or ultraviolet light increased the phenolic content of tobacco (Koeppe et al., 1970). Also, greater day length induces higher levels of phenolics and terpenes in both short- and long-day plants (Taylor, 1965; Zucker, 1969).

Mineral deficiencies also play an important role in the control of allelopathic chemical concentration. Deficiencies of boron, calcium, magnesium, nitrogen, phosphorus, or sulfur result in an increase in the concentrations of most phenolic compounds in the leaf but a decrease in the concentrations of some phenolic compounds. A potassium deficiency seems to cause more phenolic concentrations to decrease than increase. Clearly, each allelopathic chemical must be investigated to determine exactly how its concentration is affected by nutrient deficiency.

All stress conditions appear to cause an increase in phenolic concentrations in foliage. Two things are important in this regard. First, an increase in concentration does not necessarily mean an increase in amount since the plant growing under stress may be smaller. Second, data are available on only two or three of the many classes of allelopathic agents and only on a few specific allelopathic chemicals. A much more thorough investigation will be necessary before we can learn exactly how the environment influences the plant's content of allelopathic chemicals. The chronological or phenological age of the plant also can influence inhibitor content. Both have strong effects on inhibitor levels, but the literature contains very little documentation.

## VI. CHEMICAL NATURE OF ALLELOPATHIC COMPOUNDS

Although most allelopathic chemicals are secondary plant products, many of them occur widely in the Plant Kingdom. Aliphatic organic acids are common plant constituents and are present in most soils as well. These acids are common decomposition products of plant residue. Dihydroxystearic acid, crotonic acid, as well as oxalic, formic, butyric, lactric, acetic, and succinic acids can inhibit germination or growth. Under aerobic conditions, however, all of these compounds are quickly broken down in the soil. Aliphatic acids, therefore, are not a major source of allelophatic activity.

Lactones are represented by a single identified allelopathic agent, patulin. This simple lactone is produced by a *Penicillium* growing on wheat (*Triticum*) straw and is toxic to corn (*Zea maize*).

Alkaloids comprise the largest class of secondary plant compounds. They occur almost exclusively in the angiosperms. Many alkaloids, such as nicotine, are specific for a family or even a species. Little work has been done on alkaloids as allelopathic chemicals, although there is ample evidence that nicotine is a strong inhibitor of even closely related plants. Even though the chemistry may be complicated, alkaloids should receive more attention when attempts are made to isolate the active chemicals in allelopathic interactions.

Amino acids are so universal that it seems incongruous that they might be allelopathic. However, several cases of amino acids and polypeptides having allelopathic effects are known (Owens, 1969). There is also the possibility that amino acid analogs act as allelopathic compounds.

Cyanogenic glycosides were mentioned earlier. Amygdalin, or its reduced form (prunasin), and dhurrin have been shown to have allelopathic effects. Not only are these glucosides of mandelonitrile hydrolyzed to produce HCN, but the benzaldehyde or hydroxybenzaldehyde produced during hydrolysis is oxidized to benzoic acid, which itself may be toxic to some plants.

Terpenoids comprise a large group of secondary plant products that contain several allelopathic chemicals. Monoterpenes, such as α-pinene, β-pinene, camphor, and cineole, are among the primary allelopathic agents in arid zones. Sesquiterpenes, such as β-caryophyllene, bisabolene, and chamazulene, have also been shown to be allelopathic. Also, the sequiterpene lactones arbusculin A, achillin, and viscidulin C are known to be inhibitors of plant growth.

Phenolics comprise the largest group of secondary compounds in plants and are more often identified as allelopathic agents than all other chemicals put together. Of the simple phenols, hydroquinone and its glycoside, arbutin, have been identified as allelopathic inhibitors.

Simple phenolic acids are widely distributed among plants, and many have been identified as inhibitors. Benzoic, gallic, ellagic, vanillic, salicylic, and sulfosalicylic acids are all present in leaf leachates and soil, and they are inhibitory to plant growth. Among the more complex phenolic acids, the large group of cinnamic acids and coumarins contains many inhibitors. Cinnamic, coumaric, caffeic, chlorogenic, and sinapic acids are known allelopathic chemicals that apparently occur widely in nature.

Coumarins are lactones of *o*-hydroxycinnamic acid; they occur as glycosides in plants and are easily leached into the environment. Scopoletin, scopolin, esculetin, esculin, and methylesculin are coumarins that have been shown to have allelopathic potential.

Quinones comprise a large class of phenolic compounds that occur widely in plants. Only juglone, 5-hydroxynapthoquinone, has been identified as an allelopathic chemical. In contrast to other allelopathic chemicals, such as the cinnamic acids, juglone occurs only in the genus *Juglans,* but the concentration varies widely within the species of the genus.

Flavanoids comprise the largest class of phenolic chemicals, but only a few have been shown to be allelopathic. Phlorizin or its breakdown products, glycosides of kaempferol, quercetin, and myricetin are known to be inhibitors. This large class is difficult to assay and probably contains several more inhibitors. Similarly, the condensed and hydrolyzable tannins may also contain many allelopathic chemicals.

To date many of the methods used to separate and assay potential inhibitors have been crude. Because phenolic compounds are more easily isolated than many other potential allelopathic agents, the current list of known allelopathic chemicals is more closely related to the technology used to discover them than to their occurrence in nature.

## VII. MECHANISMS OF ACTION OF ALLELOPATHIC CHEMICALS

Few studies have investigated the physiological effects of allelopathic chemicals; nevertheless, a wide range of processes have been invstigated. It appears that some chemicals interfere with ontogenetic processes, some with energy transfer processes, and still others with nutrient uptake or transport.

### A. Inhibition of Ontogenetic Processes

Volatile inhibitors from *Salvia leucophylla* inhibited mitosis and root cell elongation in several plants tested by Muller and Hauge ( 1967). Jensen and Welbourne ( 1962) found that juglone inhibited mitosis in test plants, and Cornman ( 1946) found that coumarin blocked mitosis in onion roots. Growth induced by gibberellin and indoleacetic acid (IAA) is also inhibited by allelopathic chemicals. Corcoran *et al.* ( 1972) tested a wide variety of identical allelopathic chemicals and found that several tannins, benzoic acids, and cinnamic acids inhibit gibberellic acid-induced growth. They found that some allelopathic chemicals inhibited but some stimulated IAA-induced growth. Many of the effects of allelopathic chemicals may be due to inhibition of root ontogeny in the affected species.

## B. Inhibition of Energy Transfer Processes

Respiration, oxidative phosphorylation, photosynthesis, and enzyme activity may also be inhibited by allelopathic chemicals. Monoterpenes, sesquiterpenes, and coumarins have been shown to inhibit respiration (McCahon et al., 1973; Muller et al., 1968). Many phenolic compounds reduce respiration but usually only when the inhibitor is present in high concentration (Demos et al., 1975). Juglone inhibits oxidative phosphorylation even when it is applied in low concentration (Koeppe, 1972). Einhellig showed that scopoletin inhibits photosynthesis in tobacco. This is apparently caused by a reduction in stomatal aperture (Einhellig and Kuan, 1971).

Effects on enzyme activity have not been as well studied, but there are indications that allelopathic chemicals interfere with some enzymes (Owens, 1969).

## C. Inhibition of Nutrient Uptake

Seemingly the most common, but least understood, effect of allelopathic chemicals is inhibition of nutrient uptake (Buchholtz, 1971). Many studies have shown that allelopathically affected plants have much lower nutrient contents than normal, but few of these studies have shown how this occurs. Several workers have shown that phenolic acids inhibit ion uptake in cell suspensions or isolated mitochondria (Glass and Bohm, 1971; Demos et al., 1975). Either these responses were not independent of respiratory metabolism, or no information was collected on independence from respiratory metabolism. Glass and Dunlop (1974) have shown that several phenolic acids that inhibit ion uptake depolarize root membranes. Glass's studies give the best insight into how nutrient uptake is influenced by phenolic acids, but knowledge of how allelopathic chemicals affect nutrient uptake is far from complete.

## VIII. PATH OF ALLELOPATHIC CHEMICALS IN THE ENVIRONMENT

### A. Model of the Fate of an Allelopathic Chemical in the Environment

The potential fate of an allelopathic chemical in the environment is outlined schematically in Fig. 1. Donor plants produce toxins or their precursors, and these toxins are released into the environment in any

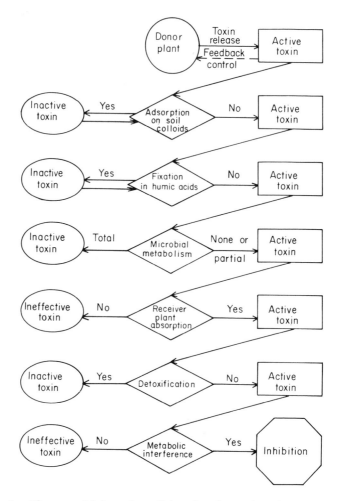

**Fig. 1.** The potential fate of an allelopathic chemical in the environment.

one of the several ways already discussed. Whether there is "feedback control" or any other form of control over this process is unknown.

Whatever quantity of toxin is released generally enters the soil, where several things may happen to it. The order of the second, third, and fourth steps in Fig. 1 is not necessarily the order in which the reactions actually occur. In fact, if the toxin is a volatile substance, these steps may be skipped entirely. In most cases, however, it appears that allelopathic chemicals enter the soil.

Soil colloids are capable of adsorbing most allelopathic chemicals.

Such adsorption would result in temporary loss of toxin activity. Chemical changes could occur during adsorption that would permanently deactivate the toxin. The adsorption reactions are usually reversible, however, so that some or all of the toxin would still be available for absorption by a receiver plant.

The toxin is also likely to be adsorbed or complexed by soil humic acids. If the reaction is a simple adsorption reaction, all or part of the toxin might later become avalable for absorption by a receiver plant. If the toxin is complexed or precipitated by its reaction with soil humic substances, then it would be deactivated.

The toxin may undergo microbial degradation either while it is free in soil solution or while it is adsorbed. This could destroy all or part of the toxin, and there is evidence that most of the natural organic chemical groups that contain allelopathic compounds can be metabolized by some microorganism. The possibility always exists that the microbial degradation product from the metabolism of an active toxin will itself be an allelopathic chemical.

The reactions that a toxin undergoes in the soil are largely controlled by edaphic factors such as moisture regime, nutrient status, or organic matter content. Soil moisture regime helps to determine whether aerobic or anaerobic decomposition takes place, which, in turn, helps to determine the quantity of toxin metabolized and the nature of the decomposition products. Soil nutrient status and soil temperature help to determine the rate of microbial activity. The nature and amount of soil organic matter determine whether simple adsorption or complexing by humic substances takes place. Microbial degradation is also controlled by the spectrum of microorganisms present in the soil. These edaphic effects mean that different things will happen to the same toxin introduced into different soils or even into the same soil at different times.

It seems unlikely that the allelopathic chemicals that may be extracted from plant material are actually those that reach the host plant, yet nearly all our information on allelopathic compounds is derived from extracts that have never been exposed to the soil. Some compounds, such as juglone, may remain unchanged in the soil under some circumstances (Fisher, 1978), but many compounds, such as ferulic or salicylic acid, are converted to other chemicals in the soil.

Toxin that is free in the soil solution is available for uptake by the receiver plant. Most, if not all, allelopathic chemicals are taken up by plants, but plants may discriminate against certain toxins on the basis of size (molecular weight) or some other factor. However, we do not know exactly which plants absorb which chemicals. It is only poorly understood why and how plants are able to discriminate against some chemicals. If the plant does not absorb the toxin, the toxin is usually ineffective.

Once the toxin is absorbed, it must be translocated to the site where it is capable of interfering with metabolism. If translocation is blocked, the toxin will be ineffective. Some plants may be capable of detoxifying an allelopathic chemical that is absorbed. The evidence for such capability is largely indirect, but this is certainly an area deserving of considerable research. If the toxin is absorbed and translocated but not detoxified within the plant, the toxin interferes with the host plant's ontogeny or its metabolism.

## B. Studies That Cross Disciplinary Boundaries

Many reactions can be inhibited. A few chemicals have been shown to inhibit specific reactions, but these same chemicals might also affect other reactions. In addition, the number of chemicals that have been studied is only a fraction of those that have been shown to have allelopathic effects. Figure 1 shows that the complete path of a toxin from its origin in the donor plant to its action in the host plant is very complex. In order to understand fully the allelopathic effects, each step along this pathway must be elucidated. Clearly, to achieve this requires an intedisciplinary effort. Physiological, ecological, microbiological, organochemical, and biochemical studies are necessary. This can seldom be accomplished by a single scientist. To date, the complete pathway in Fig. 1. has not been traced for any allelopathic chemical. Until this can be done for several, mystery will continue to cloud the credibility of allelopathy.

### References

Bonner, J .(1946). Further investigations of toxic substances which arise from guayule plants. *Bot. Gaz.* **107**, 343–351.

Bonner, J. (1965). The Isoprenoids. *In* "Plant Biochemistry" (J. Bonner and J. E. Varner, eds.), pp. 665–692. Academci Press, New York.

Bonner, J., and Galston, A. W. (1944). Toxic substances from the culture media of guayule which may inhibit growth. *Bot. Gaz.* **106**, 185–198.

Buchholtz, K. P. (1971). The influence of allelopathy on mineral nutrition. *In* "Biochemical Interactions Among Plants," pp. 86–89. Natl. Acad. Sci., Washington, D.C.

Corcoran, M. R., Geissman, T. A., and Phinney, B. O. (1972). Tannins as gibberellin antagonists. *Plant Physiol.* **49**, 323–330.

Cornman, I. (1946). Alteration of mitosis by coumarin and parascorbic acid. *Am. J. Bot.* **33**, 217.

Cowles, H. C. (1911). The causes of vegetative cycles. *Bot. Gaz.* **51**, 161–183.

DeCandolle, M. A.-P. (1832). "Physiologie Vegetale," Vol. III. Bechet Jeune, Libr. Fac. Med., Paris.

Demos, E. K., Woolwine, M., Wilson, R. H., and McMillan, C. (1975). The effects of ten phenolic compounds on hypocotyl growth and mitchondrial metabolism of mung beans. *Am. J. Bot.* **62**, 97–102.

Einhellig, F. A., and Kuan, L.-Y. (1971). Effects of scopoletin and chlorogenic acid on stomatal aperature in tobacco and sunflower. *Bull. Torrey Bot. Club* **98**, 115–162.

Fisher, R. F. (1978). Juglone inhibits pine under certain moisture regimes. *Soil Sci. Soc. Am. J.* **42**, 801–803.

Fisher, R. F., Woods, R. A., and Glavicic, M. R. (1978). Allelopathic effects of goldenrod and aster on young sugar maple. *Can. J. For. Res.* **8**, 1–9.

Glass, A. D. M., and Bohm, B. A. (1971). The uptake of simple phenols by barley roots. *Planta* **100**, 93–105.

Glass, A. D. M., and Dunlop, J. (1974). Influence of phenolic acids on ion uptake. IV. Depolarization of membrane potentials. *Plant Physiol.* **54**, 855–858.

Harborne, J. B. (1964). "Biochemistry of Phenolic Compounds." Academic Press, New York.

Horsley, S. B. (1977a). Allelopathic inhibition of black cherry by fern, grass, goldenrod, and aster. *Can. J. For. Res.* **7**, 205–216.

Horsley, S. B. (1977b). Allelopathic inhibition of black cherry. II. Inhibition by woodland grass, fern and clubmoss. *Can. J. For. Res.* **7**, 515–519.

Jensen, T. E., and Welbourne, F. (1962). The cytological effect of growth inhibitors on excised roots of *Vicia faba* and *Pisum sativum*. *Proc. S.D. Acad. Sci.* **41**, 131–136.

Koeppe, D. E. (1972). Some reactions of isolated corn mitochondria influenced by juglone. *Physiol. Plant.* **27**, 89–94.

Koeppe, D. E., Rohrbaugh, L. M., Rice, E. L., and Wender, S. H. (1970). The effect of x-radiation on the concentration of scopolin and caffeoylquinic acid in tobacco. *Radiat. Bot.* **10**, 261–265.

Lefevre, M. (1952). Auto et heteroantagonisme chez les algues d'eau douce. *Ann. Stn. Cent. Hydrobiol. Appl.* **4**, 5–197.

Linder, P. J., Craig, J. C., and Walton, T. R. (1957). Movement of $C^{14}$-tagged alpha-methoxy-phenylacetic acid out of roots. *Plant Physiol.* **32**, 572–575.

McCahon, C. B., Kelsey, R. G., Sheridan, R. P., and Shafizadeh, F. (1973). Physiological effects of compounds extracted from sagebrush. *Bull. Torrey Bot. Club* **100**, 23–28.

McCalla, T. M., and Haskins, F. A. (1964). Phytotoxic substances from soil microorganisms and crop residues. *Bacteriol. Rev.* **28**, 181–207.

Molisch, H. (1937). "Der Einfluss einer Pflanze auf die Andere-Allelopathic." Fischer, Jena.

Muller, C. H. (1966). The role of chemical inhibition (allelopathy) in vegetational compc.ition. *Bull. Torrey Bot. Club* **93**, 332–351.

Muller, C. H. (1969). Allelopathy as a factor in ecological process. *Vegetatio* **18**, 348–357.

Muller, C. H., and del Moral, R .(1966). Soil toxicity induced by terpenes from *Salvia leucophylla*. *Bull. Torrey Bot. Club* **93**, 130–137.

Muller, C. H., and Hauge, R. (1967). Volatile growth inhibitors produced by *Salvia leucophylla*: Effect on seed anatomy. *Bull. Torrey Bot. Club* **94**, 182–191.

Muller, C. H., Hanawalt, R. B., and McPherson, J. K. (1968). Allelopathic control of herb growth in the fire cycle of California chaparral. *Bull. Torrey Bot. Club* **95**, 225–231.

Nickell, L. G. (1960). Antimicrobial activity of vascular plants. *Econ. Bot.* 13, 281–318.

Norstadt, F. A., and McCalla, T. M. (1968). Phytotoxic substances from a species of *Penicillium*. *Science* 140, 410–411.

Ohman, J. H., and Kommedahl, T. (1960). Relative toxicity of extracts from vegetative organs of quackgrass to alfalfa. *Weeds* 8, 666–670.

Owens, L. D. (1969). Toxins in plant disease: Structure and mode of action. *Science* 165, 18–25.

Patrick, Z. A. (1955). The peach replant problem in Ontario. II. Toxic substances from microbial decomposition products of peach root residues. *Can. J. Bot.* 33, 461–486.

Patrick, Z. A., Toussoun, T. A., and Koch, L. W. (1964). Effect of crop residue decomposition products on plant. *Annu. Rev. Phytopathol.* 2, 267–292.

Proctor, V. W. (1957). Studies of algae antibiosis using *Maematococcus* and *Chlamydomonas*. *Limnol. Oceanogr.* 2, 125–139.

Rice, E. L. (1974). "Allelopathy." Academic Press, New York.

Rice, T. R. (1954). Biotic influences affecting population growth of planktonic algae. *U.S. Fish Wildl. Serv., Fish. Bull.* No. 54, 227–245.

Schreiner, O., and Lathrop, E .C. (1911). Examination of soils for organic constituents. *U.S. Bur. Soils, Bull.* No. 80.

Taylor, A. O. (1965). Some effects of photoperiod on the biosynthesis of phenylpropane derivatives in *Xanthium*. *Plant Physiol.* 40, 273–280.

Tukey, H. B., Jr. (1970). The leaching of substances from plants. *Annu. Rev. Plant Physiol.* 21, 305–324.

Way, J. T. (1847). On the fairy-rings of pastures, as illustrating the use of inorganic manures. *J. R. Agric. Soc. Engl.* 7, 549–552.

Zucker, M. (1969). Induction of phenylalanine ammonia-lyase in Xanthium leaf discs. Photosynthetic requirement and effect of daylength. *Plant Physiol.* 44, 912–922.

Chapter 18

# Self-Induced Disease

## H. HOESTRA

## I. INTRODUCTION

Self-induced disease (SID) in human pathology would include cases of people eating, drinking, or smoking too much, exposing skin to too much sunlight, or suicidal tendencies. Plants live quite differently, as individuals or in groups, and their SIDs should be defined in a different way, although some parallels do exist. SIDs occur as a result of the presence or activities of a plant species, the way it grows or is being grown. A plant's good life may turn against the plant, as an individual or as a species. A fine crop or a good orchard may stimulate forces, including diseases, that lead to decline. This becomes evident particularly when we want to replant or resow the same crop on the same site. Replant diseases are SIDs when they are specifically related to the repeated growing of the species concerned, i.e., when a crop grows poorly *because* the same crop has been grown (shortly) before on the same land. The word "diseases" is used here in a broad sense, including poor growth with few or no specific symptoms but excluding cases that are caused by depletion of soils, erosion, or similar factors based on poor practices. The causal factors are obviously in the soil, and these SIDs are called "site-bound." Since perennial and annual crops live very differently, their site-bound SIDs will be considered separately.

Another category of SIDs can be found among plants that "eat too much," such as when high levels of nitrogen are available in the soil

331

and taken up by the plant, leading to diseases. This subject was discussed extensively by Huber in Volume III. Also, physiological or other changes in the plant during its development may lead to a greatly increased susceptibility to diseases. These SIDs are called "plant-bound."

A cultivated plant has brought with it many characteristics from its wild stage. It is considered useful to look for aspects of SIDs in wild plants and natural communities.

## II. SITE–BOUND SELF–INDUCED DISEASES

### A. Perennial Crops

*1. Specific Apple Replant Disease*

Several factors may cause poor growth of apple. Drought, waterlogging, poor soil structure, competition by weeds, competition by other trees, and shading are among the conditions to which apple is very sensitive. In our study of apple replant problems over the years, the aforementioned factors could all be eliminated as causes of replant diseases. Nematodes were the next factor to draw our attention, especially the endoparasitic nematode *Pratylenchus penetrans,* which causes considerable damage to young trees (Hoestra and Oostenbrink, 1962) and also to rootstocks in nurseries (Oostenbrink *et al.,* 1957). Damage to young trees is more spectacular than damage to older trees, and the nematodes therefore attracted particular attention in replant situations.

Several annual crops are more "efficient host plants" for P. *penetrans* than woody crops. Cultivation of the latter usually does not lead to very high nematode populations. On the other hand, apple and many other woody crops are very susceptible to damage (Hijink and Oostenbrink, 1968). In practice, this means that damage of P. *penetrans* to apple is particularly severe when efficient, mostly annual host plants precede apple. Therefore, this is not considered a SID.

Light, sandy soils are the favored habitat of the nematode, probably because the soil pores permit quick movement of the nematode to new food sources. In The Netherlands most apple orchards are on heavier soils, where the nematode is not prevalent or damaging. However, in these areas the growers also encounter replant problems. It became clear (Hoestra, 1968) that a specific apple replant disease was present that was not caused by P. *penetrans* or other nematodes and was characterized by the following:

1. Specificity. Apple that is planted after a previous crop of apple does not grow well. Other crops, including fruit species such as cherry, do grow well after apple.

2. Persistence in time. Several years after grubbing the former orchard, the causal factor is still present in the soil and apple, when replanted, still shows clear signs of a specific apple replant disease.

3. Persistence in place. The causal factor apparently is strongly bound to the soil. In practice, the disease does not move from the former site of apple trees or rootstocks (Thompson, 1959). Under experimental conditions the causal factor is not leached out of the soil with water. When alternate layers of "apple" soil and other soil are placed on top of each other, the causal factor does not enter the layers of "fresh" soil, as indicated by the condition of the root systems of apple seedlings grown on these "sandwiched" soil profiles.

4. Recovery after changing the soil. Trees or seedlings recover if the soil is changed.

5. Nonlethality. Trees in the field or seedlings in pots may stop growth completely but will not die. The disease affects the young feeder roots only. After the first years, trees recover to some degree and start to grow.

The specificity distinguishes specific apple replant disease from all the other causes of poor growth mentioned. It also means that the disease is *self-induced*. The factor determining poor growth of apple is the former presence of apple on the same land.

According to Savory (1966), the above-mentioned characteristics also apply to several other replant diseases of perennial crops. The cause of specific apple replant disease remains obscure. There is evidence that bacteria or actinomycetes play a role, but some characteristics suggest the influence of toxic and persistent compounds in the soil. Data from the literature are contradictory on this point (Bünemann and Jensen, 1970; Bunt and Mulder, 1972; Otto and Winkler, 1977; Savory, 1969; Sewell, 1978). It is doubtful, however, whether the disease can be pinpointed to a single factor. Hoestra (1968) pointed out the effect of environmental conditions on the disease. Soils of low pH are less affected, and keeping the soil rather moist also seems to counteract the disease. Currently, growers water young trees with good results. Experimental studies in the field are presently underway to test the effect of artificially lowering the pH of soils in apple orchards (Jonkers and Hoestra, 1978).

### 2. Specific Cherry Replant Disease

Like apple, cherry suffers from a specific replant disease that is clearly not associated with *P. penetrans*. The cause of specific cherry replant disease is, however, understood. In 1965 Hoestra drew attention to the role of *Thielaviopsis basicola* in specific cherry replant disease. His observations were confirmed and extended by Sewell and Wilson (1975). They concluded that results with *T. basicola* are consistent with all the features that characterize the specific replant disease of cherry, including symptoms, immobility, and persistence of the causal agent in the soil, normal growth of affected trees after transfer to "nonreplant" soil, and specificity (including differing effects on species within the genus *Prunus*)

### 3. Replant Disease of Grape

Not all authors agree that grape is affected by a specific replant disease (Savory, 1966). This may be partly the result of the many other diseases, including various soilborne diseases, that mask the effect of "true" replant diseases. Moser (1969), with his broad knowledge of grapes, believes that there is no doubt as to the existence of replant problems in grapes and their specificity. Replant effects are obvious: Grapes grow better on fresh land than on land in which grapes were previously grown, and the differences may be visible for a period of 20 years or more. In Moser's numerous experiments, grapes suffered from the addition of grape leaves, wood, bark, etc., to soil. Leachates from grape soil transmitted the disease. Moser concluded that specific replant disease of grape is caused by host-made toxins. These conclusions are convincing, but Moser is incorrect in suggesting that all replant problems of other crops also are caused by toxins. His suggestion that replant effects have a definite meaning in natural ecosystems is very interesting.

### 4. Replant Disease and Related Problems of Peach

The peach replant problem was investigated by a multidisciplinary approach in Ontario, Canada (Mountain and Boyce, 1958). The nematode *P. penetrans* was the main factor in this case and, as with apple and cherry, caused damage mainly in orchards planted on light soils. The pathogenicity of the nematode to peach roots was associated with the breakdown of a peach glucoside, amygdalin, under the influence of the nematode. The existence of a true specific peach replant disease was demonstrated by Gilmore (1963), but the effect of this SID was not strong, and this problem is no longer considered of great importance. In the southeastern United States, peach is plagued by a number of problems, often known as "peach short life," particularly in replanted orchards. The total stress of a number of factors causes the trees to decline or die, and the elimination of any of these factors—by improved soil tillage not damaging the roots, application of nematicides or fungicides, and weed control—is highly beneficial (Hendrix and Powell, 1970). Peach short life is a clear example of a problem involving more than a single causal factor.

## B. Annual Crops

### 1. Rye

Scholte and Kupers (1977) published their experiences on the "lack of self-tolerance" of winter rye. In an 18-year rotation experiment on light, sandy soil, rye grown 2 years in succession showed a depression in seed yield of 30% and in straw yield of 10%, compared to rye grown after other crops, including oats.

In contrast to other species of small cereals, rye has always been considered to be self-tolerant. The cited observations show that replant problems of this kind may remain underestimated for a long time because they are nonlethal, without specific symptoms and with reductions in yield of the same order of magnitude as caused by "normal" fluctuations. The cause of the specific lack of self-tolerance in rye is currently unknown.

### 2. Wheat

There is a strong interest in the role of *Gaeumannomyces* (*Ophiobolus*) *graminis*, the cause of take-all disease, in association with monoculture of wheat. An interesting sequence of events has been observed. In the first years after a monoculture of wheat has been started, an increasing attack by take-all disease occurs, leading to considerable losses, especially on "virgin" land such as that found in The Netherlands in new polders (Gerlagh, 1968). The disease becomes less severe with time. The phenomenon has been widely recognized (Jepsen and Jensen, 1976; Vez, 1976) and is called "ophiobolus decline." Gerlagh (1968) demonstrated the cause of this decline to be of a biological nature. The parasite stimulates the activity of its own parasites in turn. The phenomenon has also been observed with other pathogens of wheat: "heterodera decline." Parasites of cysts are considered to be the responsible agents (Jakobson, 1974; Ohnesorge *et al.*, 1974). The identification of fungal parasites of *Heterodera avenae* cysts opens up perspectives for biological control (Kerry, 1975; Tribe, 1977). This also is the case for *Gaeumannomyces graminis* (Deacon, 1976).

### 3. Potato

The potato cyst nematode (*Globodera rostochiensis* and *G. pallida*) is the cause of a self-induced disease of potatoes. The nematode persists in the soil through its cysts. The decline phenomenon has not as yet been clearly demonstrated, although rather large oscillations of nematode populations have been observed.

When the potato cyst nematode is not present, as, for example, in new polders in The Netherlands, the existence of other factors affecting production in various rotation schemes can be studied. Hoekstra and Maenhout (1976) reported a 12-year crop rotation experiment carried out on "virgin" polder soil. There were 3-, 4-, and 6-year rotations for potato. In the 3-year rotation and later also in the 4-year cropping system, yields began to decrease compared to the 6-year rotation. The reasons for the differences in yield (and quality) have not become clear.

### 4. Sugar Beet

With sugar beet cyst nematode (*Heterodera schachtii*), the decline phenomenon has also been observed in monoculture (Van Ingen and Temme, 1973). In the rotation experiment just mentioned on new polder soil (Hoekstra and Maenhout, 1976), sugar beet yields, in contrast to potato yields, were not influenced by the frequency of cropping. This is in agreement with data in Shipton's (1977) review showing that sustainable monocropping systems of sugar beet are possible when ample organic material is supplied and when a good soil structure can be maintained.

## C. Natural Ecological Systems

To understand the background of SIDs it is useful to follow Moser's (1969) suggestion that replant diseases can be traced back to phenomena occurring in natural ecosystems. He carefully observed wild plants over the years in relation to the site they occupied and noted that decline occurred sooner or later almost universally, making the site less attractive to the species under observation but more easily accessible to other plant species. This phenomenon favors species sequences and helps to keep in check species that are otherwise too dominating. As in SIDs this may be caused by organisms and/or toxins. Although Whittaker (1970) underestimates the role of organisms, in his excellent review he correctly states that allelopathic (biochemical) effects have a significant influence on the rate and species sequence of plant succession. These are mostly cases in which plants defend themselves against other species. However, cases of autotoxicity are also quoted, such as inhibition of germination of seeds by nearby (germinating) seeds of the same species and factors toxic to seedlings of the same species in forests. The soft chaparral in California is known for its ability to prevent growth of grass and herbs in belts 1–2 m wide around patches of this shrub. This is caused by volatilization of terpenes from the leaves and subsequent adsorption of these compounds onto soil particles. Heavy concentrations of these terpenes are presumed to be the cause of reduced vigor in the centers of old shrub patches (autotoxicity), although the role of microorganisms in the soil also needs to be investigated. Fairy rings formed by sunflower species in some areas are similarly attributed to autotoxicity. DeFreitas and Fredrickson (1978) showed that production of autoinhibitors can play an important role in maintaining species diversity of microbial systems.

Roots of guayule (*Parthenium argentatum*) release *trans*-cinnamic acid, which has a stronger inhibitory effect on guayule itself than on other plants. This substance apparently acts to govern the growth of roots in such a way that they avoid one another in occupying soil space. A similar behavior of roots has been observed in the case of apple. Went (1970) also comments on the case of guayule. In the natural habitat, the desert, the plant is very evenly spaced, with a considerable distance between adjacent shrubs. When grown as a closely spaced field crop, plants along the edges grow much better than those in the center. This is a clear example of a plant species resisting man's manipulation in turning it into a field crop.

According to Ponomarenko (1975) in his study on natural stands of apple in the Caucasus mountains, the apple tree grows solitary and on the edges of open spaces. It is very sensitive to shading and competition by other trees. Young apple trees beneath parent trees were never found. Apparently the species has a good system of defense against its own offspring. It is tempting to suppose that the mechanisms responsible for specific replant disease, in natural conditions, help to weaken the young plant when the seed germinates in the immediate vicinity of the mother tree.

## III. PLANT–BOUND SELF–INDUCED DISEASES

Plants may, in other ways, induce conditions leading to SIDs. When plants reach certain stages in their development, changes in the tissues occur, rendering them susceptible to disease. Maize removes sugar from the stalk, making itself susceptible to stalk rot caused by various fungi and bacteria (see also Volume III, Chapter 17). Messiaen (1957) studied several lines and hybrids of maize with different sugar contents in the stalks. In one case he removed the ear to increase sugar content in the stalk. He showed a good correlation between sugar content of stalks and resistance to fungi such as *Gibberella zeae* and *Colletotrichum graminicola*.

A review by Horsfall and Dimond (1957) demonstrates the importance of tissue sugar and disease susceptibility. Some diseases are favored by a high sugar content, such as the rusts and the powdery mildews. Stalk diseases of maize are diseases favored by a low sugar content. Another example is *Alternaria solani* of tomato. When fruiting is prohibited, the sugar content in the leaves is high, and the plants are immune to target spot.

An infected plant, in one way or another, may become susceptible to other pathogens, Nematologists are particularly aware of this and have collected much information on complex diseases in which nematodes predispose plants to other disease agents, such as fungi and bacteria (Powell, 1971a.b). Self-interactions also occur. Jatala and Jensen (1976) reported that sugar beets are particularly sensitive to a second inoculation with *Heterodera schachtii*.

Grainger (1956) studied two host–parasite combinations with respect to the relation between host nutrition and attack by fungal parasites: *Helminthosporium avenae* of oats and *Phytophthora infestans* of potato. In both cases the hosts are not susceptible to the fungi at certain stages of their development. Periods of low disease potential in both hosts were found to be when growth was rapid and also when a low percentage of total carbohydrate was available in the whole plant.

## IV. DISCUSSION

Self-induced diseases associated with replanting or monoculture (cultivation of the same crop in the same soil year after year) become obvious when other factors causing crop failures or poor growth can be eliminated. In poorly managed monocultures, low yields may be attributed to the monoculture system, while in fact this is not correct. The bad reputation of monoculture must, at least partly, be explained historically. This can be illustrated by Craven's (1965) thorough study of the declining yields, soil depletion, and erosion in tobacco monoculture over a long period in the history of Virginia and Maryland (1606–1860). Not SIDs but ignorance, bad habits, and short-sightedness were the factors causing this large-scale destruction of soil fertility. It is now clear that under conditions of good management tobacco can be listed among the crops suitable to be grown quite well in monoculture for many areas. Shipton (1977) demonstrates this in a stimulating review in which he gives a balanced appreciation of monoculture and crop rotation in relation to soilborne diseases.

The cases of potato and rye illustrate that, in the absence of known serious pathogens, monoculture still does not provide yields as high as when these crops are grown in a favorable rotation scheme. It is obvious that many factors may contribute to some loss of yield in monocropping. It is not surprising to see that, after the identification of the more obvious and often more serious causes of damage to crops, other factors remain which reduce the yield. It is also not surprising that these "leftover" factors are often only mildly damaging, coming to the fore only in agricultural systems that have already reached a high degree of perfection.

The potato cyst nematode represents an SID that does not seem to be subject to efficient disease decline. Another, and more striking, case is *Stromatinia gladioli* in The Netherlands. Once the pathogen is present in the soil, it is virtually impossible to regrow gladiolus. In the earlier sections of this chapter, declining disease intensity with SIDs was shown to occur for *Gaeumannomyces graminis, Heterodera avenae,* and *H. schachtii.* There is increasing evidence that natural enemies, particularly fungi, are to a great extent responsible for this decline.

Replant diseases of perennial crops are very much determined by the fact that when the young tree is taken from a nursery, transported, stored, and subsequently planted in a new environment, inevitable damage occurs with each step. In addition, the naturally sparse root system of many fruit trees makes them sensitive to any factor that interferes with the functioning of the feeder roots. In contrast, grain crops, which are sown, become established in a more natural way. Their root systems are better able to support damage and can regenerate rather quickly. The significance of monoculture with these crops is that they tolerate enough of the parasite to allow the organisms responsible for decline to increase their populations. However, marked differences in behavior among the cereals grown in monoculture exist. Shipton (1975) reported that in Britain for the economically viable cultivation of cereals under monoculture the choice of spring barley is common. Growers also exploit this by using barley to build up the antagonists of *Gaeumannomyces graminis* and then grow wheat.

A cyclic behavior of pathogen and antagonist activities can be expected to occur when hyperparasites of the antagonists begin to play a role or when the number of pathogens is so far reduced that the antagonists tend to disappear due to lack of food, providing the pathogen with a new chance to cause disease. Indeed, Ohnesorge *et al.* (1974) observed that in monocultures of cereals patches with large populations of *Heterodera avenae* after some years became patches with small populations, and vice versa.

Baker and Cook (1974) drew attention to the existence of disease-suppressive soils, which prevent certain soilborne diseases including SIDs from occurring. In these cases, certain soilborne diseases never become a problem in monoculture. The basis for this may be biological (Louvet *et al.*, 1976) or physical (Stotzky and Martin, 1963). Heavy soils can be considered disease suppressive for *Pratylenchus penetrans*, due to pore size and movement of the nematodes, as was noted in the discussion of replant diseases of peach, apple, and cherry.

It is popular to advance hypotheses as to the causes of the unknown SIDs. It is doubtful, however, if devoting so much time and energy to the identification of these diseases is useful. It seems to be more appropriate to approach these problems with a broader view toward monoculture as a system, of which the various problems, as well as aspects of farm management, are closely interrelated. Indeed, these are cases in which "thought and effort is too often focused on elucidating *the* cause for a phenomenon as opposed to the complex of determining elements" (Bateman, Volume III, Chapter 3).

It should be clear why in a number of cases monoculture is feasible but also why factors leading to at least some loss of yield are difficult to avoid. Among the complex of elements determining the occurrence of SIDs, certainly man's decision to grow crops in monoculture should be mentioned. Monocultures are man-made ecosystems, and SIDSs of plants are, in a way, man-made too. Modern research techniques may show maximal yields above the yields obtained by farmers. With a long-term in mind, it becomes doubtful if maximal yields per surface area should be *the* goal of research and, if not, as in the case of well-managed and sustainable monocultures, the stability of this artificial ecosystem deserves higher priority than attempts at maximizing the yield. Resarch on the growing of multilines of cereals is based on such a long-term view rather than on maximizing yields (Kampmeijer and Zadoks, 1977). It is beyond the scope of this chapter to discuss crop rotation versus monoculture in much detail. Crop rotation is a very good system and still of great use to mankind, although less imperative than it was in the past. When it is the farmer's choice to apply crop rotation, he must tackle the problems of this system. Likewise, he has to face and live with the associated problems, including SIDs, when experiences have shown that monoculture is feasible.

### References

Baker, K. F., and Cook, R. J. (1974). "Biological Control of Plant Pathogens." Freeman, San Francisco, California.

Bünemann, G., and Jensen, A. M. (1970). Replant problem in quartz sand. *Hortic. Sci.* 5, 478–479.

Bunt, J. A., and Mulder, D. (1972). The possible role of bacteria in relation to the apple replant disease. *Meded. Rijksfac. Landbouwwet., Gent* 38, 1381–1385.

Craven, A. O. (1965). "Soil Exhaustion as a Factor in the Agricultural History of Virginia and Maryland, 1606–1860." Peter Smith, Gloucester, Massachusetts.

Deacon, J. W. (1976). Biological control of take-all by *Phialophora radicicola* Cain. *EPPO (Eur. Mediterr. Plant Prot. Organ.) Bull.* **6**, 297–308.

DeFreitas, M. J. and Fredrickson, A. G. (1978). Inhibition as a factor in the maintenance of the diversity of microbial ecosystems. *J. Gen. Microbiol.* **106**, 307–320.

Gerlagh, M. (1968). Introduction of *Ophiobolus graminis* into new polders and its decline. *Neth. J. Plant Pathol.* **73**, Suppl. 2.

Gilmore, A. E. (1963). Pot experiments related to the peach replant problem. *Hilgardia* **34**, 63–78.

Grainger, J. (1956). Host nutrition and attack by fungal parasites. *Phytopathology* **46**, 445–456.

Hendrix, F. F. and Powell, W. M. (1970). Control of root pathogens in peach decline sites. *Phytopathology* **60**, 16–19.

Hijink, M. J., and Oostenbrink, M. (1968). Vruchtwisseling ter bestrijding van planteziekten en -plagen. *Versl. Meded. Plantenziektenk. Dienst Wageningen. Separate Ser.* No. 368, 1–7.

Hoekstra, O., and Maenhout, C. A. A. A. (1976). Results of the crop rotation experiment "De Schreef" (in Dutch, English summary). *Proefst. Akkerbouw, Lelystad, Rapp.* No. 33.

Hoestra, H. (1965). *Thielaviopsis basicola*, a factor in the cherry replant problem in the Netherlands. *Neth. J. Plant Pathol.* **71**, 180–182.

Hoestra, H. (1968). Replant diseases of apple in the Netherlands. *Meded. Landbouwhogesch. Wageningen* **68** (13), 1–105.

Hoestra, H., and Oostenbrink, M. (1962). Nematodes in relation to plant growth. IV. *Pratylenchus penetrans* (Cobb) on orchard trees. *Neth. J. Agric. Sci.* **10**, 286–296.

Horsfall, J. G., and Dimond, A. E. (1957). Interactions of tissue sugar, growth substances, and disease susceptibility. *Z. Pflanzenkr. Pflanzenschutz* **64**, 415–421.

Jakobsen, J. (1974). The importance of monocultures of various host plants for the population density of *Heterodera avenae*. *Tidsskr. Planteavl* **78**, 697–700.

Jatala, P., and Jensen, H. J. (1976). Self-interactions of *Meloidogyne hapla* and *Heterodera schachtii* on *Beta vulgaris*. *J. Nematol.* **8**, 43–48.

Jepsen, H. M., and Jensen, A. (1976). Continuous cereal growing in Denmark: Experimental results. *EPPO (Eur. Mediterr. Plant. Prot. Organ.) Bull.* **6**, 371–378.

Jonkers, H., and Hoestra, H. (1978). Soil pH in fruit trees in relation to specific replant disorder of apple. I. Introduction and review of literature. *Sci. Hortic.* **8**, 113–118.

Kampmeijer, P., and Zadoks, J. C. (1977). "Epimul. A Simulator of Foci and Epidemics in Mixtures of Resistent and Susceptible Plants, Mosaics and Multilines." Pudoc, Wageningen, Netherlands.

Kerry, B. R. (1975). Fungi and the decrease of cereal cyst-nematode populations in cereal monoculture. *EPPO (Eur. Mediterr. Plant Prot. Organ.) Bull.* **5**, 353–361.

Louvet, J., Rouxel, F., and Alabouvette, C. (1976). Recherches sur la resistance des sols aux maladies. I. Mise en evidence de la nature microbiologique de la résistance d'un sol au développement de la Fusariose vasculaire du melon. *Ann. Phytopathol.* **8**, 425–436.

Messiaen, C. M. (1957). Richesse en sucre des tiges de mais et verse parasitaire. *Rev. Pathol. Veg. Entomol. Agric. Fr.* **36**, 209–213.

Moser, L. (1969). Die Hemmstoffe, ihre Aufgaben und ihre Bedeutung in den Pflanzengemeinschaften. *Rebe Wein. Obstabau Fruchtverwert.* **19**, 87–95.

Mountain, W. B., and Boyce, H. R. (1958). The peach replant problem in Ontario. VII. Pathogenicity of *Pratylenchus penetrans. Can. J. Bot.* 27, 459–470.

Ohnesorge, B., Freidel, J., and Oesterlin, U. (1974). Untersuchungen zur Dispersions-dynamik von *Heterodera avenae* Wollenw. auf einer Fläche mit Getreidedaueranbau. Z. *Pflanzenkr. Pflanzenschutz* 81, 356–363.

Oostenbrink, M., s'Jacob, J. J., and Kuiper, K. (1957). Over de waardplanten van *Pratylenchus penetrans. Tijdschr. Plantenziekten* 63, 345–360.

Otto, G., and Winkler, H. (1977). Untersuchungen über die Ursache der Bodenmüdigkeit bei Obstgehölzen. VI. Nachweis von Aktinomyzeten in Faserwurzeln von Apfelsämlingen in Böden mit verschiedenen Müdigkeitsgraden. *Zentralbl. Parasitenkd., Infektionskr. Hyg. Abt. 2* 132, 593–606.

Ponomarenko, V. V. (1975). Materials to the knowledge of apple-trees of the Caucasus (in Russian, English summary). *Bot. Zh. (Leningrad)* 60, 53–68.

Powell, N. T. (1971a). Interaction of plant parasitic nematodes with other disease complexes. *Annu. Rev. Phytopathol.* 9, 253–274.

Powell, N. T. (1971b). Interaction of plant parasitic nematodes with other disease causing agents. *In* "Plant Parasitic Nematodes" (B. M. Zuckerman, W. F. Mai, and R. A. Rohde, eds.), Vol. 2, pp. 119–136. Academic Press, New York.

Savory, B. M. (1966). "Specific Replant Diseases." Commonw. Agric. Bur., Farnham Royal, England.

Savory, B. M. (1969). Evidence that toxins are not the causal factors of the specific apple replant disease. *Ann. Appl. Biol.* 63, 225–231.

Scholte, K., and Kupers, L. J. P. (1977). The causes of the lack of self-tolerance of winter-rye, grown on light sandy soils. 1. Influences of foot rots and nematodes. *Neth. J. Agric. Sci.* 25, 255–262.

Sewell, G. W. F. (1978). Potential soil treatments for the control of apple replant disease. *Annu. Rep. East Malling Res. Stn. for 1977,* 97–99.

Sewell, G. W. F., and Wilson, J. F. (1975). The role of *Thielaviopsis basicola* in the specific replant disorders of cherry and plum. *Ann. Appl. Biol.* 79, 149–169.

Shipton, P. J. (1975). Yield trends during take-all decline in spring barley and wheat grown continuously. *EPPO (Eur. Mediterr. Plant Prot. Organ.) Bull* 5, 363–374.

Shipton, P. J. (1977). Monoculture and soilborne plant pathogens. *Annu. Rev. Phytopathol.* 15, 387–407.

Stotzky, G., and Martin, R. T. (1963). Soil mineralogy in relation to the spread of *Fusarium* wilt of banana in Central America. *Plant Soil* 18, 317–338.

Thompson, J. A. (1959). The occurrence of areas of poor growth in a fruit tree nursery. *Rep. East Malling Res. Stn. 1958* pp. 80–82.

Tribe, H. T. (1977). Pathology of cyst-nematodes. *Biol. Rev. Cambridge Philos. Soc.* 52, 477–507.

Van Ingen, C. G., and Temme, J. (1973). Aaltjesbesmetting bij een voortgezette monocultuur van suikerbieten. *Kali* 87, 221–228.

Vez, A. (1976). Possibilités et limites de la monoculture en relation avec l'application de diverses mesures culturales. *EPPO (Eur. Mediterr. Plant Prot. Organ.) Bull.* 6, 281–288.

Went, F. W. (1970). Plants and the chemical environment. *In* "Chemical ecology" (E. Sondheimer and J. B. Simeone, eds.), pp. 71–82. Academic Press, New York and London.

Whittaker, R. H. (1970). The biochemical ecology of higher plants. *In* "Chemical ecology" (E. Sondheimer and J. B. Simeone, eds.), pp. 43–70. Academic Press, New York and London.

Chapter 19

# Iatrogenic Disease: Mechanisms of Action

JAMES G. HORSFALL

## I. INTRODUCTION

The rapid rise of organic pesticides in the late 1940s was accompanied by a rapid rise in iatrogenic diseases. It seems appropriate at this time to assemble the mechanisms of action.

An iatrogenic disease of a crop is one that is induced or worsened by a plant pathologist's prescription for that crop. It is a disease generated by the "doctor," so to speak. *Iatro* is the Greek word for doctor. We find it in the word "psychiatry." Although iatrogenic diseases are fascinating, they tend to be swept under the scientific rug because we doctors do not like to think that we have been responsible for a disease. Kreutzer (1960) calls the phenomenon the "boomerang effect" and "disease exchange."

Cowling (1978) discusses agricultural and forest practices that favor epidemics. A few of the diseases he includes are made worse by the prescriptions of plant pathologists. Most of these, however, are intensified

PLANT DISEASE, VOL. IV

by the decisions of farmers and foresters to neglect certain principles of epidemiology in managing their crops.

Yarwood (1970) also discusses what he calls "man-made diseases." He shows that diseases can be worsened by grafting, plant breeding, smog, fertilizers, and tillage. Diseases also can be intensified by agricultural chemicals, herbicides, fungicides, insecticides, nematicides, and growth-promoting substances. I shall deal only with those made worse by agricultural chemicals and leave the rest to Yarwood (1970) and Cowling (1978).

An inventory (see Table I) of the 45 chemically induced iatrogenic diseases on 21 hosts discussed in this chapter shows a striking distribution. Thirty-two are due to herbicides and growth substances, 11 to fungicides, 1 to insecticides, and 2 to nematicides. They include 39 fungal diseases, 4 viral diseases, and 2 bacterial diseases.

In 1956 in Switzerland, I saw a chemically induced iatrogenic disease for the first time. Dr. S. Blumer of the Wadenswil Experiment Station showed me how zineb was controlling *Plasmopara* on his grapes but was worsening powdery mildew on the same vines and *Botrytis* on the grapes they bore. This stirred my interest in the subject, and it is for this reason that I am writing this chapter.

An expositor always worries about how to organize the data. He asks himself, "How shall I split the melon?" In this case it can be split into the kinds of disease, the kinds of chemical, or the modes of action. I have chosen to split it according to modes of action so that our understanding of disease processes will be increased.

How can a chemical exacerbate a biotic disease? It can alter one or more features of the epidemiologic cone, as discussed by Browning, Simons, and Torres in Volume I. It can reduce the resistance of the host, increase the inoculum potential of the pathogen, or alter the microclimate so that it is more favorable to the disease. And, of course, a change in any one of these will raise the rate at which an epidemic develops. Let us discuss these mechanisms in order.

## II. REDUCED RESISTANCE OF THE HOST

Since agricultural chemicals are designed to affect one living organism, we might expect that they could affect another.

### A. Changes in Sugar Relations

Horsfall and Dimond (1957) generalized the sugar story by suggesting that many diseases are "high-sugar diseases" while others are "low-sugar diseases." If, therefore, a compound increases the amount of sugar in

**TABLE I**

Summary of Iatrogenic Diseases

| Pathogen | Host | Compound | |
| | | Common Name | Chemical Name |
|---|---|---|---|
| *Agrobacterium* | Cherry | Dichlone | 2,3-Dichloro-1,4-naptho-quinone |
| | Cherry | Captan | *N*-Trichloromethylmer-capto-4-cycloheximide-1,2-carboximide |
| *Botrytis* | Cyclamen | Benomyl | Methyl 1-(butylcarba-moyl)benzimidazole-2-yl carbamate |
| | Tomato | Zineb | Zinc ethylenebis (dithio-carbamate) |
| *Erysiphe* | Bean | Maleic hydrazide | 6-Hydroxy-3(2*H*)-pyri-dazinone |
| | Currant | Simazine | 2-Chloro-4,6-bis(ethyl-amino)s-triazine |
| | Grape | Zineb | Zinc ethylenebis(dithio-carbamate) |
| | Poplar | Zineb | Zinc ethylenebis(dithio-carbamate) |
| *Fusarium* | Bean | EPTC | S-Ethyl *N,N*-dipropyl-thiocarbamate |
| | Bean | Chloramben | 3-Amino-2,5-dichloro-benzoic acid |
| | Bean | Alachlor | 2-Chloro-2',6'-dimethyl-*N*-(methoxymethyl)-acetanilide |
| | Bean | Atrazine | 2-Chloro-4-ethylamino-6-isopropylamino-*s*-triazine |
| | Cotton | Trifluralin | α,α,α-Trifluoro-2,6-di-nitro-*N,N*-dipropyl-*p*-toluidine |
| | Pea | Atrazine | See above |
| | Potato | — | Tetrachloronitrobenzene |
| *Helminthosporium* | Wheat | 2,4-D | (2,4-Dichlorophen-oxy)-acetic acid |
| *Ophiobolus* | Wheat | Mecoprop | 2-(4-Chloro-2-methyl-phenoxy)propionic acid |
| | Wheat | Oxytril | 3,4-Dibromo-4-hydroxy-benzonitrile |
| *Pratylenchus* | Strawberry | PCNB | Pentachloronitrobenzene |
| *Puccinia* | Wheat | DDT | Dichlorodiphenyltri-chloroethane |
| | Wheat | Maleic hydrazide | See above |

(*continued*)

TABLE I (*Continued*)

| Pathogen | Host | Compound | |
| | | Common Name | Chemical Name |
| --- | --- | --- | --- |
| *Pythium* | Maize | Picloram | 4-Amino-3,5,6-trichloro-picolinic acid |
| | Pine | PCNB | See above |
| *Rhizoctonia* | Bean | Gibberellic acid | — |
| | Bean | Trifluralin | See above |
| | Cotton | Trifluralin | See above |
| | Maize | Picloram | See above |
| | Rice | 2,4-D | See above |
| | Sugar beet | Pyramin | 5-Amino-4-chloro-2-phenyl-3(2*H*)-pyri-dazinone |
| | Sugar beet | Row Neet | S-Ethyl-*N*-ethylthiocy-clohexane carbamate |
| | Sugar beet | Tillam | S-Propylbutylethyl thio-carbamate |
| *Sclerotinia* | Lettuce | DD | 1,2-Dichloropropane plus 1,2-dichloropro-pene |
| | Lettuce | Nemagon | 1,2-Dibromo-3-chloro-propane |
| *Sclerotium* | Clover | EPTC | See above |
| | Peanut | Benomyl | See above |
| *Thielaviopsis* | Soybean | Chloramben | See above |
| | Soybean | Alachlor | See above |
| *Ustilago* | Maize | 2,4,5-T | (2,4,5-Trichlorophen-oxy)acetic acid |
| *Verticillium* | Strawberry | Terrachlor | Pentachloronitrobenzene |
| | Tomato | Gibberellic acid | — |
| | Tomato | Maleic hydrazide | See above |
| *Tobacco mosaic virus* | Cotton | Indoleacetic acid | Indoleacetic acid |
| | Cotton | 2,4-D | See above |
| | Cotton | Diuron | 3-(3,4-Dichlorophenyl)-1,1-dimethylurea |
| | Tobacco | 2,4-D | See above |

the host, the host should become more susceptible to high-sugar diseases, more resistant to low-sugar diseases, and vice versa. How does this work out in the practice of plant pathology?

## 1. High-Sugar Diseases

As far as I know, Johnson (1946) was the first to report an iatrogenic disease due to pesticides. He found that DDT converts a rust resistant wheat to a susceptible one. Two years later Forsyth and Samborski

(1948) showed that DDT inhibits the outflow of sugars from the leaf and thereby worsens this high-sugar disease. They then reduced the outflow of sugar from the leaves and increased the rust by "searing" with heat the lower part of the leaf and by spraying with maleic hydrazide, which damages the phloem. Both of these treatments kept the sugar concentration in the host high and thereby decreased the resistance to disease.

Since maleic hydrazide inhibits sugar transport out of the leaf, it should increase the leaf-rust phase of the disease without appreciably affecting the stem-rust phase. This is precisely what happened in the experiments of Samborski *et al.* (1960).

Some powdery mildews appear to be high-sugar diseases. Miller (1952) increased powdery mildew on snap beans by treating them with maleic hydrazide.

### 2. Low-Sugar Diseases

Again, if maleic hydrazide inhibits sugar transport, low-sugar diseases should be reduced on the foliage and increased in the lower stem or root. It reduces alternarial disease on tomato leaves (O'Brien and Bornmann, 1952) and increases the fusarial wilt in the stems of tomato (Waggoner and Dimond, 1952).

(2,4-Dichlorophenoxy)acetic acid generally decreases the sugar content in plant tissues. As expected, it also favors helminthosporial infection on leaves of maize (Hale *et al.*, 1962) and on the lower stem and roots of wheat (Hsia and Christensen, 1951).

### 3. Sugar Status Uncertain

(2,4-Dichlorophenoxy)acetic acid favors an increase in the size of lesions induced by tobacco mosaic virus on hypersensitive tobacco (Simons and Ross, 1965). It also increases the replicating capacity of cotton for tobacco mosaic virus (Cheo, 1971). Cheo proposed that the compound induces production of ethylene; ethylene induces senescence; and senescence reduces the metabolic resistance of cotton to the virus.

## B. Changes in Structural Resistance

Romig and Sasser (1972) showed that trifluralin and dinoseb, by reducing the content of cellulose and methylated pectin in cell walls, reduce the resistance of snap beans to damping-off caused by *Rhizoctonia solani*. This apparently reduces resistance to mechanical penetration. The ethylene bis series of dithiocarbamates increases powdery mildew on the leaves of many crops. Barner and Roder (1964) hold that these compounds thin the leaf surface and make the development of haustoria easier. Similarly, Cunningham (1953) showed that tetrachloronitroben-zene worsens fusarial rot of potato tubers by inhibiting the formation of the cork layer that normally forms under the harvesting wounds where *Fusarium* invades. According to Rademacher (1959), 2,4,5-T increases the brittleness in corn stems and reduces resistance to boil smut.

The scanning electron microscope shows that EPTC alters the surfaces of the exposed cells of the hypocotyl of the navy bean and presumably decreases resistance to *Fusarium solani* (Wyse *et al.*, 1976). Mecoprop induces the development of swollen and bulbous roots on treated wheat and "seems," according to Nilson (1973), to assist *Ophiobolus graminis* in penetrating the tissue. Heitefuss (1973) claims that several herbicides alter the cell wall of the host and thus reduce resistance to disease.

## C. Increase in Nitrogen Nutrition

Simazine, the weed killer, intensifies many diseases. Ries *et al.* (1963) reported that the compound accelerates the uptake of nitrogen from the soil, and so Upstone and Davis (1967) suggest that this is why it worsens powdery mildew of gooseberries.

## D. Changes in Mineral Nutrition

A dramatic and puzzling iatrogenic disease is the *Botrytis* rot of some fruits caused by the ethylenebis (dithiocarbamate) fungicides. It appears on grapes in Europe, on tomatoes and strawberries in Florida, and on tomatoes in Nova Scotia. I discuss this disease here because the best evidence suggests some disturbance in mineral nutrition.

Stall (1963) has worked on the Florida case. He noticed that the fruit-rot phase of the disease of tomato is worst on sandy soils low in pH and low in calcium. Whereupon, he ran a dosage test on liming from 1000 to 9000 kg per hectare. The *Botrytis* fruit rot declined in a typical die-away curve as calcium dose increased. This seemed a clear signal, although the data on the ghost-spot phase of the disease on the fruit introduced considerable noise into the system. That phase of the disease declined as calcium was increased from 1000 to 3000 kg per hectare and then increased again with doses up to 9000 kg.

The zinc and manganese sprays increased ghost spot only at the medium level of calcium. We must conclude that in Florida the effect of calcium depends on the symptom recorded by the investigator.

Harrison (1961) could demonstrate no effect of lime on the fruit-rot phase of the disease in Nova Scotia. Hence, the effect of calcium is not clear.

Later Stall *et al.* (1965) showed that the disease is exaggerated by high levels of phosphorus. They did not show that zinc, zineb, manganese, or maneb affects the phosphorus, however. This effect of phosphorus is contradicted by Moore's (1944) work showing that *Botrytis* on broad bean is worsened by low, not by high, levels of phosphorus.

Another possibility, I think, needs more testing. *Botrytis* loves pectin. Tomato and strawberry fruits have considerable pectin. Calcium forms salts with pectins and makes them less nutritious for the fungus, as Edgington *et al.* (1961) showed for *Fusarium*. This fits Stall's data that fruit rot decreases with calcium dose to the soil. Perhaps maneb and zineb somehow make the pectin more nutritious for *Botrytis*.

*Rhizoctonia solani* is also a pectin eater, as Bateman and Lumsden (1965) showed. Altman and Campbell (1977) reported "preliminary" evidence that calcium added to the soil counteracts the enhancing effect of herbicides on *R. solani* on sugar beet.

### E. Encouraging the Host to Excrete Nutrients

It has always puzzled me that hosts excrete nutrients that encourage pathogens. They seem to ring the dinner gong. Come and get it! This seems to have a negative survival value. No doubt several pesticides accentuate this excretion and thereby encourage the pathogen even more. This phenomenon is clearly borderline between two mechanisms of action—an effect on the host, on the one hand, and an increase in inoculum potential, on the other. We shall discuss the effect under the host.

Peterson *et al.* (1963) first reported an increase in root exudation resulting from treatment with an agricultural chemical. They found that gibberellic acid increases exudation of carbohydrates from roots of kidney bean and accentuates attack by *Rhizoctonia* on them. Similarly, Altman (1972) showed that pyramin increases exudation of glucose from sugar beets and increases damping-off by *Rhizoctonia*. Lockhart and Forsyth (1964) claim that the *Botrytis* problem on fruits as discussed in the previous section is not due to a calcium effect but to an increased exudation of carbohydrates and amino acids. This fits the principle that *Botrytis* needs exogenous nutrients in order to penetrate well. Similarly, Lai and Semeniuk (1970) hold that picloram increases carbohydrate excretion from corn seedlings and increases *R. solani* and *Pythium arrhenomanes*.

Lee and Lockwood (1977) reported that the roots of soybeans treated with chloramben secrete more electrolytes and amino acids than the controls. These excretions increase the germination of the spores of *Thielaviopsis basicola* and cause development of larger lesions on the roots.

On the other hand, Romig and Sasser (1972) hold that the exacerbation by trifluralin and dinoseb of root rot of common bean caused by *Rhizoctonia solani* is not due to an increase in root excretion.

## III. INCREASED INOCULUM POTENTIAL OF THE PATHOGEN

How do pesticides affect inoculum potential? I suggest that they do so by affecting "capacity" *sensu* Baker (1978). They improve its nutrition (discussed partially in the previous section), stimulate the pathogen (whatever that means!), reduce the parasites and pathogens of the pathogen, and encourage synergistic organisms.

### A. Changes in Nutrition of the Pathogen

In general, the better the nutrition of the pathogen in its saprophytic stage, the higher its potential for successful invasion. *Botrytis* is a classic organism for this case. It very commonly enters its host from a piece of dead tissue like a blossom. According to Newhook and Davidson (1953), New Zealand tomatoes show increased fruit rot when the vines are sprayed with $\beta$-naphthoxyacetic acid to "set" the fruit. The blossom clings to the fruit and dies. *Botrytis* colonizes the dead blossom, enters the fruit, and causes rot.

Many cases of iatrogenic soilborne diseases have been ascribed, sometimes without much data, to the killing off of the organisms that compete with the pathogen for food in its saprophytic stage. Farley and Lockwood (1969) reported good data. They showed that pentachloronitrobenzene kills off the saprophytes that compete with pathogens for food. This allows the inoculum potential to rise.

Neubauer and Avisohar-Hershenson (1973) came close to the same conclusion in accounting for the worsening effects of trifluralin on the *Rhizoctonia* disease of cotton. They treated contaminated soil and sampled it for *Rhizoctonia* by the "baiting" technique. The percentage of invaded baits increased over time. They concluded that the compound increases the "saprophytic activity" of *Rhizoctonia*. Presumably this means that the competitors for food are inhibited, and thereby, inoculum potential is increased, but they did not look for the demise of the competitors. Altman (1969) made the startling observation that 25 herbicides themselves provide substrate for *R. solani*.

Of course, the host excretions discussed in the previous section also increase the inoculum.

## B. Stimulation of the Pathogen

Probably every pesticide since Bordeaux mixture was discovered has been found to "stimulate" one organism or another. It is not surprising that the same explanation has been advanced to account for some iatrogenic diseases. For example, Partyka and Mai (1958) reported that the dichloropropene/dichloropropane mixture for nematodes increases lettuce drop induced by *Sclerotinia sclerotiorum*. The inoculum here is the sclerotium. They reported that the mixture stimulates the sclerotia to form apothecia and then ascospores. Similarly, atrazine increases the number of propagules of the *Fusarium* that is pathogenic on wheat and maize (Percich and Lockwood, 1975).

According to Kurodani *et al.* (1959), 2,4-D increases the size and number of spots of sheath blight of rice that is induced by *Rhizoctonia* sp. They ascribed this to stimulation of the pathogen because their fungus grew faster in culture on a decoction of treated rice than on the controls. Perhaps it uses the compound as a substrate for growth.

Benomyl enhances a basidiomycetous pathogen on turf, according to Smith *et al.* (1970). Here again the pathogen grew faster in media fortified by benomyl than in the control media. Maleic hydrazide increases verticillial wilt of tomato by "stimulating" the growth of the parasite in the host vessels (Sinha and Wood, 1967). Similarly, trifluralin increases the number of chlamydospores of the cotton *Fusarium* in the soil (Tang *et al.*, 1970).

## C. Pathogens and Parasites of the Pathogen Are Eliminated

Pathogens have their pathogens and parasites, too. Many authors have ascribed the increase in disease caused by prescription chemicals to the reduction of competitors. The trouble is that almost none of them has distinguished between competitors for food and hyperparasites. Deep and Young (1956) came close. They found that crown gall was increased when the roots of cherry seedlings were dipped in suspensions of dichlone and captan. They suspected competitors, took fungi from normal cherry roots, and found less crown gall than in the untreated controls. Perhaps, this treatment merely introduced competitors for food, but more likely it introduced hyperparasites for the *Agrobacterium*.

Gibson *et al.* (1961) seem to have the best evidence for inhibition of hyperparasitism by agricultural chemicals. They found that PCNB when applied to pine seedlings worsened the damping-off induced by *Pythium*. They showed that *Penicillium paxilli*, a pathogen of *Pythium*, is killed by PCNB but that *Pythium* is not.

Bollen and Scholten (1971) also seem to have a good case. They found that benomyl-tolerant strains of *Botrytis* attack cyclamen-treated plants more severely than controls when the host is treated with benomyl. They ascribed this to a reduction in the pathogenicity of a *Penicillium* on the *Botrytis*.

## D. Synergistic Organisms Are Encouraged

I find it strange that we have very few data on the effect of synergistic organisms. Nematodes are synergistic with *Verticillium* on strawberry because they create wounds for easy entrance into the host. Pentachloronitrobenzene in the soil increases the number of nematodes and thereby increases the *Verticillium*, according to Rich and Miller (1964).

## E. Saprophytes Become Pathogens

Peeples (1970) found evidence that the herbicide Eptam in soil converts *Trichoderma viride* from a soil saprophyte to a vigorous parasite on ladino clover. This discovery opens a whole new vista of disease physiology.

## IV. ALTERED MICROCLIMATE TO FAVOR DISEASE

Microclimate occupies the third corner of the magic triangle of disease that moves through time. We have very few data on its effect on iatrogenic diseases; I could find only two papers. Bachman *et al.* (1975) reported that benomyl and chlorothalonil increase white rot due to *Sclerotium rolfsii* when sprayed on peanuts to control *Cercospora* on the leaves above. Their explanation is that the control of *Cercospora* saves the leaves. These shade the soil and raise the humidity there sufficiently to encourage *Sclerotium rolfsii* on the crowns below. Wyse and Meggitt (1973) found that Eptam increases the fusarial root rot of navy bean at 22° but not at 30°C.

## V. FUTURE PROSPECTS

In preparing this chapter, I was impressed with two things: (1) the number of iatrogenic diseases that exist and (2) the way in which they could help the investigator develop new insights into the complex battle between host and parasite. Here is a device for worsening his pet disease.

He can impose two experimental variables on his experimental system—the pathogen and the iatrogenic chemical. The reactions to both may increase his understanding of each. Johnson (1946), in his first case of an iatrogenic chemical, suggested just this. I am astonished that so few have followed up his proposal.

### References

Altman, J. (1969). Predisposition of sugar beets to *Rhizoctonia solani* damping-off with herbicides. *Phytopathology* **59**, 1015 (Abstr.)

Altman, J. (1972). Increased glucose exudate and damping-off of sugar beets in soils treated with herbicides. *Phytopathology* **62**, 743. (Abstr.)

Altman, J., and Campbell, C. L. (1977). Effect of herbicides on plant diseases. *Annu. Rev. Phytopathol.* **15**, 361–385.

Bachman, P. H., Roderiguez-Kabana, R., and Williams, J. C. (1975). The effect of peanut leaf spot fungicides on the nontarget pathogen, *Sclerotium rolfsii*. *Phytopathology* **65**, 772–776.

Baker, R. (1978). Inoculum potential. *In* "Plant Disease: An Advanced Treatise" (J. G. Horsfall and E. B. Cowling, eds.), Vol. II, pp. 137–157. Academic Press, New York.

Barner, J., and Roder, K. (1964). Die Einwirkung von Fungiciden auf die Ausbildung den Blattdeckgewebes und das Verhalten einiger Pflanzenparasiten. *Z. Pflanzenkr.* **17**, 210–215.

Bateman, D. F., and Lumsden, R. D. (1965). Relation of calcium content and nature of pectin substances in bean hypocotyls of different ages to susceptibility to an isolate of *Rhizoctonia solani*. *Phytopathology* **55**, 734–738.

Bollen, G. J., and Scholten, G. (1971). Acquired resistance to benomyl and some other systemic fungicides in a strain of *Botrytis cinerea* on cyclamen. *Neth. J. Plant Pathol.* **77**, 83–90.

Cheo, P. C. (1971). The effect of plant hormones on virus-replicating capacity of cotton infected with tobacco mosaic virus. *Phytopathology* **61**, 869–872.

Cowling, E. B. (1978). Agricultural and forest practices that favor epidemics. *In* "Plant Disease: An Advanced Treatise" (J. G. Horsfall and E. B. Cowling, eds.), Vol. II, pp. 361–381. Academic Press, New York.

Cunningham, H. S. (1953). A histological study of the influence of sprout inhibitors on *Fusarium* infection in potato tubers. *Phytopathology* **43**, 95–98.

Deep, I. W., and Young, R. A. (1956). Increased incidence of crown gall on Mazzard cherry following preplanting treatment with organic fungicides. *Phytopathology* **46**, 640. (Abstr.)

Edgington, L. V., Corden, M. E., and Dimond, A. E. (1961). The role of pectic substances in chemically induced resistance to *Fusarium* wilt in tomato. *Phytopathology* **51**, 179–182.

Farley, J. D., and Lockwood, J. L. (1969). Reduced nutrient competition by soil microorganisms as a possible mechanism for pentachloronitrobenzene-induced disease accentuation. *Phytopathology* **59**, 718–724.

Forsyth, F. R., and Samborski, D. J. (1948). The effect of various methods of breaking resistance on stem rust reaction and content of soluble carbohydrate and nitrogen in wheat leaves. *Can. J. Bot.* **36**, 717–723.

Gibson, I. A. S., Ledger, M., and Boehm, E. (1961). An anomalous effect of pentachloronitrobenzene on the incidence of damping-off caused by a *Pythium* sp. *Phytopathology* 51, 531–533.

Hale, M. G., Roane, C. E., and Huang, M. R. A. (1962). Effects of growth regulators on size and number of leaf spots, and on O₂ uptake and extension growth of coleoptile sections of corn inbred lines K41 and K44. *Phytopathology* 52, 185–191.

Harrison, K. A. (1961). The control of late blight and gray mold in tomatoes in Nova Scotia. *Can. Plant Dis. Surv.* 41, 175–178.

Heitefuss, R. (1973). Der Einfluss von Herbiciden auf Bodenfurtige Pflanzenkrankheiten. *Proc. Eur. Weed Res. Control Symp., Herbic.-Soil*, pp. 99–218.

Horsfall, J. G., and Dimond, A. E. (1957). Interactions of tissue sugar, growth substances, and disease susceptibility. *Z. Pflanzenkr. (Pflanzenpathol.) Pflanzenschutz* 64, 415–421.

Hsia, Y. T., and Christensen, J. J. (1951). Effect of 2,4-D on seedling blight of wheat caused by *Helminthosporium sativum*. *Phytopathology* 41, 1011–1020.

Johnson, T. (1946). The effect of DDT on the stem rust reaction of Khapli wheat. *Can. J. Res. Sect. C* 24, 23–25.

Kreutzer, W. A. (1960). Soil treatment. *In* "Plant Pathology: An Advanced Treatise" (J. G. Horsfall and A. E. Dimond, eds.), Vol. III, Chap. 11, pp. 431–476. Academic, New York.

Kurodani, K., Yokogi, K., and Yamamoto, M. (1959). On the effect of 2,4-dichloropheny acetic acid on the mycelial growth of *Hypochnus sasakii*, Shirai. *Shokobutsu Byogai Kenkyu* 6, 132–135.

Lai, M. T., and Semeniuk, G. (1970). Pichloram-induced increase of carbohydrate exudation from corn seedlings. *Phytopathology* 60, 563–564.

Lee, M., and Lockwood, J. L. (1977). Enhanced severity of *Thielaviopsis* root rot induced in soybean by the herbicide, chloramben. *Phytopathology* 67, 1360–1367.

Lockhart, C. L., and Forsyth, F. R. (1964). Influence of fungicides on tomato and growth of *Botrytis cinerea* Pers. *Nature (London)* 204, 1107–1108.

Miller, H. J. (1952). A method of obtaining high incidence of powdery mildew on snap beans in the greenhouse for fungicide screening tests. *Phytopathology* 43, 114. (Abstr.)

Moore, W. C. (1944). Chocolate spot of beans. *Agriculture (London)* 51, 266–269.

Neubauer, R., and Avizohar-Hershenson, Z. (1973). Effect of the herbicide, trifluralin, on *Rhizoctonia* disease in cotton. *Phytopathology* 63, 651–652.

Newhook, F. J., and Davison, R. M. (1953). Combined hormone-fungicide sprays for control of *Botrytis* fruit-rot in glasshouse tomatoes. *Nature (London)* 172, 351.

Nilson, H. E. (1973). Influence of herbicides on take-all and eye-spot in winter wheat in a field trial. *Swed. J. Agric. Res.* 3, 115–118.

O'Brien, G., and Bornmann, A. (1952). Fungicidal properties of maleic hydrazide (unpublished report). *In* J. S. Zundel, "Literature Summary on Maleic Hydrazide." U.S. Rubber Co. MHIS No. 6, p. 19, Naugatuck, Connecticut.

Partyka, R. E., and Mai, W. F. (1958). Nematocides in relation to sclerotial germination in *Sclerotinia sclerotiorum*. *Phytopathology* 48, 519–520.

Peeples, J. L., Jr. (1970). Effect of the herbicide, EPTC, on growth responses, antagonism, and pathogenesis associated with *Sclerotium rolfsii* and *Trichoderma viride* Pers. Ex. Fr. Ph.D. Thesis, Auburn Univ., Auburn, Alabama.

Percich, J. A., and Lockwood, J. L. (1975). Influence of atrazine on the severity of *Fusarium* root rot in pea and corn. *Phytopathology* 65, 154–159.

Peterson, L. J., Devay, J. E., and Houston, B. R. (1963). Effect of gibberellic acid on development of hypocotyl lesions caused by *Rhizoctonia solani* in red kidney bean. *Phytopathology* **53**, 630–633.

Rademacher, B. (1959). Einige Beispiele fur Kettenwirkungen nach Anwendung von Herbiziden. *Nachrichten bl. Dtsch. Pflanzenschutzdienstes* (*Braunschweig*) **11**, 155–156.

Rich, S., and Miller, P. M. (1964). *Verticillium* wilt of strawberries made worse by soil fungicides that stimulate meadow nematode populations. *Plant Dis. Rep.* **48**, 246–248.

Ries, S. K., Larsen, R. P., and Kenworthy, A. L. (1963). The apparent influence of simazine on nitrogen nutrition of peach and apple trees. *Weeds* **11**, 270–273.

Romig, W. R., and Sasser, M. (1972). Herbicide predisposition of snap beans to *Rhizoctonia solani*. *Phytopathology* **62**, 785–786.

Samborski, D. J., Person, C., and Forsyth, F. R. (1960). Differential effects of maleic hydrazide on the growth of leaf and stem rusts of wheat. *Can. J. Bot.* **38**, 1–7.

Simons, T. J., and Ross, A. F. (1965). Effect of 2,4-dichlorophenoxy acetic acid on size of tobacco mosaic virus lesions on hypersensitive tobacco. *Phytopathology* **55**, 1076–1077. (Abstr.)

Sinha, A. K., and Wood, R. K. S. (1967). The effect of growth substances on *Verticillium* wilt of tomato plants. *Ann. Appl. Biol.* **60**, 117–128.

Smith, A. M., Stynes, B. A., and Moore, K. J. (1970). Benomyl stimulates growth of a Basidiomycete on turf. *Plant Dis. Rep.* **54**, 774–775.

Stall, R. E. (1963). Effects of lime on incidence of *Botrytis* gray mold of tomatoes. *Phytopathology* **53**, 149–151.

Stall, R. E., Hortenstein, C. C., and Iley, J. R. (1965). Incidence of *Botrytis* gray mold of tomato in relation to calcium–phosphorum balance. *Phytopathology* **55**, 447–449.

Tang, A., Curl, E. A., and Rodriguez-Kabana, R. (1970). Effect of trifluralin on inoculum density and spore germination of *Fusarium oxysporum* f. sp. *vasinfectum*. *Phytopathology* **60**, 1082–1086.

Upstone, M. E., and Davies, J. C. (1967). The effect of simazine on the incidence of American gooseberry mildew on black currants. *Plant Pathol.* **16**, 68–69.

Waggoner, P. E., and Dimond, A. E. (1952). Effect of stunting agents, *Fusarium lycopersici*, and maleic hydrazide upon phorphorus distribution in tomato. *Phytopathology* **42**, 22. (Abstr.)

Wyse, D. L., and Meggitt, W. F. (1973). The relationship between root rot and herbicide application in navy bean. *Annu. North Centr. Weed Control Conf.,* *28th*, p. 28.

Wyse, D. L., Meggitt, W. F., and Penner, D. (1976). Herbicide root rot interaction in navy bean. *Weed Sci. Soc.* **24**, 16–21.

Yarwood, C. E. (1970). Man-made diseases. *Science* **168**, 218–220.

*Chapter 20*

# Mycotoxins and Their Medical and Veterinary Effects

W. F. O. MARASAS AND S. J. van RENSBURG

## I. INTRODUCTION

Mycotoxins are secondary metabolites of fungi that are harmful to both animals and man. Mycotoxicoses are diseases caused by the ingestion of foods or feeds made toxic by these fungal metabolites. Mycotoxicology may be considered to be the branch of science dealing with the mycology and phytopathology of toxin-producing fungi, the chemistry of the mycotoxins produced, and the pathology of the mycotoxicoses caused in animals and human beings.

Mycotoxins have undoubtedly presented a hazard to human and animal health since the earliest times, and this threat can only become

PLANT DISEASE, VOL. IV

more important as the demand on the available food supply increases with the population explosion. If the food supply is limited, the mycotoxin hazard is exacerbated in at least two ways: More fungus-damaged, potentially mycotoxin-containing foodstuffs are consumed rather than discarded, and malnutrition enhances the susceptibility to lower levels of foodborne mycotoxins (Newberne, 1974).

We hope in this chapter to familiarize plant pathologists with some of the known medical and veterinary effects of mycotoxins and to point out some of the unknown effects. The discussion will be limited to some mycotoxicoses and suspected mycotoxin-related syndromes that occur in nature, and special attention will be given to those caused by plant pathogenic fungi. Several excellent books dealing with mycotoxins have been published in recent years. These are listed in the references, and the interested reader is urged to consult them for comprehensive information.

## A. Relation of Mycotoxins to Plant Pathology

Mycotoxicoses are diseases in which the interaction of multiple factors in the causal complex must be considered. Some of the factors involved in the occurrence of a field outbreak of a mycotoxicosis caused by a plant pathogenic fungus are the following: the requirements for the infection of a susceptible host plant by a pathogenic strain of the mycotoxin-producing fungus; environmental and other factors favorable for the development of the disease; genetic capability of the pathogen (or the host–pathogen interaction) to produce a metabolite or metabolites harmful to animals or man; environmental and other conditions conducive to the elaboration of the mycotoxin(s) and the accumulation of sufficient quantities of these toxic metabolites in the diseased plant to cause a toxicosis in the consumer; and, finally, the consumption of sufficient quantities of toxin-containing plant material by a genetically as well as physiologically susceptible consumer. In addition, different fungi growing in or on the same substrate can influence each other in various ways, and combinations of mycotoxins produced by one or more fungi present can act either additively, synergistically, or depressively following simultaneous ingestion by the consumer. Many of the factors in this complex situation clearly belong in the realm of plant pathology.

The very nature of mycotoxicoses requires a multidisciplinary approach in which the plant pathologist has an important role to play. In the elucidation of the causative complex of a suspected mycotoxicosis, the plant pathologist has to isolate and identify the fungi involved and prepare bulk cultures of these fungi, either on culture media or by inoculating living plants, for toxicological and chemical studies. He can make a significant contribution in defining the climatic and other environmental factors influencing the development of the plant disease that eventually causes disease in the consumer. Most importantly, however, the control of mycotoxic diseases is primarily a plant pathological problem of preventing fungal development in the field and in storage.

If plant pathologists are to develop effective methods to control the mycotoxin hazard, it is essential that they be aware of at least some of the human and animal diseases known and suspected of being mycotoxicoses and of the mechanisms by which they are caused. This chapter was written with this objective in mind.

## B. Epidemiological Features of Mycotoxicoses

Just as many experimental designs may be classed as irrelevant to the human situation, so, too, many manifestational epidemiological associations are unrelated to direct cause. Proof of etiology is best obtained by direct experimentation on the target species, which is not ethical in the case of man. Hence, assumption of etiology depends on indirect scientific experimentation corroborated by epidemiological associations.

Human mycotoxicoses classically occur in a subsistence economy in which people exist on local produce and little is purchased elsewhere. In poor socioeconomic situations, survival may depend on the success of a single staple crop, with few alternatives, leading to an unvaried diet. When the staple provides a good medium for toxic fungal growth, the situation is precarious, and critically few factors can lead to disastrous outbreaks, such as the great epidemics of ergotism and more recently alimentary toxic aleukia.

The tendency for mycotoxicoses to be confined to certain geographic areas is linked to habits, customs, and local environmental factors favoring fungal growth. The highest known rates of primary liver cancer in the world are on a coastal land mass that butts into the warm Mozambique current off Africa, resulting in intense humidity and heat, which favors the growth of *Aspergillus flavus* on a variety of foodstuffs. In Russia, alimentary toxic aleukia does not occur on sandy soils which are capable of absorbing the moisture of vernal thawing. All mycotoxicoses affect rural populations much more severely than urbanized populations for many obvious reasons.

Seasonal recurrence is particularly obvious when the putative fungus is parasitic in the fields; here, outbreaks commence soon after harvesting. When storage fungi are implicated, unseasonal wet weather or extreme moist and warm environments are usually involved. In general, human diseases found to be mycotoxicoses are decreasing with time. A major factor here is the introduction of exotic crops, leading to a more varied diet. This was most obvious in Europe, where staples such as potatoes and maize from the New World did much to eliminate ergotism. This change is still occurring in Africa, where the increasing use of Western-type vegetables and rice is reducing dependency on the hazardous mono-cereal diet.

## II. VETERINARY MYCOTOXICOSES

### A. Aflatoxicosis

It has become customary to introduce papers and reviews dealing with mycotoxins with more or less the following sentence: An outbreak of a disease that caused the death of 100,000 turkeys in Great Britain in 1960 caused a surge in the interest in mycotoxins. The tremendous interest caused by this episode is perhaps difficult to understand, because many other mycotoxins cause fulminating field outbreaks in animals, of which some had been known for centuries and many had been under active scientific investigation long before 1960. The explanation must be sought in the discovery that turkey X disease was caused by the aflatoxins, a group of highly substituted coumarins produced by *Aspergillus flavus* and *A. parasiticus*, and that aflatoxin $B_1$ is one of the most potent naturally occurring carcinogens known. In addition to the fact that aflatoxin in contaminated feedstuffs can cause liver cancer in some experimental animals at levels lower than 1 ppb, residues also occur in animal products such as meat, milk, and eggs. Consequently, it is not surprising that the aflatoxins have become the most thoroughly investigated mycotoxins and to date the only ones for which tolerance levels are prescribed by law in many countries (Wogan, 1965, pp. 153–273; Goldblatt, 1969; Ciegler *et al.*, 1971, pp. 3–178; Purchase, 1971, pp. 141–151; Purchase, 1974, pp. 1–28; Moreau, 1974; Rodricks *et al.*, 1977, pp. 6–186).

A vast amount of information has accumulated on the veterinary effects of aflatoxin. Suffice it to say that the aflatoxins occur naturally in biologically significant amounts in a variety of feedstuffs all over the world, particularly in association with insect damage and conditions of high substrate moisture. Field outbreaks of acute aflatoxicosis occur in such animals as pigs, cattle, and poultry, but much more important are the chronic effects of the long-term consumption of low levels of afla-toxin. This can result in the development of liver cancer, for instance, in hatchery-reared trout, or in reduced weight gain, which is probably the major source of economic loss in animal production caused by the aflatoxins.

## B. Diplodiosis

A neurotoxic condition characterized by ataxia and paralysis occurs sporadically in cattle and sheep grazing on corn ears infected by *Diplodia maydis* (= *D. zeae*) in harvested lands in South Africa, Rhodesia, and Zambia (Wyllie and Morehouse, 1977, pp. 119–128; Rodricks *et al.*, 1977, p. 458; Wyllie and Morehouse, 1978, pp. 163–165). In 1918, the disease was reproduced experimentally in cattle and sheep with pure cultures of D. *maydis* in South Africa, and thus diplodiosis was one of the first veterinary field problems that was proved experimentally to be a mycotoxicosis. However, the neurological signs typical of field cases have not yet been reproduced in laboratory animals (Rabie *et al.*, 1977). The only toxic metabolite of *D. maydis* that has been characterized thus far is diplodiatoxin. However, this compound could account for only a fraction of the toxicity of culture material, and the chemical nature of the mycotoxin that causes diplodiosis in the field is still unknown. It also remains to be explained why diplodiosis is a recurrent, sporadic problem only in southern Africa despite the widespread occurrence of *D. maydis* as a serious ear- and stem-rot pathogen of corn.

## C. Ergot Poisoning

Ergot poisoning in domestic animals is caused by the consumption of mature sclerotia of *Claviceps purpurea* infecting the seed heads of cereals and grasses (Kadis *et al.*, 1972, pp. 321–373; Burfenning, 1973; Purchase, 1974, pp. 69–96; Rodricks *et al.*, 1977, pp. 583–595). Field cases of ergot poisoning occur in either a nervous or a gangrenous form. The nervous form is characterized by staggering, convulsions, and posterior paralysis and the gangrenous form by lameness and the loss of extremities. Other signs of ergot poisoning include loss of milk production and reduced reproductive efficiency. The biologically active compounds in *C. purpurea* sclerotia are discussed in Section III, D.

The fact that natural outbreaks of ergot poisoning of animals are still occurring proves that the oldest known mycotoxicosis has not yet been controlled effectively in the field.

## D. Facial Eczema

Facial eczema is a photosensitizing disease characterized by severe occlusive damage to the biliary system manifested clinically as jaundice and photosensitivity (Wogan, 1965, pp. 105–110; Mateles and Wogan,

1967, pp. 69–107; Kadis *et al.*, 1971, pp. 337–376; Purchase, 1974, pp. 29–68; Moreau, 1974; Wyllie and Morehouse, 1978, pp. 73–84). The disease is caused in grazing animals by the consumption of toxic conidia of the cosmopolitan saprophytic fungus *Pithomyces chartarum* growing on pasture litter. During favorable climatic conditions large numbers of conidia are produced by the fungus growing on dead grass material. The conidia become detached and adhere to the surface of the leaves of pasture plants, where they are ingested by grazing animals. These conidia contain an epipolythiopiperazine, sporidesmin, which has been shown to cause lesions very similar to those of field cases of facial eczema in experimental animals. However, the actual levels of sporidesmin present in toxic pastures have not yet been determined. Field outbreaks occur sporadically in sheep and cattle in Australia, New Zealand, and South Africa. The disease is of major economic importance in the North Island of New Zealand, where thousands of sheep may be affected during some seasons.

In New Zealand, concentrated efforts have been made to forecast outbreaks of facial eczema on the basis of climatic conditions conducive to sporulation and actual spore numbers of *P. chartarum* in pastures. Practical preventive and control measures, such as the removal of animals from pastures when outbreaks are forecasted and the spraying of pastures with fungicides, have also been developed. These endeavors to forecast and control field outbreaks of a mycotoxicosis provide excellent models that can be applied to other veterinary mycotoxicoses.

## E. Fusariotoxicoses

### 1. Estrogenic Syndrome

An estrogenic syndrome (Mateles and Wogan, 1967, pp. 119–130; Kadis *et al.*, 1971, pp. 107–138; Purchase, 1974, pp. 129–148; Moreau, 1974; Hidy *et al.*, 1977; Rodricks *et al.*, 1977, pp. 341–416) of pigs known as vulvovaginitis and characterized by enlargement of the vulva, mammary glands, and uterine hypertrophy is caused by *Fusarium graminearum* in infected seeds of corn and barley. Field outbreaks of porcine hyperestrogenism are known to occur in many countries, but the disease and its associated infertility problems are of major importance in the midwestern United States and also in Eastern Europe.

The estrogenic metabolite zearalenone is produced by *F. graminearum* and certain other *Fusarium* species, mainly after harvest in corn stored on the cob and subjected to alternating periods of low and moderate temperatures. Zearalenone is not a mycotoxin in the true sense of the word because it is virtually nontoxic and a single dose of 20,000 mg/kg does not cause death in experimental animals. It exerts estrogenic and anabolic actions. Zearalenone is a highly active compound, however, and levels of 1–5 ppm in the feed can cause undesirable physiological effects resulting in economic loss. Zearalenone has been detected in feedstuffs implicated in porcine hyperestrogenism at levels of 0.1–2909 ppm and often occurs in association with other fusariotoxins, such as the trichothecenes deoxynivalenol and T-2 toxin (Mirocha et al., 1976). Chronic effects resulting from the simultaneous ingestion of these compounds have not yet been determined.

## 2. Feed Refusal and Emesis

Epidemics of ear rot of corn induced by *F. graminearum* occur in the Midwestern United States during some cool and wet seasons. Considerable economic loss results from the loss in productivity and the purchase of alternative feed. It plays havoc with the commercial marketing of corn. The infected corn is usually refused by swine, but if consumed it can cause vomiting. A trichothecene produced by *F. graminearum* and known as deoxynivalenol or vomitoxin has been implicated in both feed refusal and emesis and has been found to occur naturally in corn at levels from 0.1 to 12 ppm. However, deoxynivalenol often occurs together with other fusariotoxins such as zearalenone in naturally infected corn (Mirocha et al., 1976), and the refusal response is much greater for naturally infected corn than for feeds with equal amounts of deoxynivalenol added. Thus, other factors, such as a possible synergistic effect between deoxynivalenol and zearalenone and/or other unknown mycotoxins, are probably involved in the feed refusal syndrome (Purchase, 1974, pp. 199–228; Vesonder et al., 1976; Rodricks et al., 1977, pp. 189–207).

## 3. Hemorrhagic Syndrome

Outbreaks of a hemorrhagic syndrome characterized by bloody diarrhea, lack of weight gain, and death accompanied by hemorrhagic lesions in the internal organs occur sporadically in the north central United States and elsewhere (Wogan, 1965, pp. 87–104; Hertzberg, 1970, pp. 163–173; Kadis et al., 1971, pp. 207–292; Smalley, 1973; Purchase, 1974, pp. 199–228; Rodricks et al., 1977, pp. 189–207). The etiology of this disease has not yet been completely resolved but is associated with the consumption of moldy corn that is late to mature or high in moisture at the time of the first killing frosts. The most toxic fungus consistently isolated from such corn is *Fusarium tricinctum*, which should probably be known more correctly as *F. sporotrichioides*. This fungus produces several highly toxic trichothecenes, notably T-2 toxin, which is a potent dermal toxin and causes severe hemorrhagic gastroenteritis upon oral administration. Moldy corn implicated in a lethal hemorrhagic toxicosis of dairy cattle in Wisconsin has been shown to contain 2 ppm of T-2 toxin, and this toxin has also been detected chemically in mixed feed that caused bloody stools in cattle in Nebraska (Mirocha *et al.*, 1976). Trichothecenes such as T-2 toxin are, however, difficult to detect in natural products, and the development of rapid, routine analytical methods will undoubtedly lead to the incrimination of trichothecenes in more veterinary problems of unknown etiology.

### 4. Leukoencephalomalacia

A remarkable neurotoxic condition of horses and donkeys known as leukoencephalomalacia is characterized by focal liquefactive necrosis of the white matter of the brain (Purchase, 1971, pp. 223–229; Wilson *et al.*, 1973; Marasas *et al.*, 1976). Field cases of the disease have been recorded for a long time in the United States, China, South Africa, and Egypt, where outbreaks occur sporadically in association with the flooding of the Nile Delta. Although the disease was first reproduced experimentally in horses with naturally infected moldy corn as long ago as 1902, the causative role of *Fusarium moniliforme* was established only during the 1970s. The chemical nature of the leukoencephalomalacia mycotoxin is, however, still unknown. *Fusarium moniliforme* is one of the most prevalent seedborne fungi of corn in most corn-producing countries, and the leukoencephalomalacia mycotoxin obviously occurs naturally in corn under certain circumstances because of the known occurrence of field cases.

It is most disconcerting that one of the fungi most commonly associated with corn intended for animal and human consumption can produce a potent neurotoxin that cannot be detected chemically at present and that nothing is known about the environmental conditions conducive to the production of this toxin in the field.

### F. Lupinosis

Lupinosis is a hepatotoxic condition characterized by severe liver damage and jaundice and is caused by the ingestion of lupins (*Lupinus* sp.) infected by *Phomopsis leptostromiformis* (Purchase, 1971, pp. 185–193; Purchase, 1974, pp. 111–127; Wyllie and Morehouse, 1977, pp. 111–118; Wyllie and Morehouse, 1978, pp. 213–217). Field outbreaks have been reported in sheep, cattle, pigs, and horses in Europe, Australia, New Zealand, and South Africa. Lupinosis is one of the most important sheep diseases in western Australia, where lupins are cultivated extensively. Two toxic metabolites (phomopsins A and B) have recently been isolated from cultures of *P. leptostromiformis* and found to produce typical lupinosis in sheep (Culvenor *et al.*, 1977). However, the molecular structures have not yet been elucidated.

The history of lupinosis research is interesting because the plant disease (stem blight of lupins) as well as the animal disease (lupinosis) caused by *P. leptostromiformis* were first described around 1880 and subsequently extensive but separate literatures developed. The fact that the same fungus causes both the plant and the animal disease only came to light 90 years later, in 1970. It seems more than likely that similar associations still remain to be made in other similar situations.

## G. Moldy Sweet Potato Toxicosis

Enzootics of a fatal respiratory disease of cattle in the United States and Japan have long been attributed to the ingestion of mold-damaged sweet potatoes (Purchase, 1971, pp. 223–229; Purchase, 1974, pp. 327–344). Pathological lesions in cattle that die are restricted to the lungs and include edema, congestion, and hemorrhage. It is now well established that this disease is caused by compounds formed by the host in response to invasion by certain fungi that are not otherwise toxic. Although chemical or physical damage as well as many invading organisms can stimulate the sweet potato to produce furanoterpenoid stress metabolites such as ipomeamorone, the specific pulmonary toxins (4-ipomeanol and ipomeanine) are formed only when certain fungi, particularly *Fusarium solani*, are also present. It appears then that the lung edema factors are formed when a stress metabolite produced by the sweet potato is catabolized by a specific invading fungus (Burka *et al.*, 1977).

The phytoalexins and related stress metabolites have received considerable attention in the phytopathological literature, but their effects on animal and human consumers have been almost completely neglected (Kadis *et al.*, 1972, pp. 211–247). The discovery of the toxic nature of several metabolites formed in mold-damaged sweet potatoes has clearly indicated a need for more information on the toxicological properties of phytoalexins and related compounds.

## H. Salivation Syndrome

During the 1950s numerous outbreaks of profuse salivation (slobbering) occurred in cattle consuming red clover hay in the midwestern United States (Kadis *et al.*, 1971, pp. 319–333, Crump, 1973; Purchase, 1974, pp. 97–109). Affected animals refused further feed, and this resulted in loss of productivity and additional loss from the purchase of replacement forage. This problem was eventually traced to the pathogenic fungus *Rhizoctonia leguminicola,* the causative agent of black patch disease of red clover, which is favored by high humidity combined with mild temperatures. The salivation factor in red clover infected with *R. leguminicola* is an alkaloidal mycotoxin, slaframine, which is metabolized by the animal to a physiologically active quaternary amine very similar to acetylcholine.

No satisfactory control measures against or resistance to black patch disease are known, and the mycotoxicological problems associated with diseased plants have resulted in the virtual elimination of red clover as a forage crop in the Midwest.

## III. HUMAN MYCOTOXICOSES

### A. Aflatoxicosis

Aflatoxin is one of the most common toxic, naturally occurring compounds to which man is exposed. It not only is a potent carcinogen, but is so toxic to laboratory animals that the single-dose $LD_{50}$ values vary from 0.33 mg/kg for ducklings to 3.7 mg/kg for vervet monkeys. Much lower repeated doses also are fatal (Rodricks *et al.,* 1977, pp. 701–705).

If one ventures to extrapolate such animal data to the human situation, then extensive field work on the contamination of food has shown that potentially lethal intakes by human beings are possible. Considering the ubiquitous nature of *A. flavus* and its remarkable propensity to synthesize the toxic aflatoxins, it seems anomalous that so few incidents of acute toxicity are reported. Undoubtedly the current realities of clinical medical practice in vast, remote areas of the world mask the true implications of aflatoxin exposure.

The few diagnoses of aflatoxicosis recorded are invariably the consequence of accidental encounters with mycotoxicological researchers, which created the opportunity to equate disease occurrence with quantitative exposure to aflatoxin. Not only does this interpretation of the significance of exposure depend on experimental data, but in cases in which the concentration of aflatoxin in the consumed food is known, some type of "dose–response" relationship is evident.

Highest exposure of humans to aflatoxin has been recorded in Thailand, where acute deaths within 72 hours of onset of symptoms from a disease resembling Reye's syndrome in children are common. Aflatoxin is present in tissues in high concentrations, and the only incriminated food examined contained 6 mg/kg aflatoxin. An outbreak of hepatitis in western India involving 500 people, 100 of which eventually died, was ascribed to aflatoxin-contaminated maize, which, on a wet food basis, contained about 2 mg/kg. A similar death in Africa was ascribed to the consumption of cooked cassava containing 1.7 mg/kg aflatoxin. This was the lowest concentration associated with relatively rapid human deaths and interestingly is the same concentration as the minimum found necessary in monkey feed to cause deaths within a short period.

An apparent toxic hepatitis is common in Mozambique, and the measurement of aflatoxin concentrations in some 2000 meals revealed a maximal contamination of 1.5 mg/kg.

In Senegal and India, the consumption of peanut meal supplements subsequently found to contain 0.3–1.0 mg/kg was associated with subacute toxic signs, chronic hepatitis, and possibly childhood cirrhosis.

## B. Alimentary Toxic Aleukia

In addition to ergotism (Section III,D), a hemorrhagic disease known as alimentary toxic aleukia must be ranked as one of the most pernicious human diseases that has ever been associated with the consumption of fungus-infected foodstuffs (Bilay, 1960, pp. 89–110; Wogan, 1965, pp. 77–85; Kadis et al., 1971, pp. 139–189; Purchase, 1974, pp. 229–262; Moreau, 1974; Rodricks et al., 1977, pp. 323–336). Sporadic outbreaks of this lethal disease characterized by necrotic stomatitis, exhaustion of the bone marrow, extreme leukopenia, and multiple hemorrhages have been recorded in the USSR since the beginning of the nineteenth century, but the disease occurred with maximal severity during the closing years of World War II. In 1944, near-famine conditions existed in certain areas of the USSR, and people were forced to eat grain that had been left unharvested during the previous season and had over-wintered under the snow. The winter of 1943–1944 was also abnormal in that a very thick snow cover and relatively high January–February temperatures were associated with alternate freezing and thawing of the soil in spring. The consumption of the grain that had overwintered in this way had disastrous results, and although the actual number of people that died has not been disclosed, it is estimated to be hundreds of thousands. The Soviet authorities mounted a massive research effort on the disease, and this resulted in the isolation of several toxic fungi from the overwintered toxic grain. The most toxic isolates belonged to *Fusarium sporotrichioides* and *F. poae*. The essential lesions of alimentary toxic aleukia were reproduced in experimental animals with pure cultures of these two fungi, and it was also shown that the toxicity of the cultures was greatly enhanced by low temperatures and alternate freezing and thawing. Considerable controversy has surrounded the chemical nature of the mycotoxin(s) that caused this extraordinary disease. Although this is obviously impossible to prove beyond a doubt, the available evidence from various sources strongly suggests that the causative factors were trichothecene mycotoxins, primarily T-2 toxin, produced in the overwintered grain by *F. sporotrichioides* and *F. poae* (Lutsky *et al.*, 1978).

*Fusarium* species capable of producing trichothecene mycotoxins are commonly associated with cereals in temperate regions of the world, and trichothecenes such as T-2 toxin, diacetoxyscirpenol, and deoxynivalenol have been shown to occur naturally in animal feedstuffs (Mirocha *et al.*, 1976). Therefore, the need exists for continued awareness of the potential threat of these mycotoxins to human health.

## C. Celery Photodermatitis

A contact photodermatitis that occurs among celery harvesters results from sensitization of the skin to sunlight by the handling of celery plants infected by *Sclerotinia sclerotiorum*. The active compounds with phototoxic activity formed in celery tissue infected by *S. sclerotiorum* are two furocoumarins (psoralens)—xanthotoxin and bergapten (Kadis *et al.*, 1972, pp. 3–91). The active metabolism of living celery tissue as well as *S. sclerotiorum* is required for the formation of these psoralens. They are not formed by *S. sclerotiorum* infecting other vegetables, and they are not induced to form in celery infected by other rot fungi (Wu *et al.*, 1972). The mechanism by which *S. sclerotiorum* alters the chemical components of celery from an inactive to a photoreactive state may be similar to the induction of the lung edema factors in sweet potatoes by *Fusarium solani* (Section II,G).

## D. Ergotism

Classical ergotism (Purchase, 1974, pp. 69–96) undoubtedly constituted one of the greatest recurring calamities of European history for 2000 years. Innumerable epidemics, often accompanied by war and famine and causing the deaths of up to 40,000 people, reached a peak during the great epidemics of France and Germany from 1770 to 1780—a century after Dodart's classic exposition of the etiology. Gangrenous epidemics continued until 1855, and the last outbreaks were recorded in Russia and England in 1928.

Necrotic or gangrenous symptoms predominated during medieval epidemics. Early symptoms included cold or prickling sensations in the limbs, followed by severe muscular pains. Soon, affected limbs became swollen and inflamed with violent burning pains and sensations of intense heat (St. Anthony's fire). Mummification or moist gangrene ensued, and the extent of necrosis varied from the shedding of parts of fingers or toes to the loss of all four limbs.

Nervous symptoms were prominent in convulsive ergotism, which predominated during the last few epidemics. Typical were a variety of unpleasant peripheral sensations followed by acute convulsions, which alternated with periods of drowsiness. Brain and spinal lesions caused death or permanent mental impairment.

The discovery that the ergot, a 1.5- to 6-cm slightly curved, blackish structure, was in fact a single sclerotium of *Claviceps purpurea* was somewhat hampered by the belief that it arose as the result of metamorphosis of an overnourished kernel of grain during periods of excessive rain or sunshine. Each small sclerotium contains some 100 different groups of substances, including estolides, diverse amines, amino acids, glucans, pigments, enzymes, and fatty acids. Of prime interest, with respect to pharmacology and toxicology, are the alkaloids, which can be divided into the unimportant clavine alkaloids and the derivatives of lysergic acid. The subgroups of simple lysergic acid amides and derivatives of the peptide type both contain highly active alkaloids.

The mode of action of various alkaloids is variable and complex, but they are classed according to their ability to cause vasoconstriction, uterine contraction, adrenergic blockade, serotonin antagonism, and important central effects. Notable are the peripheral effects, which are mainly due to an excitant action on the contraction of all smooth muscle. Contraction of smooth muscle in arteries results in reduced blood flow to the limbs, which can result in necrosis and gangrene. Some alkaloids potentiate uterine contraction in the term-sensitized uterus and are exceedingly valuable drugs for controlling hemorrhage at childbirth. There is no evidence that ergot is abortifacient, and this mistaken belief has led to frustrated attempts and overdosage, with tragic consequences.


## IV. SUSPECTED MYCOTOXIN–RELATED SYNDROMES

### A. Veterinary Syndromes of Uncertain Etiology

#### 1. Bermuda Grass Tremors

Bermuda grass tremors of cattle (Rodricks *et al.*, 1977, pp. 583–595) are characterized by nervousness, muscular tremors, and posterior paralysis. Outbreaks have occurred sporadically for at least 20 years in the southern United States. The disease occurs in animals grazing Bermuda grass pastures during the fall and appears to be associated with environmental stress conditions such as drought or frost. An outbreak in Louisiana in 1971 affected some 25,000 animals, and more than 600 died from secondary causes. The clinical signs are very similar to the nervous form of ergot poisoning, and a *Claviceps* sp. has in fact been isolated from toxic pastures and shown to produce ergot alkaloids in culture (Porter *et al.*, 1974). Other investigators have claimed that ergot has been eliminated as a possible cause of the disease. One particular fungus that produces black sclerotia and perithecia has been found to predominate on the leaves of toxic grass (Diener *et al.*, 1976), but this fungus has not yet been identified or established in culture. The problem presents a fascinating challenge for mycologists, plant pathologists, and veterinary toxicologists.

## 2. Fescue Foot

Winter pastures of fescue grass in the United States are sometimes associated with outbreaks of lameness together with gangrene of the tail tip, ears, and hooves in grazing cattle. Although these clinical signs are very reminiscent of the gangrenous form of ergot poisoning, *Claviceps* species are apparently not involved. A search for other vasoconstrictory mycotoxins has led to the isolation of several toxic *Fusarium* strains from toxic fescue hay. Some of these produced trichothecenes as well as a toxic butenolide in culture. Intramuscular administration of the pure butenolide to a heifer caused the tip of the tail to become necrotic and drop off. However, the butenolide has not yet been shown to occur naturally, and a causative role in fescue foot has consequently not been proved (Kadis *et al.*, 1971, pp. 191–206).

A very interesting recent development has been the isolation of two systemic clavicepitaceous endophytes, *Balansia epichloë* and *Epichloë typhina*, from toxic fescue pastures (Bacon *et al.*, 1977). Microscopic examination of the pith of fescue culms revealed a much higher frequency of intercellular hyphae of *E. typhina* in toxic than in nontoxic pastures. This fungus had not previously been investigated because the infections are symptomless. The investigation of endophytes that cause symptomless infections may prove to be a rewarding approach in the elucidation of the etiology of some veterinary field problems.

## 3. Ill Thrift

It has been said that the mycotoxin problem is like an iceberg—we see only the tip in the form of the acute, fulminating, fatal episodes, some of which have been described above. The major portion is under the surface and largely obscure because the effects are undramatic and elusive. These low-level effects may be expressed mainly as a reduced feed efficiency and retarded growth rate, although the available nutrients appear to be sufficient. Affected animals may also be predisposed to secondary effects such as enhanced susceptibility to infectious diseases (Rodricks *et al.*, 1977, pp. 745–750). Ill thrift occurs in a variety of situations, such as in sheep grazing certain pastures, in feedlot cattle, in hatchery fish, or in broiler chickens, and probably represents the major economic loss caused by mycotoxins in animal production.

Although ill thrift is a known chronic effect of the consumption of low levels of certain mycotoxins such as aflatoxin, indications are that most naturally occurring problems of this nature remain undiagnosed. These problems are usually nonspecific, the feedstuffs involved are contaminated by a large and varied mycoflora, and known mycotoxins cannot be detected by routine chemical methods. Consequently, mycotoxicologists are reluctant to investigate ill thrift problems, and few papers are published about them.

The first step in approaching such a problem should be to present experimental proof that fungi and their metabolites are part of the causative complex under field conditions. Forgacs *et al.* (1962) did this by demonstrating that the poultry hemorrhagic syndrome could be prevented or precipitated simply by drying or wetting the litter in broiler houses. Hamilton (1975) proved experimentally that the weight gain, pigmentation, and carcass grade of broiler chickens could be improved under field conditions by simple sanitary measures. Once it has been established that an ill thrift problem is mycotoxic in nature, one can proceed to investigate the causal complex by chronic feeding studies with individual fungi and combinations of fungi isolated from the feed, identification of the mycotoxins produced by these fungi, and finally applying sensitive analytical techniques to detect low levels of the suspected mycotoxins in the feed. If these efforts can lead even to the partial elimination of ill thrift problems, the results will be of immense economic importance to the animal industry.

## B. Human Syndromes of Uncertain Etiology

### 1. Balkan Nephropathy

Balkan nephropathy is a fatal, chronic kidney disease that occurs endemically in some rural populations in Bulgaria, Rumania, and Yugoslavia. The disease is characterized by tubular atrophy and interstitial fibrosis and is strikingly similar to a disease of pigs in Denmark known as mycotoxic porcine nephropathy. Comparable renal lesions have been induced experimentally in pigs with ochratoxin A, a mycotoxin produced by several *Aspergillus* and *Penicillium* species. Ochratoxin A has also been shown to occur naturally in grains associated with field cases of porcine nephropathy (Purchase, 1974, pp. 419–428; Rodricks *et al.*, 1977, pp. 489–506). These considerations prompted an investigation of the possible human exposure to ochratoxin A. Preliminary surveys of home-produced cereals have revealed a higher frequency of ochratoxin A contamination in an endemic area of Balkan nephropathy than in an area where the disease is absent, thus indicating a higher exposure to foodborne ochratoxin A in the endemic area (Krogh *et al.*, 1977).

### 2. Kashin–Beck Disease

Kashin–Beck or Urov disease is a chronic, disabling, deforming osteoarthrosis involving the peripheral joints and spine (Bilay, 1960, pp. 37–50, 89–110, 117–125; Nesterov, 1964). The disease has been known for more than 100 years to occur endemically among the Cossacks in the valley of the Urov River in eastern Siberia and in certain other localities. By 1930 so many cases had been recorded in the area that a special Urov research station was founded where the disease was investigated intensively by Soviet scientists. The climate in the area is characterized by marked temperature changes during the day and by the fact that most of the precipitation occurs during the second half of the summer and beginning of the fall—that is, during the time when cereals are maturing and during the harvesting and threshing. These and other considerations led to mycotoxicological investigations and eventually to the experimental reproduction of osteodystrophy similar to the human disease in rats and puppies with pure cultures of *Fusarium poae* (= *F. sporotrichiella* var. *poae*). On the basis of these results the hypothesis was formulated that Kashin–Beck disease is a mycotoxicosis caused by certain geographically restricted strains of *F. poae* that produce a metabolite in grain capable of causing constriction of the blood vessels to the epiphyses, which results in osteodystrophy.

These astonishing findings have received little or no attention outside the USSR, and the host of interesting questions raised by them have remained unanswered. Mycotoxicological investigations of other bone and joint diseases similar to Kashin–Beck disease could be most rewarding.

### 3. Thrombocytopenic Purpura (Onyalai)

Onyalai is an acute bleeding disease with clinical features similar to those of idiopathic thrombocytopenia. Whereas the latter occurs sporadically all over the world in all races, onyalai is endemic to Africa and exceedingly common in certain localities and among particular ethnic groups of blacks. This difference in the epidemiology and the fact that hemorrhagic bullae in the mouth are reputed to be more common in onyalai indicates a distinctive disease entity. The epidemiological characteristics of onyalai are typical of those described for mycotoxicoses.

For 70 years the cause of the disease has been the subject of much work and speculation, but recently a promising lead was the observation in one area that only people who use millet as the dietary staple are prone to the disease. Subsequent mycological examination of millet and also grain sorghum obtained from patients' homes revealed the invariable presence of toxic strains of *Phoma sorghina* (Rabie *et al.*. 1975).

Cultures of *P. sorghina,* when added to the rations of chickens, results in hemorrhage from the beak and cloaca and death within 4 days. Rats develop thrombocytopenia within 14 days, which is followed by hematuria, epistaxis, melena, and death. In a fraction of the fungal culture extract, the presence of the known mycotoxin, tenuazonic acid, was established (Steyn and Rabie, 1976). However, current experimental work in rats and primates with chemically pure tenuazonic acid has failed to induce a thrombocytopenia in either of these species. Thus it remains unresolved whether either *Phoma sorghina* or tenuazonic acid may be considered as a causative factor in the etiology of onyalai.

### 4. Primary Liver Cancer

Aflatoxin in variable amounts is consumed daily by millions of people, particularly in Africa and parts of Asia. These continents are also where primary liver cancer occurs in epidemic proportions. Undoubtedly, there are numerous hepatocarcinogens in the environment, but in most countries the rate stays well below 5 per 100,000 per year (Rodricks *et al.,* 1977, pp. 705–711). In Africa, south of the Sahara, rates 100 times higher than those in Western countries have been recorded, and today there is little doubt that aflatoxin is the major determinant of this epidemic occurrence.

The vexing problem of human susceptibility to aflatoxin has stimulated an immense amount of biochemical and subcellular research, the emphasis being on comparative studies between susceptible and resistant species. Some metabolic associations with susceptibility have been claimed, but until the exact sequence of biochemical events directly associated with carcinogenesis is understood, such associations must be considered as possibly fortuitous.

All epidemiological studies designed to examine the relationship between the intake of aflatoxin and cancer rate have yielded some sort of positive results. These include associations with the cultivation and use of peanuts, apparently a good medium for A. *flavus* growth, to market and stored-food surveys. Ultimately, the association has been proved beyond reasonable doubt by means of accurate cancer registration programs and the chemical assay of hundreds of prepared "plate" food samples. Such programs have been conducted in four countries, and not only were "dose–response" relationships established in each individual case, but the pooled data are in good agreement (Fig. 1). An extended study embracing six districts of Mozambique, including the highest know liver cancer area in the world, has just been completed. An excellent correlation between mean intake of the populations and cancer rate ($r = 0.940$; $p < .005$) was found.

There seems little doubt that concurrent infection with hepatitis virus plays an important, if not essential, role in the genesis of the disease, since most patients are found to be carriers. To date, there is no real evidence that the virus is primarily oncogenic, since in Egypt, for example, there is a high carrier rate but liver cancer is rare.

## 5. Spina Bifida and Other Congenital Defects

Quite a stir was caused in 1972 with the publication of a hypothesis implicating potato tubers infected by the notorious plant pathogen, *Phytophthora infestans,* as the cause of spina bifida and certain other birth defects previously considered to be congenital dystrophic malformations

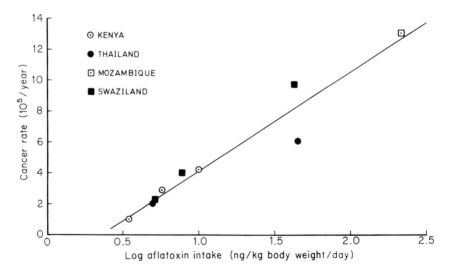

Fig. 1. Existing international data were standardized and pooled to demonstrate the relationship between the level of aflatoxin intake of populations and their rate of primary liver cancer ($r = 0.9683$; $p < .01$).

of the central nervous system (Renwick, 1972). The hypothesis was based only on statistical data, but experimental support was soon provided by the finding that blighted potatoes could induce fetal malformations involving osseous defects of the cranial vault in marmosets (Poswillo et al., 1972). At the time pregnant women were advised not to eat potatoes and even to avoid having potatoes in the house in case the teratogenic factor was volatile. The connection between P. infestans and spina bifida was, however, not supported by several subsequent epidemiological and experimental studies. According to the most recent available results, ethanol extracts of blighted potatoes are in fact embryotoxic, but so are extracts of healthy potatoes, and the responsible factor is solanine, a potato constituent (Jelinek et al., 1976). Solanine and several phytoalexins are known to accumulate in potato tubers as a result of the resistance reaction to races of P. infestans to which they are resistant, but not in the compatible reaction of host and pathogen (Kadis et al., 1972, pp. 211–247).

That was not the end of the controversy, however, and a second potato pathogen has become involved. Scott et al. (1975) found that Phoma exigua var. exigua, the cause of potato gangrene, can produce cytochalasins A and B in culture and that cytochalasin B occurs in potatoes infected by this fungus. These findings are of considerable importance because cytochalasin B causes inhibition of neural tube closure in chicken embryos (Linville and Shepard, 1972), and another cytochalasin, cytochalasin D, was recently shown to be a potent teratogen (Shepard and Greenaway, 1977).

It is a remarkable coincidence that at least two potential teratogens, solanine and cytochalasin D, have actually been found in diseased potatoes after Renwick (1972) had originally proposed such a relationship on purely epidemiological grounds. At present, there is no proof that either of these two or perhaps other compounds in diseased potatoes are related to the production of congenital defects in man, but the possibility certainly warrants further investigation.

## V. PROSPECTS FOR FUTURE RESEARCH AND CONTROL

The occurrence of natural outbreaks of mycotoxicoses in animals and man attests to the fact that the intricate interaction of the multiple factors involved in the casual complex of mycotoxicoses occurs successfully in nature, sometimes catastrophically so. The products of this successful interaction can exert harmful veterinary and medical effects by a variety of mechanisms. Morphological structures of the plant pathogens may be toxic, e.g., ergot sclerotia (Sections II,C and III,D), or the plant pathogen itself may produce mycotoxins that render the diseased plant toxic to the consumer, e.g., *Phomopsis leptostromiformis* (Section II,F). The elaboration of toxic substances may also be dependent on the active metabolism of the pathogen as well as the host, e.g., psoralen production in celery plants infected by *Sclerotinia sclerotiorum* (Section III,C), and a host–parasite interaction is also involved in the production of phytoalexins and related substances that may be toxic to the consumer, e.g., moldy sweet potato toxicosis (Section II,G). Phytotoxins produced by the pathogen are sometimes also mycotoxins, and the possibility exists that such toxins formed by root-infecting fungi may be translocated from the roots to the edible portions of the plant (Ghosal *et al.*, 1977). Saprophytic fungi can become established in living hosts in association with mechanical and/or insect damage and produce mycotoxins in the field as well as in the harvested crop during storage, e.g., aflatoxin production by *Aspergillus flavus* in groundnuts and corn (Sections II,A and III,A). Saprophyic fungi can also invade harvested crops and produce mycotoxins during storage, e.g., ochratoxin A production by *Aspergillus* and *Penicillium* species in cereal grains (Section IV,B,1). A unique mechanism by which a saprophytic fungus renders a substrate different from the one on which it is growing toxic by the production and dispersal of toxin-containing spores operates in the case of *Pithomyces chartarum* (Section II,D). Further examples of known mechanisms, as well as some unknown ones, are probably waiting to be discovered.

The multiple factors involved in the causative complex of mycotoxicoses make their etiology very difficult to elucidate. The equivalent of "Kochs postulates," consisting of the experimental reproduction of the characteristic clinical signs and pathological lesions of a naturally occurring mycotoxicosis with a chemically pure mycotoxin or combination of mycotoxins and the demonstration that this mycotoxin(s) occurs naturally in the feed at levels capable of inducing the disease, has been achieved for only a few mycotoxicoses. In most cases there are large gaps in our knowledge regarding the etiology and epidemiology of naturally occurring mycotoxicoses and suspected mycotoxicoses. The extent to which natural outbreaks of mycotoxicoses occur on a worldwide basis is largely unknown. The chemical nature of many of the mycotoxins involved is still unknown. Some known mycotoxins cannot yet be chemically detected in natural products. Some syndromes can be reproduced only in part by known mycotoxins, thus indicating that these toxins must act in association with other unknown factors in nature. Finally, the climatic and other environmental factors associated with field outbreaks of most mycotoxicoses are imperfectly known. As we saw in Section IV, the causes of several economically important animal diseases as well as certain diseases that cause untold human suffering are still unknown, although there are indications that they may be mycotoxin-related. The field for future research is wide open, and the need for such research certainly exists.

We have seen that the mycotoxin problem is a global problem. Natural outbreaks of mycotoxicoses occur in the humid tropics as well as in Siberia. Although the climate in a particular country might be favorable for the elaboration of a specific mycotoxin such as aflatoxin, the problem might be exported to another country in the form of agricultural products such as groundnuts or corn. In countries where no mycotoxin-related problems have been recorded, they have simply not been looked for. A problem cannot be controlled before it is recognized, and the positive diagnosis of a field problem as a mycotoxicosis requires a highly trained multidisciplinary team of scientists. A significant contribution to the recognition as well as control of the mycotoxin problem will be made by the introduction of compulsory courses in mycotoxicology in departments of plant pathology as well as veterinary and medical schools around the world.

Numerous possible methods for the control of mycotoxins have been proposed and discussed at length in the books cited in the references, particularly those of Krogh (1973) and Rodricks et al. (1977). Some of these are highly necessary regulatory measures which serve to protect the health of the inhabitants of some importing countries—for instance, by prescribing that only products containing less than 5 ppb of aflatoxin may be imported. Such measures are, however, not control measures in some exporting countries, where they may actually serve to exacerbate the problem if the rejected export products are consumed by the indigenous population. In the final analysis the control of the mycotoxin problem depends on the adaptation of classic as well as novel plant disease control measures to mycotoxin-producing fungi and their successful application in the field as well as in storage. This can be accomplished only by plant pathologists who are aware of the deleterious effects of mycotoxins on human and animal health.

### References

Bacon, C. W., Porter, J. K., Robbins, J. D., and Luttrell, E. S. (1977). *Epichloë typhina* from toxic tall fescue pasture. *Appl. Environ. Microbiol.* **34**, 576–581.

Bilay, V. I., ed. (1960). "Mycotoxicoses of Man and Agricultural Animals" (Engl. transl., U.S. Jt. Publ. Res. Serv., JPRS 7434). U.S. Dep. Commer., Washington, D.C.

Burfenning, P. J. (1973). Ergotism. *J. Am. Vet. Med. Assoc.* **163**, 1288–1290.

Burka, L. T., Kuhnert, L., Wilson, B. J., and Harris, T. M. (1977). Biogenesis of lung-toxic furans produced during microbial infection of sweet potatoes (*Ipomoea batatas*). *J. Am. Chem. Soc.* **99**, 2302–2305.

Ciegler, A., Kadis, S., and Ajl, S. J., eds. (1971). "Microbial Toxins," Vol. VI. Academic Press, New York.

Crump, M. H. (1973). Slaframine (slobber factor) toxicosis. *J. Am. Vet. Med. Assoc.* **163**, 1300–1302.

Culvenor, C. C. J., Beck, A. B., Clarke, M., Cockrum, P. A., Edgar, J. A., Frahn, J. L., Jago, M. V., Lanigan, G. W., Payne, A. L., Peterson, J. E., Petterson, D. S., Smith, L. W., and White, R. R. (1977). Isolation of toxic metabolites of *Phomopsis leptostromiformis* responsible for lupinosis. *Aust. J. Biol. Sci.* **30**, 269–277.

Diener, U. L., Morgan-Jones, G., Wagener, R. E., and Davis, N. D. (1976). Bermuda grass tremors in Alabama cattle in 1974. *Proc. Int. Biodegrad. Symp., 3rd,* pp. 673–678.

Forgacs, J., Koch, H., Carll, W. T., and White-Stevens, R. H. (1962). Mycotoxicoses. I. Relationship of toxic fungi to moldy feed toxicosis in poultry. *Avian Dis.* **6**, 363–380.

Ghosal, S., Chakrabarti, D. K., and Basu Chaudhary, K. C. (1977). The occurrence of 12, 13-epoxytrichothecenes in seeds of safflower infected with *Fusarium oxysporum* f. sp. *carthami. Experientia* **33**, 574–575.

Goldblatt, L. A. (1969). "Aflatoxin." Academic Press, New York.

Hamilton, P. B. (1975). Proof of mycotoxicoses being a field problem and a simple method for their control. *Poult. Sci.* **54**, 1706–1708.

Hertzberg, M., ed. (1970). "Toxic Microorganisms," Unnumbered Publ. U.S. Dept. Inter. UJNR Panels Toxic Microorg., Washington, D.C.

Hidy, P. H., Baldwin, R. S., Greasham, R. L., Keith, C. L., and McMullen, J. R. (1977). Zearalenone and some derivatives: Production and biological activities. *Adv. Appl. Microbiol.* **22**, 59–82.

Jelinek, R., Kyzlink, V., and Blattny, C. (1976). An evaluation of the embryotoxic effects of blighted potatoes on chicken embryos. *Teratology* **14**, 335–342.

Kadis, S., Ciegler, A., and Ajl, S. J., eds. (1971). "Microbial Toxins," Vol. VII. Academic Press, New York.

Kadis, S., Ciegler, A., and Ajl, S. J., eds. (1972). "Microbial Toxins," Vol. VIII. Academic Press, New York.

Krogh, P., ed. (1973). "Control of Mycotoxins." Butterworth, London.

Krogh, P., Hald, B., Plestina, R., and Ceovic, S. (1977). Balkan (endemic) nephropathy and foodborne ochratoxin A: Preliminary results of a survey of foodstuffs. *Acta Pathol. Microbiol. Scand., Sect. B* **85**, 238–240.

Linville, G. P., and Shepard, T. H. (1972). Neural tube closure defects caused by cytochalasin B. *Nature (London), New Biol.* **236**, 246–247.

Lutsky, I., Mor, N., Yagen, B., and Joffe, A. Z. (1978). The role of T-2 toxin in experimental alimentary toxic aleukia: A toxic study in cats. *Toxicol. Appl. Pharmacol.* **43**, 111–124.

Marasas, W. F. O., Kellerman, T. S., Pienaar, J. G., and Naudé, T. W. (1976). Leukoencephalomalacia: A mycotoxicosis of Equidae caused by *Fusarium moniliforme* Sheldon. *Onderstepoort J. Vet. Res.* **43**, 113–122.

Mateles, R. I., and Wogan, G. N., eds. (1967). "Biochemistry of some Foodborne Microbial Toxins." MIT Press, Cambridge, Massachusetts.

Mirocha, C. J., Pathre, S. V., Schauerhamer, B., and Christensen, C. M. (1976). Natural occurrence of *Fusarium* toxins in feedstuffs. *Appl. Environ. Microbiol.* **32**, 553–556.

Moreau, C. I. (1974). "Moisissures Toxiques dans l'Alimentation." Masson, Paris.

Nesterov, A. J. (1964). The clinical course of Kashin-Beck disease. *Arthritis Rheum.* **7,** 29–40.

Newberne, P. M. (1974). Mycotoxins: Toxicity, carcinogenicity, and the influence of various nutritional conditions. *Environ. Health Perspect.* **9,** 1–32.

Porter, J. K., Bacon, C. W., and Robbins, J. D. (1974). Major alkaloids of a *Claviceps* isolated from toxic Bermuda grass. *J. Agric. Food Chem.* **22,** 838–841.

Poswillo, D. E., Sopher, D., and Mitchell, S. (1972). Experimental induction of foetal malformation with "blighted" potatoes: A preliminary report. *Nature (London)* **239,** 462–464.

Purchase, I. F. H., ed. (1971). "Mycotoxins in Human Health." Macmillan, New York.

Purchase, I. F. H., ed. (1974). "Mycotoxins." Elsevier, Amsterdam.

Rabie, C. J., Van Rensburg, S. J., van der Watt, J. J., and Lübben, A. (1975). Onyalai—the possible involvement of a mycotoxin produced by *Phoma sorghina* in the aetiology. *S. Afr. Med. J.* **49,** 1647–1650.

Rabie, C. J., Van Rensburg, S. J., Kriek, N. P. J., and Lübben, A. (1977). Toxicity of *Diplodia maydis* to laboratory animals. *Appl. Environ. Microbiol.* **34,** 111–114.

Renwick, J. H. (1972). Hypothesis: Anencephaly and spina bifida are usually preventable by avoidance of a specific but unidentified substance present in certain potato tubers. *Br. J. Prev. Soc. Med.* **26,** 67–88.

Rodricks, J. V., Hesseltine, C. W., and Mehlman, M. A., eds. (1977). "Mycotoxins in Human and Animal Health." Pathotox Publ., Park Forest South, Illinois.

Scott, P. M., Harwig, J., Chen, Y.-K., and Kennedy, B. P. C. (1975). Cytochalasins A and B from strains of *Phoma exigua* var. *exigua* and formation of cytochalasin B in potato gangrene. *J. Gen. Microbiol.* **87,** 177–180.

Shepard, T. H., and Greenaway, J. C. (1977). Teratogenicity of cytochalasin D in the mouse. *Teratology* **16,** 131–136.

Smalley, E. B. (1973). T-2 toxin. *J. Am. Vet. Med. Assoc.* **163,** 1278–1281.

Steyn, P. S., and Rabie, C. J. (1976). Characterization of magnesium and calcium tenuazonate from *Phoma sorghina*. *Phytochemistry* **15,** 1977–1979.

Vesonder, R. F., Ciegler, A., Jensen, A. H., Rohwedder, W. K., and Weisleder, D. (1976). Co-identity of the refusal and emetic principle from *Fusarium*-infected corn. *Appl. Environ. Microbiol.* **31,** 280–285.

Wilson, B. J., Maronpot, R. R., and Hildebrandt, P. K. (1973). Equine leucoencephalomalacia. *J. Am. Vet. Med. Assoc.* **163,** 1293–1295.

Wogan, G. N., ed. (1965). "Mycotoxins in Foodstuffs." MIT Press, Cambridge, Massachusetts.

Wu, C. M., Koehler, P. E., and Ayres, J. C. (1972). Isolation and identification of xanthotoxin (8-methoxypsoralen) and bergapten (5-methoxypsoralen) from celery infected with *Sclerotinia sclerotiorum*. *Appl. Microbiol.* **23,** 852–856.

Wyllie, T. D., and Morehouse, L. G., eds. (1977). "Mycotoxic Fungi, Mycotoxins, Mycotoxicoses: An Encyclopedic Handbook," Vol. 1. Dekker, New York.

Wyllie, T. D., and Morehouse, L. G., eds. (1978). "Mycotoxic Fungi, Mycotoxins, Mycotoxicoses: An Encylopedic Handbook," Vol. 2. Dekker, New York.

*Chapter 21*

# The Epidemiology and Management of Aflatoxins and Other Mycotoxins

ROGER K. JONES

## I. INTRODUCTION

In the preceeding chapter, Marasas and van Rensberg provided a vivid introduction to many of the types of poisoning induced by mycotoxins in livestock and man. They included a brief section on the relationship of mycotoxins to plant pathology. The purpose of this chapter is to enlarge upon the subject and to offer some additional ideas about how plant pathogens produce mycotoxins and how plant pathologists can help prevent epidemics of mycotoxicoses by helping farmers prevent diseases caused by mycotoxin-producing fungi. Because most of my personal experience is with aflatoxins, they will be emphasized in this chapter, although many of the important principles of epidemiology and management discussed apply to other mycotoxin-producing fungi.

The term *mycotoxin diseases* is used to refer to those diseases in crop plants which lead to mycotoxicoses in animals. Table I lists a few of the more important mycotoxins and the organisms implicated in their accumulation in plants. The table is arranged according to whether the toxin is typically produced in the field or in the stored crop. In most cases,

PLANT DISEASE, VOL. IV

**TABLE I**
Some Major Mycotoxins and the Organisms That Have Been Implicated
in Their Accumulation in Plants

| Occurrence | Mycotoxin | Implicated organisms [a] | Reference |
|---|---|---|---|
| Only in field | Ergot alkaloids | *Claviceps purpurea* | Kadis *et al.* (1972) |
| | Diplodia toxin [b] | *Diplodia maydis* | Rabie *et al.* (1977) |
| | Phomopsis toxins | *Phomopsis leptrostromiformis* | Purchase (1974) |
| | Psoralens | *Sclerotinia sclerotiorum* | Wyllie and Morehouse (1978) |
| | Trichothecenes | *Fusarium* sp. | Kadis *et al.* (1971) |
| | T-2 Toxin | *Fusarium sporotrichioides* *Fusarium poae* | Purchase (1974) |
| Field and storage | Aflatoxin | *Aspergillus flavus* *Aspergillus parasiticus* | Goldblatt (1969) |
| | Zearalenone | *Fusarium tricinctum* | Kadis *et al.* (1971) |
| | Citrinin | *Penicillium citrinum* | Purchase (1974) |
| | Patulin | *Penicillium urticae* | Ciegler *et al.* (1971) |
| | Rubratoxin | *Penicillium rubrum* | Purchase (1974) |
| Only in storage | Ochratoxin A | *Aspergillus ochraceus* | Ciegler *et al.* (1971) |

[a] Additional organisms have been implicated in some cases (see references).
[b] The structure of this toxin is not yet known.

infection occurs in the field. The conditions for toxin production may be present in the field or they may require a storage environment. The list of implicated organisms is not exhaustive. In some cases the number of species (or genera) are too numerous to include and the taxonomy or even the toxigenic capabilities of some isolates remains in question. A more complete review of the associated fungi can be found in the references cited.

The history of man is punctuated with outbreaks of mycotoxins and mycotoxicoses. Unfortunately, investigations into the production of toxic substances by plant pathogens remained outside the realm of plant pathology until recent times.

In 1960, aflatoxins literally exploded onto the scene when 100,000 turkeys died after eating contaminated feed. Through an almost unprecedented triumph of interdisciplinary thinking, *Aspergillus flavus* and *A. parasiticus* were found to cause the production of aflatoxins. During the next 10 years, it became apparent that under the right moisture conditions *A. flavus* could colonize many substrates. Alflatoxins were found in an alarmingly wide array of food and feed products. C. M. Christensen became famous for his pioneering studies of fungi that develop in stored grain. His detailed investigations into the critical moisture conditions that permit the growth of *A. flavus* (Christensen and Kaufmann, 1968) are frequently cited in papers dealing with the occurrence of aflatoxin in stored corn. In Chapter 22 of this volume, Christensen himself tells part of this exciting story.

A further impetus for the investigation of mycotoxin diseases by plant pathologists was provided in the early 1970s. Survey results indicated that this "storage fungus," was developing in the field (Stoloff, 1976). By 1975 it was established that *A. flavus* and aflatoxin were causing serious problems in corn prior to harvest (Anderson *et al.*, 1975). Over 50 years earlier, Taubenhaus (1920) has described *A. flavus* as an ear rot of field corn, but its association with aflatoxin was not known at that time. The significance of aflatoxin accumulation in field corn prior to harvest was only beginning to be realized when disaster struck. In 1977, more than 90% of the corn crop produced in the southeastern United States was contaminated with aflatoxins (Zuber and Lillehoj, 1979). Conservative estimates of monetary losses in North Carolina alone exceeded 32 million dollars.

In recent years aflatoxins have also posed serious problems in peanuts, cottonseed, and pistachio and other tree nuts prior to harvest. Preharvest infection of many tree nuts has also been observed. But the botanical epidemiology of these diseases is just beginning to be investigated.

## II. THE CHALLENGES OF RESEARCH
## ON MYCOTOXIN DISEASES

Despite the growing importance of mycotoxin diseases, an examination of Table I will not be likely to produce any candidates for "most-important plant disease of 1980." Year by year, outbreaks of mycotoxin diseases occur sporadically. In the past, medical and veterinary scientists have been left to grapple with the consequences of mycotoxins. The challenge for plant pathologists is to apply our many tools to understand the nature and prevent the occurrence of mycotoxin diseases in the fields and granaries of the world. The needs of people and livestock for wholesome food is a compelling reason for cooperative research with other scientists to meet this challenge.

The study of mycotoxin diseases necessitates an extension beyond the traditional approaches of plant pathology. Literature on the subject is scattered widely; pertinent journals are rarely found in the same library. In addition the personal health hazards of mycotoxins must be respected. The sporadic nature of the diseases in time and space make fieldwork difficult. Mycotoxin diseases, by nature, require an interdisciplinary approach, bringing together mycologists, toxicologists, pathologists (plant, medical, and veterinary), natural products chemists, agronomists, and engineers.

Mycotoxin diseases have a historic association with postharvest physiology. This connection has been formally recognized in the journal *Phytopathology*, where Postharvest Pathology and Mycotoxins is recognized as a distinct category of publication. Yet the majority of papers published in *Phytopathology* during the last year on the subject of mycotoxins have dealt with the *field* development of aflatoxins and with ergot, which develops exclusively prior to harvest. Historic awareness of the importance of mycotoxins has focused attention on their development in storage, but this association must not limit our thinking about the spread of these diseases in the field.

Unfortunately, mycotoxins are induced by some of the most taxonomically complex genera of fungi—*Fusarium, Aspergillus,* and *Penicillium*. This has created serious problems for investigators in the field, especially since visual symptoms of disease are often entirely lacking or, if they do occur, are frequently not well correlated either with the occurrence or with the amount of mycotoxin in the crop.

The production of mycotoxins by various species of fungi is both genetically and environmentally regulated. In an experimental plot, the relationship between amount of infected tissue and toxin concentration is not linear. The genetic potential of each infection, the environmental conditions to which it has been subjected, and the length of time during which the toxin has been accumulating all affect the final concentration of toxin at harvest. Research on mycotoxin diseases too often has been concerned exclusively with toxin concentrations at harvest rather than with the dynamics of infection and toxin accumulation over time (see Volume II, Chapters 12 and 15). This lack of epidemiological thinking has muddied the distinction between environmental influences on infection versus those on toxin production.

Mycotoxins can be highly concentrated in a few infected plant parts. One infected kernel of corn among 100 healthy kernels can lead to unacceptable concentrations of aflatoxin at harvest. When the corn is milled we are faced with 10 kg of contaminated cornmeal instead of 9.9 kg of clean corn and 100 gm of highly contaminated kernels. The more the contaminated product is processed, the greater the dilution effect. Aflatoxins are carcinogens, and while dilution may hide the problem it will not eliminate it.

## III. DISEASE MANAGEMENT

The most widespread approach to regulation of mycotoxins in agricultural commodities has been through identifying and rejecting contaminated lots of produce. Suspected products are sampled and the amount of mycotoxin present is determined by a suitable analytical technique. If the material is found to contain dangerous amounts of mycotoxin the product is usually discarded. The lack of suitable screening methods in many parts of the world where mycotoxins are a problem leaves little doubt as to the eventual fate of most contaminated food or food products—they will be eaten, if not by man, at least by livestock.

Several approaches are useful in reducing the risk of human consumption of mycotoxins in agricultural commodities. Sclerotia of *Claviceps purpurea* can be separated from healthy grains by immersing them in water. The sclerotia float and can be discarded. Healthy peanuts and pistachio nuts have been separated from aflatoxin-contaminated nuts by electronic sorting. The high cash value of nut crops and the distinctive color of the contaminated produce has made electronic sorting both technically feasible and economically practical.

All of the measures mentioned above deal with the question, How can we avoid consumption of these toxic metabolites once they are produced? The question that we as plant pathologists must ask ourselves is, Are there ways we can help farmers prevent infection by mycotoxin-producing fungi in the field?

Mycotoxin diseases provide an excellent opportunity to apply the theory of disease management outlined by Apple in Chapter 5 of Volume I and the principles of epidemiology described in Volume II. Greater understanding of the ecology and epidemiology of mycotoxin-producing fungi will permit the development of tactics and strategies for their management. Knowledge is sorely lacking in this area, particularly with regard to factors that affect *in vivo* production of mycotoxins in host plants. Resistant varieties are being developed as one management tactic (Zuber *et al.*, 1978). Zuber and Lillehoj (1979) have recommended hybrids with high levels of resistance to ear-damaging insects for control of aflatoxins in corn. They also recommend the use of management practices that will either reduce stress or shift the ear development stage to miss a likely stress period. The development of early-warning systems based on spore trapping (Jones and Duncan, 1979) or degree-day temperature models (Stutz and Krumperman, 1976) may prove useful in the selection and implementation of management tactics. But, with a few exceptions, the development of effective management tactics for mycotoxin diseases is in the most primitive state.

## IV. EPIDEMIOLOGICAL CONSIDERATIONS

In the remainder of this chapter we shall discuss certain aspects of the ecology and epidemiology of A. *flavus* and aflatoxin in corn. Aflatoxin in corn is more than a veterinary problem. Ingested aflatoxin can end up in the meat, milk, and eggs of animals fed contaminated grain. Further, evidence is rapidly accumulating that both infection and most mycotoxin contamination occur prior to harvest (Stoloff, 1976).

### A. Geophytopathology

The development of aflatoxins in corn prior to harvest varies from year to year and over geographic areas within years. *Aspergillus flavus* is a high-temperature organism. Temperature is an ecological key that helps define the niche of this fungus in the field. In the United States, aflatoxin has developed more frequently in the southeastern than in the midwestern states (Lillehoj *et al.*, 1976). Caution must be used in examining aflatoxin data over the last several years for epidemiological prediction. Investigations on the problem have intensified, but available data are not always comparable. In short, we are looking for it more, therefore we will find it more.

### B. Disease Loss Assessment

Losses in crop value due to aflatoxins in corn do not follow the quantitative models for disease assessment outlined in Volume I, Chapter 4. Reductions in *yield* due to A. *flavus* usually are not significant. Very little documentation of yield loss can be attributed to the presence of the fungus alone, although decreases in viability of infected grain have been reported (Harman and Pfleger, 1974). Other, less obvious causes for loss in monetary value due to infection by A. *flavus* include the following: (1) Concentrations of aflatoxin may exceed government standards for salable feed grains; (2) problems with aflatoxins depress prices even for grain that is not contaminated; (3) aflatoxin-contaminated grain may not be accepted for storage under a government-insured warehouse loan; (4) the extra cost of testing for aflatoxin; and (5) the extra cost necessary to dry infected grain below 13% moisture content and thus halt continued accumulation of aflatoxin.

## C. Development and Dispersal of Inoculum

In the field, spores of A. flavus are exposed to exactly the same climatic and weather influences as other airborne fungi (see Volume 11, Chapter 15). In winter, temperatures well below those that are favorable for sporulation and growth of A. flavus are coupled with regular scrubbing of spores from the atmosphere by rain. The fungus lacks a sexual stage and moistened conidia lose their viability in 21–60 days (Boller and Schroeder, 1974).

Spring populations of spores must be initiated from mycelium of the fungus that overwintered in soil debris or litter. Airborne spores could also arrive on high-altitude wind currents or escape from ventilating ducts of buildings or grain storage bins. In one of our recent studies, airborne populations of A. flavus in North Carolina cornfields remained low until the end of June. Once the summer temperatures began to rise, the spore numbers increased (Jones and Duncan, 1979).

The concept of inoculum potential as described by Baker in Volume II, Chapter 7, can be adapted to the problem of aflatoxins in corn:

Aflatoxin concentration = inoculum potential × disease potential

In this instance, inoculum potential is inoculum density (number of propagules having a variable amount of toxin-producing ability) modified by environmental factors affecting the growth and development of each propagule. Disease potential can be expressed as the extent of predisposition of the host. Recent experiments in North Carolina indicate that irrigation has a major effect on disease potential, while planting date affects both inoculum density and disease potential. Early planting affords inoculum and stress escape during grain formation.

## D. Infection

Infection by A. flavus has been observed as early as the milk stage of developing ears (Taubenhaus, 1920). Grain usually is harvested before equilibrium moisture contents become too low to permit fungal growth. Thus, the process of infection appears to be governed mainly by temperatures and humidities that favor the weakly parasitic A. flavus over other ear-inhabiting fungi. On the other hand, toxin production appears to be regulated strongly by host predisposition.

The mechanism of entry of *A. flavus* into corn kernels is a subject of great controversy. Under conditions of high humidity and high temperature, we have found that *A. flavus* can colonize dying silks and invade developing kernels through the silk scar. This situation is similar to the incipient infections of *Botrytis cinerea* in cherries. Kernels with damaged pericarps are easily invaded by the fungus. Such damage can result from the feeding injuries of a wide variety of insects (Widstrom, 1979).

Wind and insects can disseminate spores from infected kernels to other plants. Insect activity and rainwater may be involved in inoculating other kernels within an infected ear. Inoculum of *A. flavus* develops in the field much like any other plant pathogen—inoculation is then followed by penetration, infection, and reproduction.

Production of aflatoxin is affected by factors which influence both *A. flavus* and its host. Like most secondary metabolites, aflatoxins are not produced under all conditions. *In vitro* studies have shown that aflatoxins develop between 18° and 35°C with an optimum of 24°C (Schindler *et al.*, 1967). *In vivo*, high concentrations of aflatoxins have been correlated with low rainfall, competition among plants for water, and insect injury (Zuber and Lillehoj, 1979).

## E. The Role of Stress in the Epidemiology of *Aspergillus flavus*

Preharvest development of aflatoxins in corn has been closely associated with drought stress. Drought stress is a confounded term. In the southeastern United States, periods of low rainfall often are associated with high temperatures and high relative humidities. The epidemiology of *A. flavus* can be influenced by "drought stress" in several ways: (1) Low rainfall increases the survival period of conidia; (2) low rainfall results in poor host growth and can inhibit the growth of mechanical barriers to infection such as the husk cover; (3) high temperatures increase spore production; and (4) high relative humidities in the absence of free water are more favorable for the germination and subsequent growth of *A. flavus* than for other ear-inhabiting fungi. In addition, drought conditions can also favor the survival and activities of many insects that provide infection courts for *A. flavus*.

## F. The Role of Nutrient Depletion in Secondary Metabolism

Mycotoxins are produced during that mysterious phase of fungal growth known as secondary metabolism. Secondary metabolites, by definition, are not essential to the growth and reproduction of the fungi;

they are thus distinct from primary metabolites. The function of most secondary metabolites is unknown. Some protect against the destructive effects of ultraviolet light. Others have antibiotic properties and may be useful in inhibiting or killing competing organisms. The selective advantages of such products are obvious, but the fact remains that the specific functions of most secondary metabolites are not known.

Bu'lock (1961, 1967) and Bu'lock et al. (1969) have suggested that it is the *process* of secondary metabolism rather than the specific effect of a given secondary metabolite that is important to fungi. Bu'lock postulated that secondary metabolism provides a mechanism for the excretion (dissimilation) of intermediates which would inhibit primary metabolism if they accumulated in fungal cells.

Bu'lock proposed two distinct phases in the growth and metabolism of a fungus in submerged culture. The tropophase is characterized by exponential growth and uptake of essential nutrients for production of primary metabolites. Secondary metabolites are initiated only after exhaustion of one of the essential noncarbon nutrients (usually nitrogen or phosphorus). The organism then enters the idiophase, during which species-specific secondary metabolites increase in abundance until the carbon source is exhausted. Such a view of secondary metabolism has been related to patulin production by *Penicillium urticae* (Bu'lock, 1967; Bu'lock et al., 1969). A similar relationship between the utilization of nitrogen and sucrose and the appearance of aflatoxin in submerged culture is demonstrated in the data of Hayes et al. (1966) and Mateles and Adye (1965).

## G. The Role of Drought Stress in Aflatoxin Production

We have seen that drought stress can affect the epidemiology of *A. flavus*. We have also seen that nutrient depletion (at least in submerged culture) can affect the onset and rate of mycotoxin production. Let us return to the field and see if we can apply these ideas to the production of aflatoxin in field corn.

The requirements of corn for water and nitrogen varies considerably at different stages of development. These requirements are minimal in the early stages, increase as the rate of growth accelerates, and reach a peak during the period between onset of flowering and grain formation. Unlike phosphorus or potassium, which reach corn roots by diffusion through the soil, over 90% of the plant's nitrogen is supplied by mass flow, carried along by soil water (Barber and Olson, 1968). This intimately links the uptake of soil water and nitrogen at a critical period during pollination. After flowering, proteolysis occurs in the vegetative organs. Approximately 65% of the nitrogen present is translocated to the ear for use in formation of grain. Water and nitrogen continue to be absorbed by the plant after initiation of the grain, but the effects of a moisture stress decrease as the grain matures (Arnon, 1975).

Low moisture supply and high temperatures during periods of stress affect the plant in many ways. The lowered activity of nitrate reductase in stressed plants results in the accumulation of nitrates and a decrease in the total nitrogen in the plant (Younis *et al.*, 1965). Under these same conditions, pollination is erratic, husk cover is incomplete, and the nutrient status (C/N ratio) of individual grains may be affected. Once seed formation is complete (black-layer formation in corn), all the rain in Olympia cannot alter the processes that have occurred. The result is a perfectly predisposed host.

In 1977, low rainfall and high temperatures in North Carolina midway in the growing season resulted in the high inoculum potential of *A. flavus*. When combined with increased insect populations and an extremely predisposed host (disease potential), an epidemic of *A. flavus* became inevitable. In 1978, nearly the same amounts of rainfall occurred, but the drought stress was largely a late season phenomenon and the aflatoxin problem was not as severe.

It was very hot early in 1977, and it was very dry. But corn will be vulnerable to another epidemic whenever such environmental conditions recur.

## V. SUMMARY

In this chapter, we have considered the history, the epidemiology, and the management of mycotoxin diseases. The need for wholesome food and feed products provides compelling justification for concern about mycotoxin diseases by plant pathologists. The concepts of inoculum potential, host predisposition, geophytopathology, and integrated management apply as well to mycotoxin diseases as to other types of plant diseases. Mycotoxin diseases may push our interdisciplinary spirit to the limit but they also provide an opportunity for important and unusually direct contributions to human and animal health.

### References

Anderson, H. W., Nehring, E. W. and Wichser, W. R. (1975). Aflatoxin contamination of corn in the field. *J. Agric. Food Chem.* **23**, 775–782.

Arnon, I. (1975). "Mineral Nutrition of Maize." Werder, A. G., Berne, Switzerland.

Barber, S. A., and Olson, R. A. (1968). Fertilizer used on corn. *In* "Changing Patterns in Fertilizer Use" (L. B. Nelson, M. H. Vickar, R. D. Munson, L. F. Seatz, S. L. Tisdale, and W. L. White, eds.). Soil Science Society of America, Madison, Wisconsin.

Boller, R. A., and Schroeder, H. W. (1974). Production of aflatoxin by cultures derived from conidia stored in the laboratory. *Mycologia* **66**, 61–66.

Bu'lock, J. D. (1961). Intermediary metabolism and antibiotic synthesis. *Adv. Appl. Microbiol.* **3**, 293–342.

Bu'lock, J. D. (1967). "Essays in Biosynthesis and Microbial Development." John Wiley, London.

Bu'lock, J. D., Shepherd, D., and Winstanley, D. J. (1969). Regulation of 6-methylsalicylate and patulin synthesis in *Penicillium urticae. Can. J. Microbiol.* **15**, 279–285.

Christensen, C. M., and Kaufmann, H. H. (1968). "Grain Storage Fungi." Univ. of Minnesota Press, Minneapolis.

Ciegler, A., Kadis, S., and Ajl, S. J. (eds.) (1971). "Microbial Toxins," Vol. VI. Academic Press, New York.

Goldblatt, L. A. (1969). "Aflatoxin." Academic Press, New York.

Harman, G. E., and Pfleger, F. L. (1974). Pathogenicity and infection sites of *Aspergillus* species in stored seeds. *Phytopathology* **64**, 1334–1339.

Hayes, A. W., Davis, N. D., and Diener, U. L. (1966). Effect of aeration on growth and aflatoxin production by *Aspergillus flavus* in submerged culture. *Appl. Microbiol.* **14**, 1019–1021.

Jones, R. K., and Duncan, H. E. (1979). Airborne populations of *Aspergillus flavus* in irrigated and non-irrigated field corn. *Phytopathology* **69**, 1–A5.

Kadis, S., Ciegler, A., and Ajl, S. J. (eds.) (1971). "Microbial Toxins," Vol. VII. Academic Press, New York.

Kadis, S., Ciegler, A., and Ajl, S. J. (eds.) (1972). "Microbial Toxins," Vol. VIII. Academic Press, New York.

Lillehoj, E. B., Fennell, D. I., and Kwolek, W. F. (1976). *Aspergillus flavus* and aflatoxin in Iowa corn before harvest. *Science* **193**, 495–496.

Mateles, R. I., and Adye, J. C. (1965). Production of aflatoxins in submerged culture. *Appl. Microbiol.* **13**, 208–210.

Purchase, I. F. H. (ed.) (1974). "Mycotoxins." Elsevier, Amsterdam.

Rabie, C. J., van Rensburg, S. J., Kriek, N. P. J., and Lübben, A. (1977). Toxicity of *Diplodia maydis* to laboratory animals. *Appl. Environ. Microbiol.* **34**, 111–114.

Schindler, A. F., Palmer, J. G., and Eisenberg, W. V. (1967). Aflatoxin production by *Aspergillus flavus* as related to various temperatures. *Appl. Microbiol.* **15**, 1006–1009.

Stoloff, L. (1976). Incidence, distribution, and disposition of products containing aflatoxins. *Proc. Am. Phytopathol. Soc.* **3**, 156–172.

Stutz, H. K., and Krumperman, P. H. (1976). Effect of temperature cycling on the production of aflatoxin by *Aspergillus parasiticus. Appl. Environ. Microbiol.* **32**, 327–332.

Taubenhaus, J. J. (1920). A study of the black and yellow molds of ear corn. *Tex. Agric. Exp. Stn., Bull.* **270.**

Widstrom, N. W. (1979). The role of insects and other plant pests in aflatoxin contamination of corn, cotton, and peanuts—A review. *J. Environ. Qual.* **8**, 5–11.

Wyllie, T. D., and Morehouse, L. G. (eds.) (1978). "Mycotoxicoses: An Encyclopedic Handbook," Vol. I. Marcel Dekker, New York.

Younis, M. A., Pauli, A. W., Mitchell, H. L., and Stickler, S. C. (1965). Temperature and its interaction with light and moisture in nitrogen metabolism of corn (*Zea mays* L.) seedlings. *Crop Sci.* **5**, 321–326.

Zuber, M. S., and Lillehoj, E. B. (1979). Status of the aflatoxin problem in corn. *J. Environ. Qual.* **8**, 1–5.

Zuber, M. S., Calvert, O. H., Kwolek, W. F., and Lillehoj, E. B. (1978). Aflatoxin $B_1$ production in an eight line diallele of *Zea mays* L. infected with *Aspergillus flavus*. *Phytopathology* **68**, 1346–1349.

*Chapter 22*

# The Effects of Plant Parasitic and Other Fungi on Man

## C. M. CHRISTENSEN

Chapter 20, of this volume, by Marasas and van Rensburg, deals with the effects on man of toxins from fungi. This chapter discusses some of the effects on man of storage fungi in grains and grain products, of poisonous and hallucinogenic mushroom, and of fungus allergens.

## I. FIELD AND STORAGE FUNGI

The terms "field fungi" and "storage fungi" were coined in the 1940s, when work first got underway on the relation of fungi to the deterioration of stored grains, to distinguish between two ecologically distinct groups of fungi.

PLANT DISEASE, VOL. IV

## A. Field Fungi

Seeds of many kinds of cultivated plants harbor a great variety of microflora, including bacteria, protozoans, slime molds, yeasts, and filamentous fungi. This is particularly true of seeds that are borne more or less exposed, such as those of the cereal plants. In humid growing seasons, developing and mature seeds of these and other kinds of crop plants may be invaded by fungi in the field, resulting in various kinds of blemishes, blights, and discolorations that may affect the quality of the seeds for various purposes, including foods and feeds. Among the most common field fungi are *Alternaria, Cladosporium, Fusarium,* and *Helminthosporium.* Malone and Muskett (1964) described 77 species of seed-borne fungi, in 60 genera. It has long been known that barley, wheat, rye, and millet invaded moderately to heavily by *Fusarium* might be toxic when consumed. This is discussed in this volume, Chapter 20. In order to grow, all field fungi require a moisture content in equilibrium with a relative humidity of 95–100%. In the starchy cereal seeds, this means a moisture content of 22–23%, on a wet weight basis, or 30–33%, on a dry weight basis. Because most seeds are harvested at moisture contents well below this, field fungi do not ordinarily continue to grow after harvest; the damage that they do is done by the time of harvest. One exception to this is ear corn, which sometimes is harvested and stored with a high enough moisture content for the growth of *Alternaria, Cladosporium, Fusarium,* and other fungi that require a high moisture content.

## B. Storage Fungi

The storage fungi, on the other hand, are capable of growing in seeds, in other plant parts, and in many other kinds of materials, at moisture contents in equilibrium with relative humidities of about 68–90%. They do not need free water. Some of them, in fact, require a high osmotic pressure to grow; they are adapted to an environment of limited water supply. The major storage fungi are *Aspergillus halophilicus, A. restrictus, A. glaucus, A. candidus, A. ochraceus,* and *A. flavus* (named in order of increasing moisture requirements), a few species of *Penicillium,* and, under special circumstances, *Wallemia sebi* and *Chrysosporium inops.* With the exception of *Aspergillus flavus,* none of these fungi invade seeds to any serious degree or extent before harvest.

Inoculum of storage fungi, like that of many other fungi, evidently is always present. Whether a given lot of grain or other substrate will or will not be invaded sufficiently to be damaged by any of these fungi depends on the moisture content of the material, its temperature, and the length of time it is held. The rate of development of these fungi also is influenced by the amount, nature, and distribution of dirt and debris in seeds lots, by whether the seed lot or food product already has been invaded by storage fungi at some point during its storage life, and by the presence and activities of insects and mites (Christensen and Kaufmann, 1974).

## 1. Effect of Moisture Content

The lower limits of relative humidity that permit growth of the common storage fungi vary, depending on species, from 68 to 90%. The moisture contents of various grains and seeds in equilibrium with different relative humidities are summarized in Table I; the data are approximations, because the equilibrium moisture contents of grains at different relative humidities vary over a range of several percentage points, depending on the history of the particular lot of grain tested and on various unknown factors. Soybeans, for example, because of their high oil content, normally have a lower equilibrium moisture content at a given relative humidity than starchy cereal seeds, such as wheat or corn. Yet I exposed samples of soybeans and corn above saturated salt solutions in the same desiccators, until equilibrium was reached, and the soybeans ended up with a *higher* moisture content than the corn.

Since the moisture contents at which the storage fungi can invade and cause damage in various grains and seeds and their products are fairly precisely known, it would seem that prevention of such damage could be easily prevented by keeping the grain or the processed material at a moisture content too low for the fungi to grow. In practice, this is sometimes difficult. The moisture content of grains and similar materials

### TABLE I

Moisture Content of Various Grains and Seeds in Equilibrium
with Different Relative Humidities at 25°–30°C

| Relative humidity (%) | Wheat, corn sorghum (%) | Rice (%) | | Soybeans (%) | Sunflower (%) | |
|---|---|---|---|---|---|---|
| | | Rough | Polished | | Seeds | Meats |
| 65 | 12.5–13.5 | 12.5 | 14.0 | 12.5 | 8.5 | 5.0 |
| 70 | 13.5–14.5 | 13.5 | 15.0 | 13.0 | 9.5 | 6.0 |
| 75 | 14.5–15.5 | 14.5 | 15.5 | 14.0 | 10.5 | 7.0 |
| 80 | 15.5–16.5 | 15.0 | 16.5 | 16.0 | 11.5 | 8.0 |
| 85 | 18.0–18.5 | 16.5 | 17.5 | 18.0 | 13.5 | 9.0 |

stored in bulk usually is determined by meters of one sort or another or by rapid infrared drying of small samples. Some of these machines or methods are of questionable accuracy. Even if the meter or method has a high degree of accuracy, it can only indicate the moisture content of the specific sample tested at the time it is tested. When a ship is loaded with maize for export, samples usually are taken automatically at specified intervals from the transfer belt, and the moisture content is determined of a representative sample of every 40,000–50,000 bushels. If the grain has been very well mixed, this sampling and testing procedure is adequate. In corn, however, a difference of only 0.5% in moisture content, in the range between 14.5 and 15.5% (the range encountered in almost all export corn), can mean the difference between safe carriage in a voyage of about a month through tropical or subtropical seas and considerable spoilage. Also, differences in temperature between different portions of bulk-stored grains can result in fairly rapid transfer of moisture from the warmer to the cooler portions. Aeration systems and schedules have been developed to enable those in charge of stored grains to maintain a uniform and moderately low temperature throughout the bulk and thus reduce moisture transfer and also reduce risk of spoilage.

Some processing procedures almost inevitably contribute to moderate contamination of the final product by storage fungi. Wheat to be milled into flour is conditioned to about 16.0% moisture to toughen the bran coat and make it easier to remove. The wheat is gradually reduced to flour by passing through a series of rapidly revolving corrugated rolls. The friction of grinding heats up the stock and raises the humidity of the air within the casings that cover the rolls. The flour that adheres to the inner walls of the roll casings acquires a moisture content high enough for storage fungi to develop. Christensen and Cohen (1950) reported that wheat just before it entered the first break rolls had only a few hundred colonies of storage fungi per gram, whereas flour scraped from the inside of roll housings had up to several million colonies of storage fungi per gram. Some samples of the best grades of white flour are contaminated with spores of *Aspergillus restrictus* and other storage fungi that seldom are found in large numbers in outdoor air. This contamination is from fungi growing within the milling machinery itself.

Macaroni, spaghetti, and other pasta products are made by mixing coarsely ground durum wheat flour with water into a thick dough and extruding this through dies. The resulting product is dried slowly over a period of several days, and during this time is subject to invasion by a variety of microflora, including storage fungi. Christensen and Kennedy

(1971) plated out samples of pasta products from 12 cities in nine states. Fungi grew from all of them, commonly from almost the entire surface of each piece. *Aspergillus flavus-oryzae* grew from 44 of the 47 samples tested, and in most of these it was the predominant fungus. Nine isolates of *A. flavus-oryzae* from these pasta products were grown in autoclaved moist corn and fed to ducklings, and five of these isolates resulted in the death of one or more of the ducklings. Van Walbeck *et al.* (1968) found 12.5 ppb of aflatoxin in a sample of dry spaghetti that was suspected of having been involved in illness of children who had consumed it.

## 2. Effect of Temperature

Within limits, low temperature can be as effective as low moisture content in preventing damage to grains and seeds and their products by storage fungi. Sound corn for example, of 18.5% moisture can be held at 5°C for 6 months or more without much invasion by or damage from storage fungi. Stored at 18.5% moisture and 8°C, however, it will be at least moderately invaded by *Aspergillus glaucus* within 6 months. In the range of moisture content between 18.0 and 22% and in the range of temperature between 5° and 10°C, a very small increase in either moisture content or temperature can make a great difference in degree of invasion by and damage from storage fungi within a given time. Aeration systems and schedules have been developed for maintaining relatively low and uniform temperature throughout bulk-stored grain. An aeration rate of 0.1 ft³ air per minute per bushel of grain will establish a uniform temperature throughout the grain in about 100 hours. Once the desired temperature has been attained, aeration can be discontinued until such time as inequalities in temperature arise. Even in such a humid, warm, and high-storage-risk site as San Juan, Puerto Rico, it is possible to select periods for aeration that combine moderate temperatures with relative humidities below 75%. Devices to monitor temperatures in stored grains were developed more than 50 years ago and have been in use throughout the world wherever grain is stored. Most processes that result in reduction of quality or in spoilage in stored grains are accompanied by a rise in temperature. Relatively dry grain is a good insulator, and so a "hot spot" may develop without this being immediately detectable by a temperature rise at the nearest thermocouple, but no extensive spoilage will develop without some temperature rise being detectable. Any temperature rise of more than a few degrees means that advanced spoilage is likely to be underway in the portion where the heat is being generated.

## C. Effects of Invasion of Grains, Seeds, and Their Products by Storage Fungi

Invasion of seeds by storage fungi can result in (1) decrease in germinability, important in seeds used for planting, for malting, and for edible sprouts; (2) discoloration of part or all of the seed or kernel, which reduces the quality of the seed for some uses; (3) heating and mustiness, which if it proceeds far enough can make the seeds or their products unusable for food or feed; (4) total spoilage. No precise figures are available on losses resulting from damage by storage fungi, and the magnitude of such losses, of course, varies from country to country and year to year. It has been estimated that in some of the tropical countries as much as 30% of the harvested food grains is lost in storage. The losses are likely to be higher in the countries that can least afford them—in part because of climates favorable to deterioration of all kinds of stored products and in part because of lack of knowledge and facilities necessary to reduce or prevent such losses. In many of the countries with chronically short food supplies, great efforts and resources are devoted to increasing the production of edible grains and seeds, but little or no effort is devoted to preserving them once they are produced or even to studying the problem. In the developed countries, storage losses overall may not amount to more than a few percent of the total harvested and so have relatively little impact on the economy in general or on the food supply. However, these losses are not uniformly distributed, and a given farmer who loses tens of thousands of bushels of soybeans or maize or a grain merchandising firm or poultry, beef, or dairy operation that loses hundreds of thousands of bushels of grain in storage may be very severely hurt. Most of such losses are needless, a result of ignorance, indifference, or carelessness. Figure 1 shows storage fungi growing from seeds, and Fig. 2 shows what sometimes is the final result of such invasion.

## D. Control of Losses Caused by Storage Fungi

We now know, without question, the agents that cause spoilage in grains and seeds and their products and the conditions under which they cause spoilage. We can detect in various ways the presence of these agents and can measure their increase long before they cause serious spoilage. We also have the facilities and the knowledge to establish and maintain conditions in stored products that will greatly reduce or eliminate storage losses. The principles and practices of good grain storage are known, and wherever these are applied losses are held to a minimum.

Fig. 1. *Aspergillus flavus* growing from a damaged or sick germ of wheat. The presence of this fungus in the wheat proves that the grain was stored for a time with a moisture content of at least 18.5%. This fungus can cause a rapid rise in temperature in stored grain and in some circumstances will produce potent toxins.

## II. TOXIC MUSHROOMS

The Hymenomycetes and Gastromycetes, which include most of the fleshly fungi, together comprise about 5000 species, of which probably several hundred are at one time or another gathered for food. No statistics are available on the number of cases of poisoning resulting

**Fig. 2.** Soybeans damaged by fungi in storage. Left, heavily caked; right, bin-burned.

from ingestion of toxic wild mushrooms in the United States, but Pilat and Usak (1954) give such figures for Switzerland in 1943 and 1944. In the two years combined there were four deaths from mushroom poisoning—two from *Amanita phalloides* and two from *Inocybe patouillardi*. Of the 356 cases of illness associated with the consumption of wild mushrooms in those years, 74 were from eating partly spoiled specimens of edible species. According to Fischer (1918), at least 90% of the fatalities caused by eating wild mushrooms were due to *Amanita phalloides* and *A. verna*. A half-century later, Simons (1971) attributed 95% of the cases of fatal poisoning from mushrooms to the *A. phalloides* complex. Fischer also stated that in France 100 people died from eating *A. phalloides* in 1900, and 153 died in 1912. Simons mentioned an outbreak of mass poisoning near Poznan, Poland, in 1918, in which 31 school children died from eating a prepared luncheon dish containing this mushroom.

### A. *Amanita phalloides–A. verna* Complex

Kaufmann (1918) described 22 species of *Amanita*, of which 8 were listed as "deadly," 7 as "suspected," 2 as "probably poisonous," and the rest as "edible but use caution." *Amanita phalloides* evidently is more common in Europe, and *A. verna* in the United States, although both occur in both countries. They are not ordinarily abundant, but in good mushroom weather in early fall it is not at all unusual to encounter them frequently and in quantity. Representative specimens of *A. verna* are shown in Fig. 3. Freshly expanded specimens are most attractive in appearance and have a pleasant mushroomy or fungusy odor, with nothing to suggest their extreme toxicity. Güssow and Odell (1928) described in detail the frightful symptoms of poisoning by these fungi.

## 1. Toxins

Two groups of toxins are present in these mushrooms—the phallotoxins, composed of phallin and phalloidin, and the amanitins, made up of $\alpha$-amanitin and $\beta$-amanitin. According to Florsheim (1972), the gastro-

**Fig. 3.** *Amanita verna.* This mushroom, along with its close relative *A. phalloides,* is responsible for the great majority of deaths from mushroom poisoning.

intestinal symptoms, which appear a few hours after the mushrooms are eaten, are produced by the phallotoxins, and the amanitins cause the liver and kidney damage, the symptoms of which appear later and the effects of which may result in death. Others attribute all of the effects to amanitins. Simons (1971) quotes evidence from other workers to the effect that the concentration of amanitins in the fresh mushroom is of the order of 5000 ppm, which is a fairly high concentration as fungus toxins go. The amount present probably varies somewhat from time to time and place to place. Consumption of 5–10 mg of the toxin evidently is enough to cause death. In a case of poisoning by A. *verna* in Minneapolis, in 1972, a husband and wife consumed sautéed specimens in a salad on a Sunday evening. Sixteen hours later they developed the characteristic gastrointestinal symptoms—nausea, vomiting, and diarrhea. They were hospitalized on Tuesday and were given intravenous fluid replacement to prevent dehydration. The woman's symptoms persisted for 3 days, the husband's for 6, but both recovered with no evidence of any liver or kidney damage. The physician who treated them supposed that the mushrooms they ate contained phalloidin toxins but not enough amanitin toxins to cause detectable harm. They were fortunate.

## 2. Treatment

According to Simons (1971), one Kubicka, in Bohemia, reported that 39 of 40 victims of *Amanita* poisoning were saved by administration of an infusion of thioctic acid, along with salts and dextrose. In St. Paul, in 1971, three men were poisoned by *Amanita verna*. Thioctic acid was administered to two of them, and they recovered. The third victim refused, on religious grounds, to receive thioctic acid treatment, and he also recovered. Florsheim (1972) stated that cytochrome c provided antidotal effects against a lethal dose of α-amanitin in mice. A French savant claimed that rabbits were able to eat "considerable quantities" of A. *phalloides* without harm, and from this he supposed that the stomachs of the rabbits contained some substance that inactivated the toxins. He devised a supposed curative paste, consisting of three hashed raw stomachs and seven hashed brains of rabbits, to be administered to the victim. There is no evidence to support the efficacy of this revolting mush.

## B. Amanita muscaria

Many books on mushrooms list A. *muscaria* as "deadly," and if consumed in sufficient quantity it unquestionably is deadly. If consumed in lesser quantity, it causes illness of more or less severity and various sorts of hallucinations (see below).

## C. *Lepiota morgani*

Cases of poisoning by *L. morgani* are fairly common but rarely fatal. Some people can eat it without harm, either because some specimens of it do not contain appreciable amounts of whatever toxins are involved, or because these people are not affected by the toxins.

## D. *Gyromitra* or *Helvella esculenta*

*Gyromitra* or *Helvella esculenta* is, for most people at most times, an edible and delicious mushroom, and some people have eaten it regularly and in quantity for years with nothing but enjoyment. Güssow and Odell state that *Gyromitra* is not only safe to eat, but a very desirable fungus for the table, but they warn against the use of old or even slightly deteriorated specimens. According to Ramsbottom (1953), large quantities of this fungus are gathered in Poland and exported to Germany, presumably after they are dried. It seems unlikely that *all* of the specimens in these large quantities would be picked at the peak of perfection or that none of them would be even slightly deteriorated. Simons (1971) states that the toxin in this species is monomethylhydrazine, whereas others believe that some people who eat this mushroom just happen to have an idiosyncratic sensitivity to it. In that regard, Simons (1971) mentions an acquaintance who, if he consumed even a very small quantity of the common cultivated mushroom in sauces or dressings, in a few hours became exceedingly ill. There are several authentic reports of illness following consumption of morels, although these generally are considered to be among the choicest of the edible fungi.

## E. *Coprinus atramentarius* and Alcohol

Some people are sensitive to a combination of *Coprinus atramentarius* and alcohol, so that if they consume even a small amount of alcohol after having partaken of the fungus they develop a rapid pulse, profuse sweating, and prostration. The symptoms abate after a few hours. This sort of poisoning is relatively rare, and Pilat and Usak (1954) mentioned only eight cases, described by seven authors in several different countries and dating back to 1906. In one case, which Pilat and Usak characterized as "specially heavy," the patient had consumed 6 pounds (at least 6 pints) of *C. atramentarius*, most likely along with other but unspecified comestibles and accompanied by "plenty of wine." No wonder he became ill.

## III. HALLUCINOGENIC MUSHROOMS

Since the early 1950s hallucinogenic mushrooms and some of the hallucinatory compounds derived from ergot alkaloids have received a great deal of publicity, some of it rather sensational and even fantastic. Included are publications by "religious" groups interested in mind-altering drugs. One of these publications describes the characteristics of the common hallucinogenic fungi, so that one presumably can go out and pick them (some of them are of common occurrence). Another gives detailed directions for growing some of these mushrooms on agar or in compost (some of them can be grown very readily in that way). Others discuss the "sacred" nature of these mushrooms and their relation to the development of religion or other sociopsychological phenomena, and still others are devoted to the effect of these psychotropic mushrooms on the mind. According to some of these accounts, consumption of these mushrooms liberates the mind and enables one to get in tune with the universe, experience religious ecstasy, and achieve Nirvana. The major fungi involved are *Amanita muscaria* and various species of the *Panaeolus–Psilocybe–Stropharia* group.

### A. *Amanita muscaria*

Fisher (1918) quoted a Russian, Krasheninnikoff, who traveled through Siberia in the early 1700s, on the use of *A. muscaria* by the Siberian peasants to produce intoxication. Those who consumed the mushroom became merry and active (maybe irrational and frenzied would be a better description) and had visions accompanied by a "horrible kind of delirium." Later, they became unconscious and remained so for some time, after which they awakened in a state of exhaustion. A good drunk required the consumption of ten specimens of the mushroom (they must have been small specimens, because the cap of a large specimen of *A. muscaria* may be a foot across, and no one, not even the gourmand who ate 6 lb of *Coprinus atramentarius*, could eat ten of them). The drunk could be passed on and shared by others, economically if not very aesthetically, by drinking the urine of the one who consumed the mushrooms. One wonders what sort of person it was who first made this momentous discovery!

Wasson and Wasson (1957), in a two-volume book on these and other "sacred" mushrooms, attribute the origin of some religions to revelations experienced by priests of ancient times who consumed *A. muscaria*. This is supported by Allegro (1970), mostly on the basis of etymology—

similarity of root words of some of the ancient languages. However, it is difficult to reconcile, for example, some of the timeless precepts of the Old Testament with the picture of frenzy, "horrid delirium," and brutal drunkenness described by Fischer. In the 1970s there were occasional outbreaks, on the West Coast and also in the humid South, of poisonings among teenagers from consumption of *A. muscaria* for "kicks." Whether their minds were liberated, and, if so, from what, is not known.

## B. *Panaeolus, Psilocybe, Stropharia*

Various species of these genera occur fairly commonly throughout much of the Northern Hemisphere, coming up in lawns and grassy places after every rain. In southern Mexico and Central America several species of them for many centuries have been consumed by divinators or witch doctors to induce a trance. According to Wasson (1958–1959), the medicine men, after consuming these mushrooms, could locate lost or stolen property, communicate with friends at a distance, and predict the future with astonishing accuracy. No specific examples were given on the kind of property located, the distance or content of communication, or the nature of the predictions. Soothsayers in many lands have been doing these things for millenia, with or without the aid of drugs. Also according to Wasson, native Americans feared and adored these mushrooms, regarded them as sacred, did not traffic in them openly, and consumed them only on special occasions in the darkness of the night and behind closed doors. Schultes (1940), an ethnobotanist of high repute who lived with some of these tribes in the 1930s, readily obtained specimens of these mushrooms in exchange for quinine pills, openly and in broad daylight. The medicine men evidently knew a good medicine when they saw it. This was 10 years or so before Wasson visited the same region. Maybe mysticism, like beauty, is partly in the eye of the beholder. Mycologists who visited some of these areas later found that the hallucinogenic mushrooms were being gathered by the natives to sell in some of the cities of the region, presumably for use by people who wanted to experience hallucinatory effects. There were complaints that the halluginogenic species were being adulterated with nonhallucinogenic kinds.

Schultes (1940) wrote that eating these hallucinogenic mushrooms induced in the divinator visions of color, some of them in kaleidoscopic patterns, accompanied by "incoherent utterances." This can hardly be taken as evidence that these medicine men were in tune with any celestial orchestration. They were temporarily deranged or, in other words, drunk.

According to Sanford (1972), "laughing mushrooms" were mentioned in a delightful Japanese folktale dating from the eleventh century. No religious overtones of any sort were involved. Sanford thinks that the mushrooms were *Paneolus papilionaceus* (or a closely related species), which is common in the United States and which he says is eaten by some thrifty New England Yankees to give them a free drunk. He describes a case of intoxication by this fungus, which involved various hallucinatory effects, none of them even remotely of a religious nature.

## IV. FUNGUS SPORES AND RESPIRATORY ALLERGY

The spores of many fungi are adapted to dissemination by air; with their large surface-to-volume ratio they are, so to speak, designed for far travel by air. Since the days of Marshall Ward, a century ago, plant pathologists have known that dissemination of airborne spores of plant pathogenic fungi is important in epidemiology. For many years, beginning in the early 1920s, Stakman and his co-workers at the University of Minnesota monitored the spread of airborne uredospores of *Puccinia graminis tritici* from Texas northward through the Great Plains to Minnesota and the Dakotas. In the 1925 epidemic in that region, Stakman (1955) calculated that in early June, 1925, there were 4000 tons of urediospores (150 billion spores per pound) on 4 million acres of wheat in Oklahoma and Kansas and that winds carried the spores to Minnesota and the Dakotas where they were deposited at the rate of 3.5 million an acre.

In the 1930s allergists began to recognize fungus spores as second only to pollens as causes of respiratory allergy in man, and this led to extensive sampling of airborne fungus spores. Enough work was devoted to this and to related problems to justify its designation as a separate discipline, aerobiology. Those interested are referred to the book by Gregory (1973).

Most of the work on airborne fungus spores has been done out-of-doors, which is reasonable enough since that is where most of the spores originate and where most of the problems are, both in the epidemiology of plant diseases and in respiratory allergy. Spores of some of the fungi involved in respiratory allergy, however, originate mainly within the buildings where the people sensitive to them live or work.

### A. Allergenic Fungus Spores in Outdoor Air

Durham (1938) mentioned a cloud consisting of "thousands of tons" of spores of *Alternaria* and *Hormodendrum* that originated probably in the stubble of grain fields of the upper Midwest in October 1936 and that was carried eastward across the country. Both of these genera are among the most important fungus allergens. Few plant pathogenic fungi are listed among the allergens, the major exception being smut (probably corn smut, *Ustilago maydis*). An allergist in Canada in 1924 reported that men exposed to a heavy concentration of rust spores (probably uredospores of stem rust) had an attack of asthma. However, Feinberg (1946) questioned whether the rust fungus was responsible for the attack, because he never had observed any reaction in patients given skin tests with extracts of pure rust spores but frequently did observe reactions in such tests to extracts from rust spores contaminated with *Alternaria, Hormodendrum,* and *Penicillium.* He also examined collections of rust spores from fields and found them to be contaminated with spores of these other fungi.

The "farmer's lung" syndrome is common enough to warrant its designation as a distinct disease. It evidently is due mostly to spores of the thermophilic actinomycetes *Thermopolyspora* and *Micromonospora* (Festenstein *et al.,* 1965).

### B. Allergenic Fungus Spores in Indoor Air

The numbers and kinds of spores in the air within some kinds of buildings, including many homes, differ greatly from those in the air out-of-doors. According to Swaebly and Christensen (1952), the predominant fungus in most samples of house dust that they studied was *Aspergillus restrictus,* but other species of *Aspergillus* and some species of *Penicillium* were also common. *Aspergillus restrictus* has not been detected in outdoor air, probably in part because the agar media used in most aerobiological studies of outdoor air would not reveal it (most strains of it require a medium of rather high osmotic pressure to grow) but also because it is prevalent only in special environments, including many homes. It may grow in various places within homes, but especially in the stuffing of chairs, sofas, mattresses, and pillows. In humid weather these pick up moisture from bodies in contact with them and become invaded by fungi that can grow without free water, but especially by *A. restrictus.* When one sits or lies down on a chair or bed with such moldy stuffing, an invisible cloud of spores is puffed into the air. Even foam rubber cushions can be invaded by these fungi.

Another favorite site for the growth of *A. restrictus* is the flour that adheres to the inner walls of roll housings in flour mills, as mentioned above. If such flour is kept in a humid place for a few weeks, it will become musty.

The air in grain elevators bears a heavy load of fungus spores, especially those of storage fungi; dilution cultures of such dust have yielded millions of colonies of these fungi per gram of dust. At times when grain cars are being unloaded at a terminal, the dust is so thick that the workers disappear as they walk into it. Some of those who work for years in such environments develop "elevator operator's lung."

Essentially the same symptoms as those of farmer's lung, and probably from approximately the same cause, are experienced by some workers in potato cellars at sorting time in the spring, where spores of various species of *Streptomyces* are likely to be abundant in the air.

## References

Allegro, J. M. (1970). "The Sacred Mushroom and the Cross." Doubleday, Garden City, New York.

Christensen, C. M., and Cohen, M. (1950). Numbers, kinds, and sources of molds in flour. *Cereal Chem.* **27,** 178–185.

Christensen, C. M., and Kaufmann, H. H. (1974). Microflora. *In* "Storage of Cereal Grains and Their Products" (C. M. Christensen, ed.), pp. 158–192. Am. Assoc. Cereal Chem., St. Paul, Minnesota.

Christensen, C. M., and Kennedy, B. W. (1971). Filamentous fungi and bacteria in macaroni and spaghetti products. *Appl. Microbiol.* **21,** 144–146.

Durham, O. C. (1938). An unusual shower of fungus spores. *J. Am. Med. Assoc.* **111,** 24–25.

Feinberg, S. M. (1946). "Allergy in Practice." Yearbook Publ., Chicago, Illinois.

Festenstein, J. L., Skinner, F. A., Jenkins, P. A., and Pepys, J. (1965). Self-heating of hay and grain in Dewar flasks and the development of farmer's lung antigens. *J. Gen. Microbiol.* **41,** 389–407.

Fischer, O. E. (1918). Mushroom poisoning. Cited in Kauffman (1918), pp. 125–164.

Florsheim, G. L. (1972). Curative potencies against A-amanitin poisoning by cytochrome C. *Science* **177,** 808–809.

Gregory, P. H. (1973). "The Microbiology of the Atmosphere," 2nd Ed. Wiley, New York.

Güssow, H. T., and Odell, W. S. (1928). "Mushrooms and Toadstools." Div. Bot., Dom. Exp. Farms, Ottawa.

Kauffman, C. H. (1918). "The Agaricaceae of Michigan." *Mich. Geol. Biol. Surv.,* Publ. No. 26, Biol. Ser. No. 5. (Reprinted, Dover, New York, 1971.)

Malone, J. P., and Muskett, A. E. (1964). Seed-borne fungi—description of 77 fungus species. *Proc. Int. Seed Test. Assoc.* **29,** 179–384.

Pilat, A., and Usak, O. (1954). "Mushrooms." H. W. Bijl, Amsterdam.

Ramsbottom, J. (1953). "Mushrooms and Toadstools." Collins, London.

Sanford, J. H. (1972). Japan's laughing mushrooms. *Econ. Bot.* **26,** 174–181.

Schultes, R. E. (1940). Teonanacatl: The narcotic mushroom of the Aztecs. *Am. Anthropol.,* **42,** 429–443.

Simons, D. M. (1971). The mushroom toxins. *Del. Med. J.* **43,** 177–187.

Stakman, E. C. (1955). Progress and problems in plant pathology. *Ann. Appl. Biol.* **42,** 22–33.

Swaebly, M. A., and Christensen, C. M. (1952). Molds in house dust, furniture stuffing, and in the air within homes. *J. Allergy* **23,** 370–374.

Van Walbeck, W., Scott, P. M., and Thatcher, F. S. (1968). Mycotoxins from food-borne fungi. *Can. J. Microbiol.* **14,** 131–137.

Wasson, R. G. (1958–1959). The hallucinogenic mushrooms of Mexico: An adventure in ethnomycological exploration. *Trans. N.Y. Acad. Sci.* **21,** 325–339.

Wasson, R. G., and Wasson, V. P. (1957). "Mushrooms, Russia, and History," 2 vols. Pantheon Books, New York.

# Author Index

Numbers in italics refer to the pages in which the complete references are listed.

411

## M

McAuslan, B. R., 258, *270*
McCahon, C. B., 325, *329*
McCalla, T. M., 315, 321, *329, 330*
McComb, A. J., 144, *152*
McCullogh, T. E., 281, *290*
McCune, D. C., 285, *289*
MacDonald, P. W., 57, *72*
Macdowall, F. D., 284, *290*
McDowell, L. L., 302, *311*
McEwen, F. L., 244, *255*
McGee, E. E. M., 175, *178*
McIntosh, A. H., 210, 211, *215, 216*
McIntyre, G. A., 87, *93*
McKeen, C. D., 102, *110*
McKeen, W. E., 140, 148, *152*
Maclean, D. J., 12, *21*
MacLeod, D. G., 304, 306, *311*
McManus, T. T., 281, *290*
McMeans, J. L., 102, *109*
McMillan, C., 325, *328*
McMullen, J. R., 362, *378*
McNulty, I. B., 285, *290*
MacPherson, H. T., 306, *310*
McPherson, J. K., 317, 325, *329*
Madden, P. A., 224, *238*
Maenhout, C. A. A. A., 336, *341*
Mai, W. F., 232, *238*, 351, *354*
Maino, A. L., 127, *132*, 186, *201*
Majernik, O., 282, *292*
Majima, R., 63, *70*
Malajczuk, N., 144, *152*
Malca, I., 60, *72*
Malone, C. P., 62, *72*
Malone, J. P., 394, *408*
Mamiya, Y., 224, *238*
Mancini, J. F., 285, *289*
Maniloff, J., 209, *215*
Mansfield, J. W., 185, *201*
Mansfield, T. A., 276, *290*
Maramorosch, K., 203, 204, 205, 206, 207, 210, 211, 212, 213, *215, 216, 217*, 251, *255*
Marasas, W. F. O., 364, *378*
Marchant, R., 106, *110*
Mark, W. R., 304, *311*
Markham, P. G., 208, 211, *216*
Maronpot, R. R., 364, *379*
Martin, J. T., 106, *110*, 280, *290*

Martin, R. T., 340, *342*
Mason, T. L., 63, *73*
Mateles, R. I., 361, 362, *378*, 388, *391*
Matile, P., 50, 57, *72*
Matta, A., 150, *152*
Matthews, R. E. F., 259, 260, 262, *270*
Maxwell, D. P., 169, *178*
Maxwell, M. D., 169, *178*
Mayama, S., 148, *152*
Meddins, B. M., 205, 208, 209, *215*
Meekison, D. M., 99, *110*
Meggitt, W. F., 348, 352, *355*
Mehlman, M. A., 360, 361, 362, 363, 366, 367, 369, 371, 372, 373, 377, *379*
Meléndéz, P. L., 117, 120, 121, 123, *132, 133*
Menzies, S. A., 88, *94*
Merlo, D. J., 197, 198, *201*
Messiaen, C. M., 337, *341*
Metzler, J. T., 277, 278, *290*
Michaelson, M. E., 119, *133*
Mijatovic, K., 294, *311*
Milbcrrow, B. V., 276, 277, *290*
Miles, P. W., 241, 242, *255*
Millar, R. L., 55, *71, 72*
Miller, G. W., 282, 284, *291, 292*
Miller, H. J., 347, *354*
Miller, J. R., 303, 304, *311*
Miller, P. M., 352, *355*
Miller, R. J., 61, 62, *71, 72*
Millerd, A., 64, *72*
Mims, C. A., 258, *270*
Mirocha, C. J., 56, 57, 58, *72, 73*, 363, 364, 368, *378*
Mishke, I. V., 107, *110*
Mishra, J. N., 85, *94*
Mitchell, H. L., 390, *391*
Mitchell, J. E., 227, *238*
Mitchell, R. E., 193, 194, *201*
Mitchell, S., 375, *379*
Mizuno, K., 300, *311*
Moczydlowski, E. G., 53, *73*
Molisch, H., 314, *329*
Moller, W. J., 145, 146, *152*
Moore, K. J., 351, *355*
Moore, R. E., 193, *201*
Moore, W. C., 88, *94*, 349, *354*
Moore, W. D., 86, *94*
Mor, N., 368, *378*
Moreau, C. I., 360, 362, 367, *378*

# Subject Index

## A

*Abies balsamea*, effects of allelopathy on, 319

Abiotic factors, role in pathogenesis, 4, 7, 8, 13, 273–288

Abnormal growth, 3, 13, 14, 16, 119, 159, 324 (*see also* Growth regulators, role in pathogenesis)

Abnormal metabolites, production of, 56, 59, 69

Abscissic acid, 13
role in stomatal closure, 276

Abscission, 160

Acarina, effects on plants, 244

*Acer saccharum*, effects of allelopathy on, 319

*Aceria ficus*, effects on plants, 244

*Achromobacter*, on bean, 186
synergism with *Pseudomonas phaseolicola*, 127, 186

Acid rain, 7, 8, 14
effect on mesophyll cells, 279
effect on plants, 278

Acidity, of plants, effect on disease, 88

Acquired disposition, 79, 81

Acquired susceptibility, 79

Actinomycetes, protect plants from soil microorganisms and nematodes, 229

ADP, role in virus synthesis, 48

*Adelges picae*, on balsam fir, 248

*Adelphocorus lineolatus*, effects on plants, 243, 245

Adhesion, of bacteria to plant surfaces, 184

Adhesorium, 171, 172

Adsorption, of allelopathic compounds, in soil, 324, 333

*Aegineta*, on sugarcane, 301
hosts of, 297

Aeration systems, for maintaining temperature in stored grain, 397

Aerosol particles, injurious, 7

Affinity transport system, of pathogens, 53

Aflatoxicosis, in animals, 360–366
in humans, 366, 367, 373, 374, 376, 377

Aflatoxin, as a carcinogen, 385
in dry spaghetti, 397
epidemiology and management of, 377, 381–390
induces liver cancer, 374
$LD_{50}$ of, 366

Agglutination, of bacteria, by lectins, 184

Aggressiveness, of pathogen populations, 27

Aging, of plants, 80

Agricultural chemicals that favor disease, 343–353

Agricultural commodities, regulation of mycotoxins in, 385, 386

427

Agricultural practices, that favor disease, 42, 43, 343–353
Agriocorpus, concept of, 115
*Agrobacterium*, 184, 185
    favored by Dichlone and Captan, 345
    hyperparasites of, 351
*Agrobacterium tumefaciens*, 185
    carbon metabolism of, 55
    plasmid of, 6, 7, 176, 197
Air pollutants, 5, 7, 12, 13, 16
    effects on cell organelles, 281, 285, 286
    alter physiological processes, 275
    effects on cells, 278–281, 284, 285
    effects on photosynthesis, 283, 284
    effects on respiration, 284, 285
    impact on yield and quality of crops, 274
    mechanisms of pathogenesis, 5, 16, 273–288
    as pathogens, 5, 273–288
    predispose plants to biotic pathogens, 100, 114, 274
    responses of plants to, 273–288
    synergism among, 100
Air, temperature, effect on disease, 83
Alachlor, diseases favored by, 345, 346
*Albugo* spp., rupture host epidermis, 172
Albertus Magnus, 294
*Alectra*, hosts of, 297
Alfalfa, *Adelphocorus lineolatus* on, 243
    *Corynebacterium insidiosm* on, 190
    spittle bugs on, 243, 245
Alfalfa wilt, 190
Algae, produce antitoxins, 316
    bacterial diseases of, 181
    similarity to fungi, 37
Alimentary toxic aleukia, 367, 368
    epidemiology of, 359
Alkaloids, as allelopathic compounds, 323
    in crop plants, 322
    produced by plant pathogens, 361, 366, 368–370
Allelopathic compounds, chemical nature of, 322–324
    degradation of, in soil, 327
    detoxification of, 328
    mechanisms of action of, 324, 325
    model of fate, in the environment, 325, 328

plant parts that contain, 320
published lists of, 320
quantities produced, 322
release of, into the environment, 320–322
Allelopathy, 16, 313–328
    in agriculture, 318
    as distinct from competition, 314
    definition of, 313, 314
    in forestry, 319
    history of, 314
    in horticulture, 318, 319
    in managed ecosystems, 318
    in natural ecosystems, 316
    role in pathogenesis, 5, 16, 313–328
Allergy, respiratory, induced by fungus spores, 406–408
Almond, twig dieback of, induced by *Rhizopus* spp., 56
*Alternaria*, on cereal grains, 394
    on grain stubble, 406, 407
    on sugar beets, 103
    on tomato, favored by low sugar content in leaves, 347
*Alternaria alternata*, 108
*Alternaria brassicicola*, on cabbage, 104
*Alternaria solani*, on tomato, 119, 338, 347
    synergism with potato virus Y, 126
*Amanita muscaria*, a poisonous mushroom, 402
    symptoms after eating, 404, 405
*Amanita phalloides*, deaths from eating, 400
*Amanita phalloides-A. verna* complex, description of, 400, 401
Amanitins, 401
*Amblypelta*, effects on plants, 245
*Ambrosia*, inhibition of, by volatile compounds, 320
Ambrosia beetles, 249, 250
*Ambrosia psilostachya*, production of allelopathic chemicals by, 316
Amino acid theory, of phytoxemia, 241
    accumulation, during stress of plants, 306
    in virus-infected plants, 260
    in giant cells, 121
    as allelopathic compounds, 323
    as sources of energy, 51

*Erysiphe polygoni* on, 82
*Fusarium solani* f. sp. *phaseoli* on, 138, 146
*Orobanche* on, 302, 304
*Pratylenchus scribneri* on, 225, 226
*Pseudomonas phaseolicola* on, 193, 194
*Pseudomonas putida* on, 185
*Rhizoctonia solani* on, 83, 84
*Uromyces appendiculatus* on, 82
*U. phaseoli* on, 82, 142
insects of, 245
Bean rust, 58
Bearberry, allelopathic effects of, 319
Beech, *Myzodendron* on, 296
*Belonolaimus*, feeding habits of, 221, 222
role in biopredisposition, 125
Beneficial plant diseases, 205, 206
Benomyl, diseases favored by, 345, 346
favors *Sclerotium rolfsii*, 352
favors turf pathogens, 351
inhibits symptom production by viruses, 264
protects plants against O$_3$ injury, 280
Benzoic acid, an allelopathic compound, 323
Bermuda grass tremors, of cattle, 369, 370
Biases, resulting from Koch's postulates, 11, 12
Biochemistry, of hosts, effect on disease, 97
Biocides, use of, in experimental analysis of multiple pathogen complexes, 129
Bioenergetic status, of cells, effects of air pollutants on, 284, 285
Biological control, 98, 99, 109
Biological fitness, definition of, 25, 26
Biophysics, of disease, 167, 168
Biopredisposition, by *Belonolaimus* spp., 125
definition of, 117
by nematodes, 122, 124, 125, 220, 227–231, 338
by nonpathogens, 89–92
by organisms external to the host, 97–109
by organisms internal to the host, 113–133
role of bacteria in, 127

role of fungi in, 126
role of *Heterodera* spp. in, 124
role of viruses in, 126
theoretical mechanisms of, 118
by *Trichodorum* spp., 125
Biosynthetic machinery, diversion of, 14
Biotic pathogens, definition of, 5
influence plant response to air pollutants, 274
Biotic stress factors, definition of, 10, 11
Biotrophs, 165, 174, 175
Black cherry, allelopathy of, 319, 321
Black path disease, of red clover, 366
Bore-holes, formed by wood-destroying fungi, 53
Botanists, plant pathologists as, 6
*Botrytis*, effects on phosphorus, 349
favored by Benomyl and Zineb, 345
favored by naphthoxyacetic acid, 350
synergism with *Meloidogyne*, 123
tolerant to benomyl, 352
*Botrytis allii*, on apples, 84
*Botrytis cinerea*, 108
conversion to pathogens on tobacco, 120
in cherries, 386
on grape, 137
on lettuce, 87
synergism with air pollutants, 100
*B. fabae*, on beans, 88
*Botrytis* rot of fruits, dithiocarbamate fungicides, 348, 349
Bracken fern, allelopathic to Douglas-fir, 319
Breeders, role in coevolution, 42, 43
Breeding, for disease resistance, 33, 42
for resistance to vectors, 207
*Brevipalpus phoenicis*, effects on plants, 244
Brood galleries, of insects, 249
Broomrapes, hosts of, 297
on sunflower, 294
Brown rust, of brome grasses, 78
Buildings, airborne spores in, 406, 408
Bunt, of wheat, 76, 105
*Bursaphelenchus*, feeding habits of, 221–223
*B. lignicolus*, on pines, 224
Butenolide, as a mycotoxin, 370

## C